INTERMEDIA POLLUTANT TRANSPORT

Modeling and Field Measurements

INTERMEDIA POLLUTANT TRANSPORT

Modeling and Field Measurements

Edited by
David T. Allen
Yoram Cohen
Isaac R. Kaplan

University of California, Los Angeles
Los Angeles, California

With the assistance of
Donato A. Kusuanco

National Center for
Intermedia Transport Research at UCLA
Los Angeles, California

PLENUM PRESS • NEW YORK AND LONDON

Library of Congress Cataloging in Publication Data

Intermedia pollutant transport: modeling and field measurements / edited by David T. Allen, Yoram Cohen, and Isaac R. Kaplan, with the assistance of Donato A. Kusuanco.
p. cm.
Papers from a conference sponsored by the National Center for Intermedia Transport Research, held Aug. 24–26, 1988 in Santa Monica, Calif.
Bibliography: p.
Includes index.
ISBN 0-306-43257-9
1. Pollutants—Environmental aspects—Congresses. 2. Transport theory—Congresses. I. Allen, David T. II. Cohen, Yoram. III. Kaplan, Isaac R. IV. National Center for Intermedia Transport Research (U.S.)
TD172.5.I54 1989 89-16035
628.5—dc20 CIP

Proceedings of a conference on Intermedia Pollutant Transport: Modeling and Field Measurements, held August 24–26, 1988, in Santa Monica, California

A Division of Plenum Publishing Corporation
233 Spring Street, New York, N.Y. 10013

Printed in the United States of America

PREFACE

The National Center for Intermedia Transport Research (NCITR) was established at UCLA in 1982 by EPA as one of six Centers of Excellence for the study of environmental pollution problems. One of the functions undertaken by the NCITR has been to hold periodic workshops and to provide a forum for the discussion of current topics in the environmental pollution arena. To this end, two other workshops have previously been held. The first, held in November 1982, was chaired by H. R. Pruppacher, R. G. Semonin and W. G. N. Slinn on Precipitation Scavenging, Dry Deposition and Resuspension. The second, held in January 1986, was chaired by Y. Cohen on Pollution Transport and Accumulation in a Multimedia Environment.

The present workshop, chaired by D. T. Allen, Y. Cohen and I. R. Kaplan, was held on August 24-26, 1988 in Santa Monica, California. The title of the workshop was Intermedia Pollutant Transport: Modeling and Field Measurements. Approximately one hundred individuals participated and twenty five papers were given, mostly by invitation. The workshop was divided into the following four broad topics: 1) Transport of Pollutants from the Atmosphere, 2) Transport of Pollutants from Soils and Groundwaters, 3) Transport of Pollutants from Lakes and Oceans, and 4) Multimedia Transport of Pollutants. The last afternoon was reserved for a Panel Discussion.

It is evident from the session headings that there was considerable diversity in the content of the contributions. Added to the broad topical distribution is the fact that the participants came from Canada, as well as different geographic locations and institutions in the USA. Hence, the intent of the organizers to assemble a group of specialists who represent different profession backgrounds and interest was accomplished.

The four main headings of the workshop were intended to cover the major geospheres which become impacted by man's activity. The topics discussed in the symposia on atmospheric pollutant transport processes covered the range of problems from acid rain chemistry to the transport of both organic and metal pollutants as aerosols. It was particularly interesting to learn that the dominant transfer process of pesticides to the Great Lakes was by atmospheric entrainment and not by river runoff (Eisenreich and Willford).

The symposium on soils and groundwater largely dealt with modeling of organic pollutants in the soil and their transport mechanisms into the air as vapors or into the groundwater as solutes. Particularly interesting was the different approaches taken in the modeling and the assumptions which had to be made concerning diffusion coefficients and phase partitioning problems.

The symposium on lakes and oceans covered the areas of chemical pollutant transport in the ocean associated with sanitary treatment wastewater discharge, both for solutes and particles

(Eganhouse), transport of pollutants from the atmosphere (microlayers) and into the sediments, and mobilization of organic pollutants associated with dredging of an estuary (Thibodeaux).

The last session dealt with modeling of pollutant transport and partitioning but was particularly concerned with evaluations of human exposure to a variety of potential toxins.

The workshop ended with a short panel discussion involving the audience, on information needed to plan realistic multimedia programs. It became very evident from the discussion that a large gap in our knowledge is the interaction of toxins with humans and animals, and how to evaluate risk assessment based on inadequate information.

The volume presents 18 manuscripts that were submitted to be published in this Proceedings. The volume opens with an introduction to intermedia transport of pollutants by the Director of the NCITR, Yoram Cohen. Following this introduction are four major sections describing transport of pollutants from the atmosphere, transport of pollutants from soils to groundwaters, transport of pollutants from soils, lakes and oceans to the atmosphere, and multimedia transport of pollutants.

The editors want to express their gratitude to EPA for its continued support of the NCITR at UCLA and for its support in sponsoring this workshop.

February 1989 D.T. Allen, Y. Cohen and I. R. Kaplan

CONTENTS

INTRODUCTION

Multimedia and Intermedia Transport Modeling Concepts in Environmental Monitoring 3
Yoram Cohen

TRANSPORT OF POLLUTANTS FROM THE ATMOSPHERE

The Role of Atmospheric Deposition in Organic Contaminant Cycling in the Great Lakes 19
Steven J. Eisenreich, Wayne A. Wilford and William M. J. Strachan

Impact of Acid Rain on Lake Water Quality 41
Dean S. Jeffries

Distribution Pattern of Atmospheric H_2O_2 in Los Angeles and its Vicinity and its Controlling Role as An Oxidant of SO_2 53
Hiroshi Sakugawa and Isaac R. Kaplan

Determination of Dry Deposition of Atmospheric Aerosols for Point and Area Sources by Dual Tracer Method 73
Hilary H. Main and Sheldon K. Friedlander

Transport of Trace Elements to Man by Atmospheric Aerosols 85
Walter John

TRANSPORT OF POLLUTANTS FROM SOILS, LAKES AND OCEANS TO THE ATMOSPHERE

Multiphase Chemical Transport in Porous Media 93
Patrick A. Ryan and Yoram Cohen

Mathematical Modeling of Air Toxic Emissions from Landfill Sites 105
Christian Seigneur, Anthony Wegrecki, Shyam Nair and Douglas Longwell

Theoretical Chemodynamics Models for Predicting Volatile Emissions to Air from Dredged Material Disposal 121
Louis J. Thibodeaux

Diffusion Experiments in Soils and Their Implications on Modeling Transport ... 161
David R. Shonnard and Richard L. Bell

Emission Estimates for a High Viscosity Crude Oil Surface Impoundments: 1. Field Measurements for Heat Transfer Model Validation ... 167
Barbara J. Morrison and Richard L. Bell

TRANSPORT OF POLLUTANTS FROM SOILS TO GROUNDWATER

Solute Transport in Heterogenous Field Soils ... 177
Martinus Th. Van Genuchten and Peter J. Strouse

Contaminant Transport in the Subsurface: Sorption Equilibrium and the Role of Nonaqueous Phase Liquids ... 189
Dermont C. Bouchard

Transport and Degradation of Water-Soluble Creosote-Derived Compounds ... 213
Edward M. Godsy, Donald F. Goerlitz and Dunja Grbic-Galic

Subsurface Processes of Nonaqueous Phase Contaminants ... 237
Danny D. Reible and Tissa H. Illangasekare

MULTIMEDIA TRANSPORT OF POLLUTANTS

Multimedia Partitioning of Dioxin ... 257
Curtis C. Travis and Holly A. Hattemer-Frey

Modeling of the Uptake and Distribution of Organic Chemicals in Plants ... 269
Sally Paterson and Donald Mackay

Multiple Pathway Exposure Factors (PEFs) Associated with Multimedia Pollutants ... 283
Thomas E. McKone

Summary ... 293

Contributors ... 295

Index ... 297

INTRODUCTION

MULTIMEDIA AND INTERMEDIA TRANSPORT MODELING CONCEPTS IN ENVIRONMENTAL MONITORING

Yoram Cohen

Department of Chemical Engineering
and
National Center for Intermedia Transport Research
University of California, Los Angeles
Los Angeles, California 90024

ABSTRACT

Environmental pollution is a multimedia problem. Pollutants do not stay where they originate but are able to migrate across environmental phase boundaries and to disperse throughout our ecosystem. Consequently, environmental exposure and risk assessments must be multimedia based analyses. In this paper, some aspects of intermedia transport are reviewed in relation to multimedia transport modeling, environmental monitoring and exposure assessment. Finally, an assessment is presented of the key research needs in the area of pollutant fate and transport modeling.

INTRODUCTION

One of the ultimate purposes of environmental pollution control is to reduce potential or existing risks. This calls for the determination of health risks which are directly related to the exposure of biological receptors to environmental chemicals. Exposure, in turn, is a function of contaminant concentrations in various environmental media that affect both the receptor directly, as well as various components of the food chain. Pollutant concentrations can in principle be obtained by either pollutant fate and transport modeling or field measurements. The above tasks require an understanding of the complex physical, chemical, and biological processes that govern the movement of pollutants among the different environmental media.

It is now well accepted that pollutants which are released to the environment as the result of a variety of human-related activities (see Table 1) move across environmental boundaries and are therefore found in most media. For example, based on simple thermodynamic partitioning calculations one can demonstrate that about 30%-50% of known synthetic organic chemicals will likely be found in all media (i.e., air, water, soil) if released to the environment (Travis et al., 1987). The realization of the multimedia nature of environmental pollution is not new (Draggan et al., 1987; Cohen, 1986a, 1986d; and references therein). It has been recognized as the original motivation for the establishment of the United States Environmental Protection Agency (EPA). The Presidential Message at the founding of the Environmental Protection Agency (EPA) in 1970 recognized that, "Despite its complexity, for pollution control purposes the environment must be perceived as a single interrelated system....some pollutants--chemicals, radiation, pesticides --- appear in all media."

Although the multimedia approach to analyzing environmental risk and pollution control strategies is attractive, its implementation is hampered by major obstacles. First, the lack of specific multimedia data bases of pollutant concentrations prevents a clear demonstration of the multimedia approach and the validation of transport and exposure models used in multimedia analyses. Second, there is still a serious deficiency in our understanding of various intermedia transport processes (Cohen, 1987, 1986a).

TABLE 1

RELEASE OF CHEMICALS TO THE ENVIRONMENT+

To the atmosphere:
• Stack emissions during manufacturing and incineration processes
• Fugitive emissions form storage tanks, leaks, and waste disposal and treatment sites
• Losses during use and disposal
To water:
• Treatment of effluents at manufacturing and formulating plants
• Spills during manufacturing and distribution
• Losses during transportation
• Disposal after use
To soil:
• Direct applications of agricultural chemicals such as pesticides or fertilizers
• Land disposal for landfilling or cultivation operations
• Spills

+ adapted from Cohen (1986d).

Consequently, there are uncertainties in the predictions of many of the existing multimedia approaches that employ highly approximate treatment of intermedia transport processes.

Despite the above difficulties, the multimedia approach is the logical approach for a comprehensive and integrated environmental management (Chemical Engineering News, 1984; Conway, 1982; Cohen, 1986b, 1986d, 1987; Draggan et al., 1987; and references therein). At the research level, the multimedia approach is important in guiding research efforts on intermedia transport, exposure assessment and risk analysis. For example, multimedia analyses can be used to determine pollutant transport pathways that are least significant based on mass balance, exposure and risk assessment considerations. Also, multimedia analysis may establish which intermedia transport parameters are known to a sufficient accuracy for environmental applications and which parameters require further refinement.

SOME MULTIMEDIA AND INTERMEDIA EXAMPLES

It is now becoming apparent that intermedia transport of pollutants can lead to pollution problems and raise serious questions about monitoring strategies as well as the enforcement of environmental regulations. For example, sulfur dioxide can be transported over great distances, and can be chemically transformed to its acidic form when combined with water vapor and deposit to terrestrial and aquatic environments via rain scavenging (Walcek and Pruppacher, 1984a, 1984b). Polychlorinated biphenyls (PCBs) and polycyclic aromatic hydrocarbons are dispersed through the environment via wet and dry deposition even though the majority of the PCBs' mass is in the soil environment (Neely, 1977; Bjorseth and Lunde, 1979; Nibset and Sarofin, 1972). The above examples illustrate that in order to track the movement of sulfur dioxide (and its transformed products) and PCBs, a multimedia monitoring and modeling effort is required. Other examples of organic chemicals that have been detected in multiple environmental media include chemicals such as DDT (Cramer, 1973; Harrison et al., 1970; Woodwell et al., 1971), trichloroethylene (Cohen and Ryan, 1985), chloroform (Khalil et al., 1983), dioxins (Travis and Hattemer-Frey, 1988b), and lead (Huntzicker et al., 1975).

Intermedia transport problems can also result when a compound that is present in one medium or is even being treated to reduce its level in a given medium, may be released or transferred inadvertently in the course of the treatment process to other media. For example, it is now well documented that air stripping of organics from biological treatment units in Publicly Owned Treatment Works plants results in air emissions of compounds such as toluene, MIBK, chlorobenzene, benzene, perchloroethylene, TCE, chloroform, 1,1,1-trichloroethane (Dunovant et al., 1986). The IEMP Philadelphia study (EPA, 1986) in particular has clearly shown that POTWs can be a significant source of emissions of chloroform, percholoroethylene and other chlorinated halocarbons. In contrast with the above volatile compounds, chemicals such as anthracene tend to accumulate in the sludge of municipal water treatment plants. The incineration of sludges from POTW can lead to the emission of metals such as zinc, cadmium and lead. Considering that about $6\text{-}7 \times 10^6$ dry tons of sludge are produced in the U.S. annually with the projected increase to $10\text{-}14 \times 10^6$ dry tons/year, sludge incineration may present a serious air pollution problem (Loehr and Ward, 1987). The subsequent wet and dry deposition of particulates will then lead to the contamination of the aquatic and terrestrial environments. Thus, the question of which media should be regulated, whether it be the air or the water has been raised as well as the question of where should monitoring stations be set in order to best allow for an appropriate pollution abatement.

An example of the multimedia partitioning of particle bound organics was demonstrated by the Southeast Ohio River Valley Study (Ryan and Cohen, 1986). It was demonstrated that benzo(a)pyrene, which originates from automobile emissions and coal fired power plants, is found in the air, water and soil compartments. Both dry and wet deposition play an important role in affecting both the atmospheric concentrations of B(a)P as well as the accumulation of B(a)P in the soil and aquatic compartments. Unfortunately, a detailed understanding of the cycling of many particle-bound organics is hindered by the lack of fundamental information on gas-particle partitioning, heterogeneous atmospheric reactions of such chemicals, and precipitation scavenging, as well as dry deposition on to complex rough terrains.

The above examples serve to illustrate that when intermedia transport is significant (from the viewpoint of exposure and risk assessments), the determination of pollutant concentrations, via monitoring, in all media of interest is a formidable task. Therefore, monitoring activities must be coupled with models of transport and exposure as discussed in the following sections.

INTERMEDIA TRANSFERS AND EXPOSURE

The assessment of risks due to the exposure of a receptor (usually a biological receptor) to pollutants is generally determined from appropriate dose-response relations. In order to utilize dose-response relations to predict the expected response of a target receptor, one must determine the dose. The dose in turn is directly related to the exposure of the receptor to the given agent. The exposure, as discussed below, is a function, among other factors, of the pollutant concentration in various environmental media that affect the receptor directly or indirectly (Ott, 1985; Cothern et al., 1986; McKone, 1986a; McKone and Kastenberg, 1986).

Exposure is defined as the contact of a receptor with a chemical or physical agent. The measure of exposure is the amount of the agent (i.e., chemical contaminant) available at the exchange boundaries (i.e., lungs, skin, intestinal tract) during a specified period of time. Exposure via a specific pathway, during a specific time interval Δt, can be defined by the following expression:

by $$E_i = \int_{t_o}^{t_o+\Delta t} I_i(t)dt \qquad (1)$$

in which I_i is the agent intake rate (or intensity of contact) of the given agent by the receptor. The intake rate I_i is expressed by

$$I_i = L_i C_i \qquad (2)$$

in which C_i is the concentration of the agent in compartment i in contact with the receptor and L_i is the extent of contact (e.g., inhalation rate is given as volume of air/unit time). The extent of contact, L_i, is obviously characteristic of the behavior of the receptor (i.e., its dynamics in the environment). For example, exposure that occurs through inhalation is a function of the time that the individual spends at various locations (indoors and outdoors) and the rate of inhalation, L_i. Such information can be obtained, for example, from population activity pattern studies. The concentration C_i in compartment i can be determined from either monitoring studies or from appropriate fate and transport models. Recently, personal monitoring has been developed to the state where exposure via inhalation, for example, can be determined by the direct determination of I_i (Wallace et al., 1986; Wallace, 1987). Once the exposure is known the actual dose delivered to particular organs (defined as the amount of the agent available for interaction with metabolic processes in a specific organ following exposure and absorption into the receptor) can be determined using an appropriate pharmacokinetic model. The dose is often related to exposure by a simple relationship

$$D_j = \sum_{i=1}^{N} E_i F_{ij} \qquad (3)$$

in which F_{ij} is an absorption factor associated with the absorption by organ j attributed to exposure pathway i. Detailed evaluation of the absorption factors requires either experimental data or prediction using the appropriate pharmacokinetic model. Clearly, as Eq. 3 indicates, the dose through its dependence on exposure is affected by dynamic processes of pollutant fate and transport, population activity patterns, and physiological characteristics of the receptor.

The exposure of the population to various pollutants can occur via three major exposure routes, namely; inhalation; dermal absorption; and ingestion. The ingestion pathway refers to both the consumption of food and drinking of water and other liquids. The intake of contaminants via food consumption is particularly significant since contaminants may accumulate in the food chain (Travis and Arms, 1987, 1988; Vaughn, 1984; Bell and MacLeod, 1983; McKone and Ryan, 1988). Thus, exposure can be strongly affected by multimedia transfers. Monitoring of the most significant exposure pathways may, for example, require monitoring the environmental compartment of lowest concentration but in which the exposure of the human population and subsequent bioaccumulation are significant.

While direct exposure of the human population (e.g., via inhalation of outdoor air) is often of prime concern (Abott et al., 1978), the exposure to contaminants via the food chain necessitates the evaluation of contaminant accumulation in the food chain. For example, the ingestion of beef may lead to exposure of chemical contaminants that have bioaccumulated within cattle. A simple illustration of the above process is possible if one regards the receptor as an environmental compartment; then a simple mass balance on the receptor (e.g., fish or cattle) compartments can be written as

$$\frac{dV_jC_j}{dt} = \sum_{i=1}^{N} L_iC_{in} - \sum_{k=1}^{M} Ex_kC_k - k_jC_jV_j, \qquad j=1,......m \tag{4}$$

in which V_j is the volume of the receptor or a component of the receptor (i.e., the cow or its milk) and C_j and C_{in} are the concentrations of the contaminant in the receptor (e.g., fish) and the media directly associated with the exposure route (e.g., concentration in the water phase), respectively. L_i represents the intake rate (or extent of contact associated with a given medium, e.g., the rate of water intake or inhalation rate) and Ex_k is the outflow stream associated with a particular elimination pathway (e.g., water outflow from the gills of a fish, urine, or fecal excretion). Finally, the last term represents the overall biochemical transformation of the given contaminant via first order reaction kinetics with a rate constant given by k_j. The above is only meant for an illustrative purpose since, in general, the metabolic rate may be a much more complex function compared to the last term in Eq. 4. It should be recognized that the parameters L_i, Ex_k, and k_j are time dependent. In order to determine the total body burden of a given contaminant, one has to resort to appropriate pharmacokinetic models. Thus, we could write Eq. 4 for j=1,...m where j would represent the particular body organ in which the contaminant accumulates. Alternatively, in order to estimate the degree of bioaccumulation (and avoid the use of pharmacokinetic models), one can make use of the common assumption that "equilibrium" exists between the outflow streams (e.g., urine) and the receptor's bulk. Accordingly, one may express this partitioning by

$$C_k = H_{kj}C_j \tag{5}$$

in which H_{kj} is a partition coefficient between the receptor j and outflow (or elimination) stream k. Thus, Eq. 4 reduces to

$$\frac{dV_jC_j}{dt} = \sum_{i=1}^{N} L_iC_{in} - C_j\left[\sum_{k=1}^{M} Ex_kH_{kj} - k_jV_j\right] \tag{6}$$

At steady state, Eq. 6 reduces to give the following expression for the so called biotransfer factor (Travis, 1988; McKone and Ryan, 1988),

$$B_t = C_j/\left(\sum_{i=1}^{N} L_iC_{in}\right) = \left[\sum_{k=1}^{M} Ex_kH_{kj} - k_jV_j\right]^{-1} \tag{7}$$

For example, one can define a biotransfer factor for the concentration of a contaminant in cow's milk where ΣL_iC_{in} is the contaminant intake via grazing, water drinking, and inhalation. In general, B_t is not a constant since V_j, the volume of the receptor, varies with time, and the partition coefficients H_{kj}, and k_j are also probably functions of the activity of the receptor, temperature, etc.; thus, H_{kj}, and the reaction rate constant k_j are time dependent. While the use of the concept of the biotransformation factor may be adequate, under some conditions, for very rough approximate calculations (to within an order of magnitude accuracy or worse; see McKone and Ryan, 1988), it may lead to significant errors in risk calculations, especially during periods where the receptor is undergoing rapid physiological changes (e.g., children whose metabolic activity and weight changes rapidly with age). The danger of using the concept of the biotransformation factor, which should be regarded as a transport parameter, is apparent in the recent attempts to correlate it with the octanol-water partition coefficient, K_{ow}, as if it was a physicochemical property of the contaminant under consideration (Travis and Arms, 1988; McKone and Ryan, 1988).

In an effort to obtain exposure estimates the concept of a bioaccumulation partition coefficient has also emerged. The bioaccumulation factor K_b is defined as (Kenaga and Goring, 1980; Kenega, 1980; MacKay, 1982)

$$K_b = C_b/C_i = B_t[\sum_{i=1}^{N} L_i C_{in}]/C_i \tag{8}$$

in which C_i is the concentration in the environmental compartment in which the exposure takes place (e.g., for fish it is the water compartment), and C_b is the concentration of the contaminant in the receptor. The above approach assumes implicitly that K_b is time invariant. For biota, for example, K_b is often correlated with the octanol-water partition coefficient, K_{ow} (Kenega and Goring, 1980; Kenega, 1980; MacKay, 1982; Veith and Rosian, 1983). The foundation for the above K_b-K_{ow}-type correlations is in the assumption that bioaccumulation is restricted primarily to the lipid phase of the receptor and the partitioning between water and octanol is characteristic of the partitioning between water and the lipid phase. The above concept is most valuable for receptors that are exposed to a single environmental compartment or that their exposure to other compartments can be easily linked to the major compartment where the receptor is found. The above approach also presents an interesting aspect of monitoring. It suggests that if K_b is indeed time invariant, then, by monitoring the contaminant concentration in the receptor, one can infer via appropriate models of multimedia partitioning as to the concentrations of the contaminant in the compartment where exposure has taken place. Such an approach, based on wildlife monitoring, was recently reviewed by Clark, et al. (1985), and Sandhu and Lower (1989).

The above discussion of exposure assessment suggests that in order to properly determine exposure one must know the contaminant concentration in the media which are in contact with the receptor (including food), population activity patterns, as well as the specific characteristic of the receptor (age, rate of growth during the exposure period, metabolic activity, etc.). While pollutant concentrations can be assessed via both monitoring and modeling, the coupling with population activity patterns especially when indoor pollution is considered is a difficult but necessary task (Ott, 1985).

The complexity of determining exposure has suggested the use of personal monitoring as demonstrated in the TEAMS study (Wallace et al., 1986; Wallace, 1987). Other approaches are based on combining environmental monitoring or modeling with information on population activity patterns to determine exposure (Ott, 1985). Regardless of the approach used to estimate exposure to the chemical of interest one must consider the fact that contaminants can cross environmental phase boundaries. Thus, in recent years, significant effort has been devoted to the development of intermedia and multimedia transport models designed to provide an integrated view of pollutant partitioning in the environment.

More recently, there have also been suggestions to use plants as sensors of the level of contamination in the environment (Gaggi et. al., 1985; Calamari, 1987). Although the process of pollutant uptake and accumulation by plants is a complex process (Mackay and Paterson, 1988), future effort in this area will undoubtedly add to our understanding of exposure and environmental monitoring.

MULTIMEDIA TRANSPORT MODELS

Multimedia fate and transport models vary in their level of complexity. Tables 2 and 3 list some of the major multimedia models that were developed over the last several years. There are three major classes of models as listed below:

i. Multimedia-compartmental (MCM) models
ii. Spatial-multimedia (SM) models.
iii. Spatial-multimedia-compartmental (SMCM) models

The need for simple multimedia screening level models led to the development of the compartmental multimedia modeling approach (Cohen, 1981, 1986c, 1987; Gillet at al., 1974). In this approach, all the environmental media under consideration are assumed to be well mixed. In these models, estimated concentrations in the air and water compartments can be reasonably approximated for

TABLE 2

MULTIMEDIA TRANSPORT MODELS - COMPARTMENTAL MODELS

Model	Reference
* Kinetic	Wiersma, 1979
* Fugacity	MacKay and co-workers 1979-1987
* MCM	Cohen and co-workers 1981-1987
* GEOTOX	McKone, 1985; McKone and Kastenberg, 1986
* RATE	Bolten et al., 1983

TABLE 3

SPATIAL MULTIMEDIA MODELS

Model	Reference
* UTM-TOX	Browman et al., 1982; Patterson et al., 1984; Patterson, 1986
* TOX-SCREEN	McDowell and Hetrick, 1982
* RAPS	Whelan et al., 1987
* MCEA	Onishi et al., 1982
* ALWAS	Tucker et al., 1982, 1984

cases involving area or dispersed atmospheric emissions (Cohen and Ryan, 1985; Ryan and Cohen, 1986). The soil environment, however, is highly non-uniform, so estimates from a compartmental model are likely to be inaccurate. Compartmental models have been introduced in different levels of complexity depending on whether convection (advection) streams among compartments, transformations, time dependent sources are included in the model formulation (Cohen, 1986c, 1987).

There are two categories of spatial multimedia models: (i) partially coupled, integrated multimedia models; and, (ii) composite multimedia models. The partially coupled models consist of single-medium models that are solved sequentially. Output from different sub-modules (e.g., air, water) is shared and managed usually by a central executive program. Since the modules are fixed by the model developer, the general applicability of the integrated model is limited. The partially coupled modeling approach as implemented by the RAPS methodology (Whelan, 1987) is useful for screening purposes and prioritization of chemical waste sites. The UTM-TOX model (Patterson, 1986) and the ALWAS model (Tucker et al., 1982) are examples of comprehensive multimedia models which consist of integrated single-medium models.

Composite multimedia models consist of individual pathway models selected by the user. The individual components are not connected. The user may, however, run the models sequentially and use the output file from a given pathway as part of the output file for the subsequent pathway. The above approach does not allow for specific feedback among the different environmental media. The user may, however, analyze the individual modules and allow for manual feedback, at least as a first level of an approximation. The composite multimedia models require extensive input data, user expertise, and computer resources. They can, however, be adapted to handle site specific problems. Examples of composite models include the Land Disposal Restriction (LDR) methodology (EPA, 1986), the chemical migration risk assessment (CMRA) methodology (Onishi et al., 1981a, 1985; Whelan and Parkhurst, 1983), the multimedia contaminant environmental exposure assessment (MCEA) methodology (Onishi et al., 1982a,b; Whelan and Onishi, 1983). Detailed reviews of the above multimedia models have been published by Cohen (1986c, 1987, 1988) and Onishi et al. (1988).

More recently, Cohen and co-workers have introduced the spatial-multimedia-compartmental (SMCM) model (Cohen, 1988; Mayer, 1988). The SMCM model treats the soil and sediment compartments as non-uniform compartments. Additionally, this model includes runoff, precipitation scavenging, soil drying, and temperature dependence of various model parameters. In the SMCM model partial and ordinary differential equations that describe the different compartments are solved simultaneously with the appropriate boundary conditions. The SMCM model has an improved resolution compared to the simple compartmental and Fugacity-type models as well as a capability of describing a variety of intermedia processes.

The variety of models described above can serve for a multitude of objectives. They can be used to aid in exposure assessment and risk analyses by providing information regarding pollutant concentrations. Another area of importance is the role of models in relation to pollutant monitoring.

ENVIRONMENTAL MONITORING

Environmental monitoring is carried out for a variety of reasons. Monitoring can help us understand the fate and transport of pollutants, to evaluate the state of the environment, to assist in exposure and risk assessment (Behar, 1979), and to test various theoretical transport models. Monitoring is often accomplished as required by environmental regulations. Most of the existing regulations however, are intended to regulate a single medium without explicit regard to the interaction between media.

The integration of monitoring and modeling activities is imperative for successful exposure and risk assessment programs. The potential interaction between model output/input and its relation to monitoring is outlined in Table 4. For example, predicted concentration levels can aid in the selection of the appropriate analytical technique, while predictions of spatial and temporal concentration profiles may suggest the appropriate monitoring strategy regarding the location and number of sampling sites as well as the sampling frequency. Another important interaction is that of pathway analysis. Undoubtedly, the recurrsive application of monitoring and modeling can aid in the design of monitoring strategies and the improvement of theoretical models of transport and exposure.

TABLE 4

THE POTENTIAL CONNECTION BETWEEN MODELING AND MONITORING

Model Output	Relation to Monitoring
Concentration levels	Selection of analytical technique
Concentration field	Location and number of sampling sites
Temporal concentration profiles	Frequency of sampling
Fluxes	Sampling method and frequency of sampling
Pathway analysis	Design of monitoring strategy

CONCLUSIONS

Various aspects of the need for a national pollution control program which account for problems arising from multimedia transfers have been described in the recent NRC report on "Cross Media Approaches to Pollution Control" (National Academy Press, 1987) and the recent report on "Geochemical and Hydrologic Processes and their Protection" (Draggan et al., 1987). While there are some major legislative and administrative issues facing the establishment of a multimedia pollution control program, there are still some scientific questions that have to be answered. It is imperative that multimedia pathways should be clearly identified and quantified and monitoring protocols and standard models are established. The proper design of such a program would have to rely on our ability to properly model and monitor the multimedia transport of chemicals in the environment and the associated multimedia exposures and the associated risk.

It is unlikely that a single comprehensive multimedia model (or a modeling approach) will be developed in the near future to predict the transport, transformations, and accumulation of every toxic chemical that has been introduced or will be introduced into the environment. Presently, the user may select from among the various classes of models described earlier in this paper. One should note that the spatial models require considerable meteorological and hydrologic input data bases, while the simpler compartmental and spatial-compartmental models have a relatively modest input requirement. The choice of the particular model for a particular application requires a careful demonstration of the required temporal and spatial scales, the complexity of the model in relation to available model input parameters, model interpretation, and validation by laboratory and field measurements.

Despite intensifying efforts in the areas of multimedia fate and transport modeling, and exposure assessment, little attention has been devoted to the verification of models through laboratory and/or field studies. This deficiency stems principally from the lack of adequate multimedia monitoring data. The coordination and amplification of both multimedia monitoring and modeling activities is essential in order to progress towards an improved assessment of the environmental impact of emerging new technologies, and hazardous waste treatment and disposal practices.

ACKNOWLEDGEMENTS

This completion of this paper was funded in part by the United States Environmental Protection Agency Grant CR-812771 and the University of California Toxic Substances Research and Teaching program.

REFERENCES

Abbott, T. J., M. J. Barcelona, W. H. White, S. K. Friedlander, and J. J. Morgan, "Human Dosage/Emission Source Relationships for Benzo(a)pyrene and Chloroform in the Los Angeles Basin", Special Report to US/EPA, Keck Laboratories, Caltech (1978).

Behar, J. V., E. A. Schuck, R. E. Stanley, and G. B. Morgan, "Integrated Exposure Assessment Monitoring," Environmental Science and Technology 13, 34-39 (1979).

Bell, D. and A. F. McLeod, "Dieldrin Pollution of a Human Food Chain", Human Toxicology, 2, 75 (1983).

Bjorseth, A., and G. Lunde, "Long-range Transport of Polycyclic Aromatic Hydrocarbons," Atmos. Environ. 13, 45 (1979).

Bolten, J. G., P. F. Morrison and K. A. Solomon, "Risk-Cost Assessment Methodology for Toxic Pollutants from Fossil Fuel Power Plants", Report R-2993-EPRI, The Rand Publication Series, Rand, Santa Monica, CA (June, 1983).

Bolten, J. G. , P. F. Morrison, K. A. Solomon and K. Wolf, "Alternative Models for Risk Assessment of Toxic Emissions", Report N-2261-EPRI, The Rand Publication Series, Rand Santa Monica, CA (April 1983).

Browman, M. G., M. R. Patterson and T. J. Sworski, "Formulation of the Physicochemical Processes in the ORNL Unified Transport Model for Toxicants (UTM-TOX) Interim Report", ORNL/TM-8013, Oak Ridge National Laboratory, Oak Ridge, Tennessee (1982).

Calamari, D., M. Vighi and E. Bacci, Chemosphere, 16, 2359-2364 (1987).

Chemical Engineering News, editorial, "Panel Suggests Management Changes at E.P.A.," June 4, 33 (1984).

Clark, T. K. Clark, S. Paterson, D. Mackay and R. J. Norstrom", Wildlife Monitoring, Modeling and Fugacity", Environ.Sci.Technol., 19, 880 (1985).

Cohen, Yoram, "Compartmental Modeling of Environmental Transport," Report 81-1, National Center for Intermedia Transport Research, University of California, Los Angeles, November (1981).

Cohen, Yoram and P. A. Ryan, "Multimedia Modeling of Environmental Transport: Trichloroethylene Test Case", Environ.Sci.Technol., 9, 412 (1985).

Cohen, Yoram, and P. A. Ryan, "Environmental Partition Coefficients," NCITR Report No. 84-2, National Center for Intermedia Transport Research, University of California, Los Angeles (1984).

Cohen, Yoram (Editor), Pollutants in A Multimedia Environment Plenum Press (1986a).

Cohen, Yoram, "Intermedia Transport Modeling in Multimedia Systems", in Pollutants in a Multimedia Environment, Cohen, Y. (Editor), Plenum Press (1986b).

Cohen, Yoram, "Organic Pollutant Transport", Environ.Sci.Technol., 20, 538 (1986c).

Cohen, Yoram, "Modeling of Pollutant Transport and Accumulation in a Multimedia Environment", in "Geochemical and Hydrologic Processes and Their Protection: The Agenda for Long Term Research and Development", Draggan, S., J. J. Cohrssen, and R. E. Morrison (Eds), Praeger Publishing Company, New York (1987).

Conway, R. A. (ed), "Environmental Risk Analysis for Chemicals", Van Norstand Reinhold Co., New York (1982).

Cothern, R., W. A. Coniglio and W. L. Marcus, "Estimating Risk to Human Health", Environ.Sci.Technol. 20, 111 (1986).

Cramer, J., "Model of the Circulation of DDT on Earth," Atmos. Environ. 7, (1973).

Draggan, S., J. J. Cohrssen, and R. E. Morrison (Eds), "Geochemical and Hydrologic Processes and Their Protection: The Agenda for Long Term Research and Development", Praeger Publishing Company, New York (1987).

Dunovant, V. S., C. S. Clark, S. S. Que Hee, V. S. Hertzberg and J. H. Trapys, "Volatile Organics in the Waste Water and Airspaces of Three Wastewater Treatment Plants", J.Water.Pollut.Control.Fed., 58, 886-895 (1986).

Gaggi, C., E. Bacci, D. Calamari, and R. Fanelli, "Chlorinated Hydrocarbons in Plant Foliage: An Indication of the Tropospheric Contamination Level", Chemosphere 14 (11/12), 1673-1686 (1985).

Gillett, J. W, J. Hill, A. Jarrinen and W. Schoor, "A Conceptual Model for the Movement of Pesticides through the Environment", EPA-600/3-74-024, U.S. Environmental Protection Agency, Ecological Research Series (1974).

Harrison, H. L., O. L. Loucks, J. W. Mitchell, D. F. Parkhurst, C. R. Tracy, D. G. Watts, and Y. L. Yannacone, Jr., "Systems Studies of DDT Transport," Science 170, 502 (1970).

Huntzicker, J. J., S. K. Friedlander, and C. I. Davidson, "Material Balance for Automobile Emitted Lead in the Los Angeles Basin," Environ. Sci. Technol. 9, 448 (1975).

Kenaga, G. E., and C. A. I. Goring, "Relationship Between Water Solubility, Soil Sorption, Octanol-Water Partitioning and Concentration of Chemicals in Biota," in Aquatic Toxicology, ASTM Stp 707, J. G. Eaton, P. R. Parrish and H. C. Hendrick (eds.), pp. 78-115, American Society for Testing and Materials, Philadelphia, PA (1980).

Kenega, E. E., "Correlation of Bioconcentration Factors of Chemicals in Aquatic and Terrestrial Organisms with their Physical and Chemical Properties", Environ.Sci.Technol., 14, 553 (1980).

Khalil, M. A. K., R. A. Rasmussen and S. D. Hoyt, "Atmospheric Chloroform ($CHCl_3$): Ocean-air Exchange and Global Mass Balance," Tellus 35B, 266-74 (1983).

Loehr, R. and C. H. Ward, "Waste Treatment and Cross-Media Transfer of Pollutants", in "Cross-Media Approaches to Pollution Control", National Academy Press, Washington, D. C. (1987).

Mackay, D. "Correlation of Bioconcentration Factors", Environ.Sci.Technol., 6, 274 (1982).

Mackay, D., S. Paterson and M. Joy, "Applications of Fugacity Models to the Estimation of Chemical Distribution and Persistence in the Environment," in Fate of Chemicals in the Environment, R. L. Swann and A. Eshenroeder (eds.), ACS Symposium Series, No. 225, ACS, Washington, D.C. (1982).

Mackay, D. and Paterson, S. "Calculating Fugacity", Environ.Sci.Tecnol., 15, 106-113 (1981).

Mayer, J. G., "The Spatial Multimedia Compartmental Model (SMCM)", M.S. Thesis, Department of Chemical Engineering, University of California, Los Angeles (1988).

McDowell-Boyer, L. M. and D. M. Hetrick, "A Multimedia Screening-Level Model for Assessing the Potential Fate of Chemical Released to the Environment", Oak Ridge National Laboratory, O-RNL/TM8334 (1982), Oak Ridge, Tennessee.

McKone, T. E. and W. E. Kastenberg, "Application of Multimedia Pollutant Transport Models to Risk Analysis", in Pollutants in a Multimedia Environment, Y. Cohen (ed.), Plenum Press, New York (1986).

McKone, E. M. and D. W. Layton, "Screening the Potential Risks of Toxic Substances Using a Multimedia Compartment Model, Estimation of Human Exposure", Regulatory Toxicology and Pharmacology, 6, 359-380 (1986).

McKone, T. E. and P. B. Ryan, "Human Exposure to Chemicals through Food Chains: An Uncertainty Analysis", UCRL -99290, preprint, Lawrence Livermore National Laboratory, submitted to J.Environ.Sci.Technol. (1988).

National Academy Press, "Cross-Media Approaches to Pollution Control", Washington, D.C. (1987).

Neely, W. B., "A Material Balance Study of Polychlorinated Biphenyls in Lake Michigan," Science of the Total Environment 1, 117 (1977).

Nibset, C. T. and A. F. Sarofin, "Rates and Routes of Transport of PCBs in the Environment", Environ. Health Perspect. 1, 21-38 (1972).

Onishi, Y., S. M. Brown, A. R. Olsen and M. A. Parkhurst, "Chemical Migration and Assessment Methodology", in "Proceedings of the Conference on Environmental Engineering", American Society of Civil Engineers, Atlanta, Georgia, 1981.

Onishi, Y., G. Whelan, and R. L. Skaggs, "Development of a Multimedia Radionuclide Exposure Assessment Methodology for Low-Level Waste Management", Battelle Pacific Northwest Laboratory, PNL-3370, Richland, Washington (1982a).

Onishi, Y., G. Whelan and R. L. Skaggs, "Development of a Multimedia Radionuclide Exposure Model for Low-Level Management", PNL-3370, Pacific Northwest Laboratory, Richland, Washington (1982b).

Onishi, Y., S. B. Yabusaki, C. R. Cole, W. E. Davis and G. Whelan, "Multimedia Contaminant Environmental Exposure Assessment (MCEA) Methodology for Coal-Fired Power Plants", Vols. I and II, Battele, Pacific Northwest Laboratory, Richland, Washington (1982c).

Onishi, Y., A. R. Olsen, M. A. Parkhurst and G. Whelan, "Computer-Based Environmental Exposure and Risk Assessment Methodology for Hazardous Materials", J.Hazard.Mater. 10, 389-417 (1985).

Onishi, Y., L. Shuyler and Y. Cohen, "Multimedia Modeling of Toxic Chemicals", paper presented at the International Symposium on Water Quality Modeling of Agricultural Non-Point Sources, Utah State University, Logan, Utah, June 20-23, 1988.

Ott, W. R., "Total Human Exposure", Environ.Sci.Tech., 19, 880 (1985).

Paterson, S. and D. Mackay, "Modeling the Distribution of Organic Chemicals in Plants", paper presented at the NCITR Workshop on "Intermedia Pollutant Transport: Modeling and Field Measurements", Santa Monica, California, August 23-26, 1988.

Patterson, M., "Unified Transport Model for Organics", in "Pollutants in a Multimedia Environment", Cohen, Y. (Ed.), Plenum Press, New York (1986).

Patterson, M. R., T. J. Sworski, A. L. Sjoreen, M. G. Browman, C. C. Coutant, D. M. Hetrick, E. D. Murphy and R. J. Raridon, "A User's Manual for UTM-TOX: A Unified Transport Model", ORNL-6064, (1984). Oak Ridge National Laboratory, Oak Ridge, Tennessee.

Ryan, P. A. and Y. Cohen, "Multimedia Distribution of Particle-Bound Pollutants:Benzo(a)Pyrene Test Case", Chemosphere 15, 21-47 (1986).

Sandhu, S. S. and W. R. Lower, "In Situ Assessment of Genotoxic Hazards of Environmental Pollution", Toxic.Ind.Health 5, 73 (1989).

Travis, C. C., and A. D. Arms, "The Food Chain as a Source of Toxic Chemical Exposure", in Toxic Chemicals, Health and the Environment, L. B. Lave and A. C. Upton (eds.), The John Hopkins University Press, Baltimore, MD, 5, 95-113 (1987).

Travis, C. C., J. W. Dennison, and A. D. Arms, "The Nature and Extent of Multimedia Partitioning of Chemicals", Unpublished Report, Health and Safety Research Division, Oak Ridge National Laboratory, Oak Ridge Tennessee (1987).

Travis, C.C. and A. D. Arms, "Bioconcentration of Organics in Beef, Milk, and Vegetation", Environ.Sci.Technol., 22, 271 (1988a).

Travis, C. C. and H. A. Hattemer-Frey, "Multimedia Partitioning of Dioxin", presented at the NCITR workshop on "Intermedia Pollutant Transport: Modeling and Field Measurements", Santa Monica, August 23-26, 1988b.

Tucker, W. A., A. Q. Eschenroeder and G. A. Magill, "Air Land Water Analysis System (ALWAS): A Multi Media Model for Toxic Substances", EPA 600/S3 84 052/NTIS PB 84 171 743, U. S. Environmental Protection Agency, Athens, Georgia (1984).

Tucker, W. A., A. G. Eschenroeder and G. C. Magill, "Air, Land, Water Analysis System (ALWAS): A Multimedia Model for Assessing the Effect of Airborne Toxic Substances on Surface Quality", first draft report, prepared by Arthur D. Little, Inc. for Environmental Research Laboratory, EPA, Athens, G.A. (1982).

United States Environmental Protection Agency, "Guidelines for Estimating Exposures", Federal Register, 51, 34092-34054 (1988).

United States Environmental Protection Agency, "Hazardous Waste Managements System Land Disposal Restrictions; Proposed Rule.", Part III, 40 CFR (260) Fed.Reg. 1601-1766 (January 14, 1986).

United States Environmental Protection Agency, Report of the Philadelphia Integrated Environmental Management Project, 1986.

Vaughan, B. E., "State of Research: Environmental Pathways and Food Chain Transfer", Environ. Health Perspect. 54, 353-371 (1984).

Veith, G. and P. Kosian, "Estimating Bioconcentration Potential From Octanol/Water Partition Coefficient", in Physical Behaviour of PCBs in the Great Lakes, M. Simmons (ed.), Ann Arbor Science, Ann Arbor, Michigan 269-282 (1983).

Walcek, C. J. and H. R. Pruppacher, J. Atm.Chem., 1, 307 (1984a).

Walcek, C. J. and H. R. Pruppacher, J. Atm.Chem., 1, 269 (1984b).

Wallace, L., E. Pellizzari, L. Sheldon, T. Harwell, C. Sparacino, and H. Zelon, "The Total Exposure Methodology (TEAM) Study: Direct Measurements of Personal Exposures through Air and Water for 600 Residents of Several U. S. Cities", in "Pollution in a Multimedia Environment", Cohen, Y. (Ed), Plenum Press (1986).

Wallace, L. A., "The Total Exposure Assessment Methodology (TEAM) Study: Project Summary", U.S. Environmental Protection Agency, Washington, D. C., EPA/600/S6-87/002.

Whelan, G. and Y. Onishi, "In-Stream Contaminant Interaction and Transport", in "Proceedings of the Tenth International Symposium on Urban Hydrology, Hydraulics, and Sediment Control", University of Kentucky, Lexington, Kentucky, July 25-28, 1983.

Whelan, G. and M. A. Parkhurst, "Simulation of the Migration, Fate and Effects of Diazinon In-Stream", in "Proceedings of the D. B. Simons Symposium on Erosion and Sedimentation", Colorado State University, Fort Collins, Colorado, July 27-29, 1983.

Whelan, G., D. L. Strenge, J. G. Droppo, Jr., B. L. Steelman and J. W. Buck, "The Remedial Ection Priority System (RAPS): Mathematical Formulation", DOE/RL/87-09/PNL-6200, Pacific Northwest Laboratory, Richland, Washington (1987).

Wiersma, G. B. "Kinetics and Exposure Commitment Analyses of Lead Behavior in a Biosphere Reserve", Technical Report (1979) Monitoring and Assessment Research Center, Chelsea College, University of London.

Woodwell, G., P. Craig and H. Johnson, "DDT in the Biosphere: Where Does it Go?" Science 174, 1, 101 (1971).

TRANSPORT OF POLLUTANTS FROM THE ATMOSPHERE

THE ROLE OF ATMOSPHERIC DEPOSITION IN ORGANIC CONTAMINANT CYCLING IN LARGE LAKES

Steve J. Eisenreich, Wayne A. Willford and William M.J. Strachan

Environmental Engineering
Department of Civil and Mineral Engineering
122 CivMinE Bldg.
University of Minnesota
Minneapolis, MN 55455

GLNPO
U.S. Environmental Protection Agency
230 S. Dearborn St.
Chicago, IL 60604

NWRI
Canada Center for Inland Waters
P.O. Box 5050
Burlington, Onatrio
L7R 4A6

INTRODUCTION

The atmosphere is now recognized as an important contributor of anthropogenic organic compounds to oceanic[1-4] and freshwater ecosystems[5-10]. The Laurentian Great Lakes are recognized as being particularly susceptible to atmospheric inputs of organic contaminants because they are near and downwind of major urban and industrial centers, have large lake surface to basin area ratios, and have long water residence times[7]. The compounds studied most thus far are semivolatile organic chemicals (SOCs) including chlorinated hydrocarbons represented by polychlorinated biphenyls (PCBs), DDT and its metabolites, hexachlorobenzene (HCB), toxaphene, chlorinated cyclohexanes (HCHs), polychlorinated dioxins (PCDDs) and furans (PCDFs), and combustion related chemicals such as polycyclic aromatic hydrocarbons (PAHs). These compounds have vapor pressures, Henry's Law constants, emission sources and environmental persistence sufficient to guarantee an atmospheric pathway. Estimates of PCB input to Lake Superior[7,11] and Lake Michigan[12] indicate that atmospheric deposition represented the dominant input pathway compared to all sources for both lakes. Thus, atmospheric transport of SOCs from source regions and subsequent deposition to receptors is the most important pathway for distributing toxic chemicals worldwide. There is a growing body of evidence[3,13-18] that SOCs emitted in one location may be transported hundreds or thousands of km away before being removed from the atmosphere. Many of these chemicals enter the Great Lakes ecosystem, and are concentrated at the top of the food chain - fish, birds and humans. It is this concern that has moved the International Joint Commission (IJC), the body governing transboundary issues between the U.S. and Canada, and the U.S. Environmental Protection Agency and Environment Canada to address the air toxics issue relative to the Great Lakes.

In November, 1986, a workshop was held in Scarborough, Ontario attended by atmospheric experts from several countries to address the following questions:

1. What is the magnitude of atmospheric deposition to the Great Lakes for a list of 14 "critical" pollutants?
2. How important is atmospheric deposition relative to all sources?
3. What are the emission sources and strengths of important atmospheric contaminants?

Strachan and Eisenreich[19] prepared a working paper for the workshop which used the mass balance paradigm to assess the magnitude and importance of atmospheric deposition of the 14 "critical" pollutants to the Great Lakes. The workshop report was revised and published in 1988. Based on the results of this exercise, and the established need to modify the Great Lakes Atmospheric Deposition (GLAD) Network[20,21], the Great Lakes National Program (GLNP) of US EPA upgraded GLAD, and it is now being implemented. This paper will discuss the physical characteristics of the Great Lakes, briefly summarize atmospheric removal processes, present examples of the mass balance modeling effort performed by Strachan and Eisenreich[19], and describe the evolving research and monitoring network of GLNP aimed at assessing atmospheric inputs of toxic chemicals.

PHYSICAL CHARACTERISTICS OF THE LAURENTIAN GREAT LAKES

Table 1 lists the hydrologic and morphometric features of the Great Lakes. The Great Lakes and connecting channels/bays of the United States and Canada encompass more than 150,000 km^2 and represent nearly 20% of the earth's surface freshwater reserves. Lake Superior alone is the largest freshwater body on the planet in terms of surface area, and second to Lake Baikal in the Soviet Union in terms of volume. The Great Lakes provide drinking water, water for industrial purposes, and recreational/commercial fishing to nearly 60 million people. The surface areas of the lakes are sufficiently large compared to the drainage area of the basins that a significant if not dominant fraction of water is derived from precipitation falling directly on the lake surface. Water residence times range from 172 years for Lake Superior to 2.3 years for Lake Erie. However, 90% exchange times range up to several hundred years. Lake Superior is the deepest of the lakes having a maximum depth of 406 m and a mean depth of 149 m. Lake Erie is the shallowest of the lakes having a maximum depth of 60 m and a mean depth of 19 m. These differences illustrate that the processes occurring in deep, oligotrophic Lake Superior may be quantitatively different that those in shallow, warm, eutrophic Lake Erie.

ATMOSPHERIC DEPOSITION PROCESSES

Vapor - Particle Partitioning

Atmospheric organic chemicals exist in the vapor and particulate phases. The processes by which trace organic contaminants are removed from the atmosphere and the quantity ultimately deposited on the water or land surface depend on the distribution between the gas and particle phases[22]. Partitioning between the gas and particle phases depends on chemical vapor pressure, size, surface area and organic carbon (OC) content of the aerosol, and air temperature. The less volatile the compound, the higher the affinity for atmospheric particulate matter. Theoretical considerations[23,24] and laboratory and field measurements[7,11,25-27] indicate that PCBs, DDT, low molecular weight hydrocarbons and PAHs occur in the gas phase in "clean" or rural airsheds, while high MW PCBs, PCDDs and PCDFs[28] occur in the particle phases. Yamasaki et al.[25], Bidleman and Foreman[26] and McVeety and Hites[27] show that PAHs and many organochlorines are distributed on aerosol inversely proportional to temperature and sub-cooled liquid vapor pressure:

$$\log A(TSP)/F = m/T + b \quad (1)$$

where A and F are adsorbed (i.e., gas) and filter-bound (i.e., particle) concentrations, respectively, m and b are regression constants, TSP is total suspended particulates, and T is temperature (°K). Bidleman and Foreman[26] and Pankow[24] show that the above equation is identical to that derived by Junge[23] for physical adsorption of gases on aerosol.

TABLE 1. Hydrologic and morphometric features of the Great Lakes.

	Superior	Michigan	Huron	Erie	Ontario
Drainage Area (km^2)	127,700	118,100	133,900	58,790	70,700
Surface Area (km^2)	82,100	57,800	59,700	25,700	19,520
Mean Depth (m)	149	85	59	19	86
Maximum Depth (m)	406	281	228	60	244
Volume (km^3)	12,230	4,920	3,537	483	1,636
Mean Connecting Channel Flow (10^{10} m^3/yr)	------	-----	12	19	21
Tributary Inflow (10^{10} m^3/yr)	5.4	2.9	5.1	2.2	3.0
Mean Outflow (10^{10} m^3/yr)	7.1[a]	4.9[b]	18[c]	21[d]	25[e]
Annual Precipitation (m/yr)	0.76	0.79	0.76	0.84	0.89
Hydraulic Residence Time (yr)	172	100	20	2.3	6.5

[a]St. Mary's River [b]Mackinac Straits [c]St. Clair River/Detroit River
[d]Niagara River [e]St. Lawrence River

The gas-particle distribution is also related to the sub-cooled liquid vapor pressure for SOCs. At TSP concentrations of about 100 ug/m^3 and 20 °C, organic compounds having a vapor pressure $<10^{-5}$ torr occur predominantely in the particle phase. At colder temperatures or in higher TSP environments, a greater fraction of the atmospheric burden will likely occur in the particle phase.

Wet Deposition Processes

The mechanisms of chemical removal from the atmosphere are different for particle associated compounds than for gas phase compounds. The total extent of organic compound scavenging by falling raindrops may be given as:[29]

$$W_T = W_g(1-\phi) + W_p\phi \qquad (2)$$

where W_T, W_g and W_p are the total, gas and particle scavenging coefficients, respectively, and ϕ is the fraction of the total atmospheric concentration occurring in the particle phase.

An atmospheric gas attaining equilibrium with a falling raindrop is scavenged from the atmosphere inversely proportional to Henry's Law:

$$W_g = RT/H \qquad (3)$$

where R is the universal gas constant (atm m^3/mol °K), T is absolute temperature (°K), and H is Henry's Law constant (atm m^3/mol). Values of W_g vary with H and temperature. Field-measured values[29] varied from 3 to 10^5 demonstrating the dependence on H. In the absence of chemical reactions in the droplet, an atmospheric gas should attain equilibrium with a falling raindrop in about 10 m of fall.[29,31] The position of equilibrium defined by H is a function of temperature as it increases by about 2x for each 10 °C increase in temperature. Tateya et al.[32] reported that the temperature dependence on H for PCBs is ln H = 18.53 - 7868/T, where T is in °K. This corresponds to H = 3.8×10^{-4} atm m^3/mol at 25 °C to 9.4×10^{-5} atm m^3/mol at 10 °K. The temperature dependency of H suggets a significant difference in removal efficiency between summer and winter seasons above 20° N.

Precipitation scavenging of aerosols containing sorbed or imbedded organic compounds permit the calculation of surface fluxes:

$$F_P = W_P \times P \times C_P = W_P \times P \times C_T(\phi) \qquad (4)$$

where W_p is the particle scavenging coefficient, C_p is the concentration in the particle phase (mass/m^3), P is precipitation intensity or annual precipitation (m/yr), and C_T is the total atmospheric concentration of chemical (mass/m^3). In-cloud scavenging may produce W_p values of about 10^6 [33,34]. Depending on particle size, precipitation intensity and type of precipitation event, W_p may range from 10^3 to 10^6. Ligocki et al.[30] list-field-measured values of 10^4 to 10^5 for organic chemicals.

The total wet surface flux of organic compounds in the atmosphere may be estimated from:

$$F_{T,w} = W_g(1-\phi)C_T + W_pC_T \qquad (5)$$

For PCBs, many organochlorines and low MW PAHs, most of the atmospheric burden is in the gas phase (i.e., ϕ<0.5), but wet deposition may be dominated by the more efficient scavenging of the particle-bound chemical.

Dry Particle Deposition

The dry deposition of particle-bound species onto a receptor surface depends on the deposition layer, particle-size, and macro- and micrometeorology. In the simplest case, the flux of particles to a receptor surface may be given as:

$$F_p = v_{d,h} \times C_p \qquad (6)$$

where F_p is the surface flux of particle-bound chemical (mass m^2/t), $v_{d,h}$ is the deposition velocity (m/t) at a reference height, h, often taken as 10 m, and C_p is the chemical concentration in the atmosphere in the particle phase (mass/m^3). The $v_{d,h}$ is analogous to the inverse of the resistance to particle transfer to a surface through the turbulent and deposition layer. The resistance model is not unlike the two-layer dry deposition models presented by Slinn[34], Slinn and Slinn[35], and Georgi[36]. The two-layer model consists of a constant flux layer where particle transfer is dominated by turbulence, and a deposition layer where deposition is dominated by diffusion, interception, impaction and sedimentation. Deposition velocities estimated for oceanic conditions for 0.1 to 1.0 um particles are 0.01 to 0.1 cm/s at wind speeds of 5 to 20 m/s[36]. Deposition velocities for particles in the fine particle mode (0.1 to 2 um) are about 0.05 to 1.0 cm/s based on wind tunnel data. Field meteorological data suggest values for fine particles to be about 0.1 to 1.0 cm/s. Coarse particles (> 2 um) have deposition velocities > 2 cm/s. Field data suggest that $v_{d,h}$'s are about 0.1 to 1.0 cm/s.[19,22,27]

Precipitation scavenging of fine particles is usually more important than dry particle deposition away from emissions sources for many SOCs. Slinn[34] argues that wet and dry deposition velocities are about equal (1 cm/s) over inland lakes averaged annually for 0.1 to 2 um size particles. Over the ocean and perhaps large lakes, wet deposition is greater. For the Great Lakes where contaminants are associated with fine particles and in the gas phase, wet deposition is probably more important than dry particle deposition.[11,37]

Vapor Exchange at the Air-Water Interface

Transfer of organic gases across the air-water interface is often predicted from a two-film diffusion model.[38-40] In this model, the rate of gas transfer between the well-mixed air and water reservoirs across the stagnant gas and liquid films at the interface is assumed to be governed by molecular diffusion and is driven by the concentration gradient between the equilibrium concentrations at the interface and bulk reservoirs. For steady-state transfer, the gas flux is given by:

$$F_v = K_{OL}(C_{w,diss} - P_v/H) \qquad (7)$$

$$1/K_{OL} = 1/k_l + RT/Hk_g \qquad (8)$$

where K_{OL}, k_l and k_g are the overall, liquid and gas side mass transfer coefficients (m/d), H is Henry's Law constant (atm m^3/mol), R is the universal gas constant (82 x 10^{-6} atm m^3/mol °K), T is absolute temperature (°K), $C_{w,diss}$ is the dissolved solute concentration in water (mol/m^3), and P_v is the solute partial pressure in the atmosphere (atm). The volatilization/absorption rate may be controlled by resistance to mass transfer in the liquid phase, gas phase, or a combination of both. At typical values of k_l and k_g (20 and 2000 cm/hr, respectively), resistance to mass transfer occurs > 95% in the liquid phase for compounds having H >4.4 x 10^{-3} atm m^3/mol, and >95% in the gas phase for H <1.2 x 10^{-5} atm m^3/mol. For the range of compounds considered here, H ranges from 10^{-3} to 10^{-7} atm m^3/mol implying a full range of liquid and gas phase resistances.

Values of K_{OL} calculated for a suite of PCB congeners and PAHs by the technique of Mackay and Yuen[39] are 0.03 to 0.30 m/d. It is critical to determine the fugacity of the organic solute in the water and atmosphere to determine the direction of transfer and the magnitude of flux. The aqueous concentration of trace organic substances occurring in the ng/L range is extremely difficult. SOCs may also associate with colloidal organic matter in natural waters[41] complicating the operational separation. Our ability to estimate organic compound fluxes across the air-water interface in the Great Lakes and elsewhere remains limited.

ROLE OF ATMOSPHERIC DEPOSITION OF TOXIC CHEMICALS IN THE GREAT LAKES

Numerous attempts have been made to construct dynamic steady-state models SOCs in large lakes based on the mass balance construct[11,42-45]. Others have constructed input-output budgets for SOCs in large lakes to access the importance of external inputs and outputs and internal transformations and losses[11,12,27]. Recently, Strachan and Eisenreich[19] used the mass balance paradigm to construct input-output budgets for 14 critical pollutants in the Great Lakes. The questions addressed were: 1). What was the magnitude of atmospheric deposition of selected SOCs and trace metals to each of the lakes? and 2). How do atmospheric inputs compare to non-atmospheric inputs? The methodology and results of this study are presented in the next section.

Environmental Concentration Data

The "critical" pollutants selected for study by the International Joint Commission (IJC) included four trace elements - lead, mercury, cadmium and arsenic, and eleven SOCs including benzo[a]pyrene (B[a]P), polychlorinated biphenyls (PCBs), hexachlorbenzene (HCB), mirex, dieldrin, Lindane (-hexachlorcyclohexane), DDT, toxaphene, and polychlorinated dioxins and furans (PCdds and PCDFs). The model requires concentrations of each of these chemicals in each atmospheric, aquatic and sedimentary compartment in each of the Great Lakes, in addition to concentrations in each input and output source. A careful review of the published and "gray" literature indicated there were insufficient data to construct a mass balance. Even for chemicals that have received considerable attention in research and monitoring such as PCBs, many compartments for some lakes had no reliable concentration data. Input - output budgets were calculated only for PCBs, t-DDT, B[a]P, Mirex (Lake Ontario only) and lead. The concentrations of these compounds from which estimates of current burdens were calculated may be found in Strachan and Eisenreich[19].

Mass balance constructs will be presented here for PCBs, B[a]P, and lead. These chemicals or classes of chemicals (e.g., PCBs) were selected because they represent types of elements or chemicals thought to have significant atmospheric pathways, have sufficient environmental data available, and for which mass transfer coefficients could be reasonably estimated. Tables 2 through 4 present the concentrations of PCBs, B[a]P, and Pb in each environmental compartment for each lake based on a thorough evaluation of the literature that are used in the mass balance calculations. A question mark (?) next to a specific datum means that that value is an estimate based on similar data available elsewhere. Nonetheless, uncertainty in many of these data may be 50 to 100%.

TABLE 2. Polychlorinated Biphenyls (PCBs) in the Great Lakes.

LAKE/TRIBUTARY		CONCENTRATIONS WATER (ng/L)	SUSPENDED SOLIDS (ng/L)	SEDIMENT (μg/g)	AIR (ng/m^3)	RAIN (ng/L)
ONTARIO:	Lake	0.6	0.3	0.1	0.5	5.0
	Niagara River	----- 10	--------	---	---	---
	Tributaries	----- 10	--------	---	---	---
ERIE:	Lake	0.7 ?	0.3 ?	0.06	0.5	5.0
	Detroit River	----- 10	--------	---	---	---
	Tributaries	----- 20	--------	---	---	---
HURON:	Lake	0.7	0.3	0.1	0.5 ?	5.0
	St. Marys River	----- 0.6	--------	---	---	---
	Mackinaw Strait	----- 2.0	--------	---	---	---
	Tributaries	----- 1.0	--------	---	---	---
MICHIGAN:	Lake	1.4	0.6	0.2	0.5	5.0
	Tributaries	---- 10.0	--------	---	---	---
SUPERIOR:	Lake	0.4	0.2	0.03	0.5	5.0
	Tributaries	----- 1.0	--------	---	---	---

TABLE 3. Benzo[a]pyrene in the Great Lakes.

LAKE/TRIBUTARY		CONCENTRATIONS WATER (ng/L)	SUSPENDED SOLIDS (ng/L)	SEDIMENT (μg/g)	A I R (ng/m³)	RAIN (ng/L)
ONTARIO:	Lake	0.2 ?	0.1 ?	0.3	0.1 ?	1.0 ?
	Niagara River	----- 0.3 ? ------		---	---	---
	Tributaries	----- 1.0 ? ------		---	---	---
ERIE:	Lake	0.2	0.1	0.2	0.1 ?	1.0 ?
	Detroit River	----- 0.1 ? ------		---	---	---
	Tributaries	----- 1.0 ? ------		---	---	---
HURON:	Lake	0.07 ?	0.03?	0.2	0.1 ?	1.0 ?
	St. Marys River	----- 0.1 --------		---	---	---
	Mackinaw Strait	----- 1.0 --------		---	---	---
	Tributaries	----- 1.0 --------		---	---	---
MICHIGAN:	Lake	0.7	0.3	0.5	0.1	1.0
	Tributaries	----- 1.0 --------		---	---	---
SUPERIOR:	Lake	0.07	0.03	0.03	0.02	0.5
	Tributaries	----- 0.05 -------		---	---	---

TABLE 4. Total Lead in the Great Lakes.

LAKE/TRIBUTARY		CONCENTRATIONS WATER (ng/L)	SUSPENDED SOLIDS (ng/L)	SEDIMENT (μg/g)	A I R (ng/m³)	RAIN (ng/L)
ONTARIO:	Lake	300	100	100	75 ?	10000
	Niagara River	-----1000 --------		---	---	---
	Tributaries	------ 11 --------		---	---	---
ERIE:	Lake	750	250	100 ?	70 ?	8000
	Detroit River	-----1800 --------		---	---	---
	Tributaries	------ 10 ? ------		---	---	---
HURON:	Lake	150	50	70	50 ?	7000
	St. Marys River	----- 100 ? ------		---	---	---
	Mackinaw Strait	----- 200 ? ------		---	---	---
	Tributaries	----- 24 ? ------		---	---	---
MICHIGAN:	Lake	150	50	40	50	10000 ?
	Tributaries	----- 100 ? ------		---	---	---
SUPERIOR:	Lake	75	25	100	20	3000
	Tributaries	----- 50 ? ------		---	---	---

Input - Output (Mass Balance) Calculations

Inputs to the Great Lakes include tributary and connecting channel flows, wet and dry atmospheric deposition, direct discharge to the lakes from municipal and industrial sources, and groundwater inflows. No groundwater flow data and only limited recent direct discharge data are available for the compounds of interest; the latter, however, were included where available. Figure 1 portrays the contaminant flux pathways considered for the construction of a chemical mass budget for the lakes. Table 5 lists the simple arithmetic expressions used to calculate inputs and outputs. Inflow flux (F_i) was calculated as the product of tributary or connecting channel concentrations (C_i) and annual average volumetric flows (Q_i) listed in Table 1. Precipitation fluxes (F_w) were calculated as the product of total solute concentrations in rain ($C_{T,rain}$), annual precipitation and lake surface area. Dry particle deposition (F_d) was estimated as the product of pollutant concentration in the particle phase in the atmosphere ($C_{p,air}$), dry particle deposition velocity ($v_{d,h}$), lake surface area and the fraction of the year not raining (f_d; assumed = 0.9). The $v_{d,h}$ used in the flux calculations was 0.1 cm/s typical of values for 0.1 to 1.0 um size particles with wind speeds of about 2 to 10 m/s.[22,34,36]

Outputs from the Great Lakes include tributary and connecting channel flow, sedimentation, volatilization, and biological or chemical degradation. For this study, the latter degradation pathways are not explicitly noted as outputs and are considered to be insignificant compared to other pathways for the chemicals of interest. They are, however, implicitly included in other loss terms.

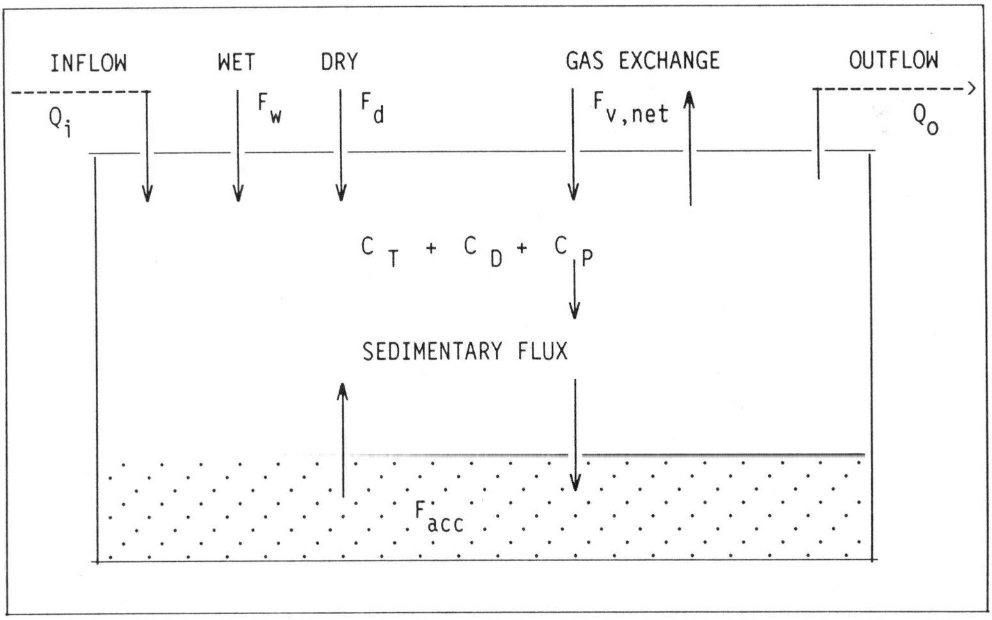

FIGURE 1. Mass balance paradigm (net atmospheric deposition).

TABLE 5. Mass balance framework: input-output calculations.

INPUTS	INFLOW:	$F_i = \Sigma\, C_i \bullet Q_i$ (Tributaries; connecting channels)
	ATMOSPHERE:	$F_w = C_{T,\ rain} \bullet P \bullet SA$
		$F_d = C_{p,\ air} \bullet V_{d,h} \bullet SA \bullet f_d$
OUTFLOW	OUTFLOW:	$F_o = C_{T,w} \bullet Q_o$ (Connecting channels)
	SEDIMENTATION:	$F_{acc} = C_{p,\ sed} \bullet W_{sed} \bullet SA \bullet f_{sed}$
	VOLATILIZATION:	$F_{v,\ net} = K_{OL} \{ C_{diss,\ w} - (C_{v,\ air} RT/H)\} \bullet SA \bullet f_v$
INFORMATION REQUIRED FOR INPUT - OUTPUT CALCULATIONS		
		Q_i; Q_o; SA; f_d; f_{sed}; f_v
		P; $V_{d,h}$; W_{sed}; K_{OL}
		$C_{T,\ rain}$; $C_{p,\ air}$; $C_{T,w}$; $C_{p,\ sed}$; $C_{diss,\ w}$; $C_{v,\ air}$

Outflow fluxes (F_o) were calculated as the total chemical concentration in the lake ($C_{T,w}$) multiplied by the annual average volumetric outflow ($Q_{o;}$ Table 1). Sedimentation fluxes due to incorporation in bottom sediments (F_{accum}) were calcualted from the concentration in surficial sediment ($C_{p,sed}$), the mass sedimentation rate (W_{sed}) and the fraction of the lake bottom accumulating sediment (f_{sed}). The last term is intended to account for the focusing of sediment into the deeper depositional basins for each lake. Table 6 lists the mass sedimentation rates used the mass balance calculations. The range of W_{sed} rates was 200 and 1000 $g/m^2/yr$ for Lake Superior and Lake Erie, respectively. These rates are lakewide averages based on tens of dated sediment cores taken mostly from the depositional basins. Chemical accumulation in sediments does not consider short-term or episodic resuspension events caused by seasonal forces.

TABLE 6. Sedimentation rates in the Great Lakes.

Lake	W_{sed} (g/m²/yr)	
Superior	200 (f_{sed} = 0.5)	(56) Kemp et al. 1978 (57) Bruland et al. 1975 (58) Evans et al. 1981
Michigan	400 (f_{sed} = 0.5)	(59) Edgington and Robbins, 1976 (60) Weininger et al. 1983
Huron	220 (f_{sed} = 0.5)	(61) Robbins 198 (62) Kemp and Harper, 1977 (63) Kemp et al. 1974
Erie	1000 (f_{sed} = 0.7)	(63) ibid. (64) Nriagu et al. 1979 (65) Thomas et al. 1976 Robbins, unpublished data
Ontario	400 (f_{sed} = 0.5)	(66) Kemp and Harper, 1976 Robbins, unpublished data (67) Durham and Oliver, 1984

f_{sed} = fractional area of lake bottom accumulating sediment.

TABLE 7. Volatilization parameters.

Compound	H atm m^3•mol (x 10^6)	K_{OL}(m/d)	$f_{diss,w}$	$f_{v,air}$
BaP	7.6	0.032	0.7	0.2
PCB (A-1252)	200	0.24	0.7	0.8
HCB	1300	0.31	0.7	0.7
Dieldrin	0.25	0.0012	0.7	0.9
δ-HCH	16	0.06	0.9	0.9
α-HCH	60	0.15	0.9	0.8
Toxaphene	1.7	0.0079	0.7	0.7
DDT	120	0.21	0.7	0.7
Mirex	790	0.30	0.5	0.2

$f_{diss,w}$ = fraction of total aqueous concentration in dissolved phase.
$f_{v,air}$ = fraction of total atmospheric concentration in vapour phase.

Net gas flux across the air-water interface ($F_{v,net}$) incorporates both the process of gas absorption and volatilization as discussed earlier. $F_{v,net}$ was calculated as the product of the overall mass transfer coefficient (K_{OL}), the concentration gradient, lake surface area and the fraction of total atmospheric concentration in the gas phase ($f_{v,air}$). Volatilization parameters are given in Table 7. The values of K_{OL} were estiamted from the parameterizations of Mackay and Yuen[39] which consider wind speed (5 m/s), temperature (15 °C), Henry's Law constant and Schmidt number. The fraction of chemical dissolved in the water phase and thus available for gas exchange ($f_{diss,w}$) was estimated from field studies[11,46-49] and laboratory experiments[50]. The fraction of the total atmospheric concentration in the gas phase (f_v) was estimated from studies by Junge[23], Yamasaki et al.[25], Pankow[24], and Bidleman[22].

The calculations of mass inputs and outputs needed to construct a mass budget for each chemical and lake was performed on a LOTUS 123 spreadsheet. A MACRO was written to obtain the individual mass fluxes into and out of the lakes. Table 8 portrays the format of the data output as it appears for all chemicals and lakes. The upper left-hand portion of the table lists the physical characteristics of the lake required to estimate inputs and outputs. The upper right-hand portion lists the chemical concentration in the various environmental compartments of the lake and the mass transfer coefficients used in flux calculations. The lower left-hand portion provides the individual input fluxes, and the middle bottom provides output fluxes. The lower right-hand columns present the estimated percentage of total lake inputs derived from wet and dry atmospheric deposition. For lakes Huron, Erie and Ontario, the percentage atmospheric contribution to total inputs is the sum of direct atmospheric wet and dry deposition to the lake surface and undirect atmospheric inputs derived from "upstream" deposition entering the lower lake via the connecting channel.

The residence time T_R of each chemical in each of the lakes is calculated as :

$$T_R = \text{Mass in Lake//Flux Out} \qquad (9)$$

$$= (C_{T,w} \times Vol_{lake})/\text{Total Flux Out}$$

Such residence times are rough estimates of the time in years required for the lake to recover from existing pollutant burdens given the processes depicted in the mass balance paradigm. At steady-state, T_R's calculated using input or output flux are equal.

Magnitude and Importance of Atmospheric Deposition

The review of the literature indicated that data on the concentrations of all but a few elements or compounds in compartments of the Great Lakes are dramatically lacking. As important as concentrations are to the mass balance construct, there is an even greater lack of information on mass transfer both between and within the aquatic and atmospheric compartments. The model is set up to easily change all parameters to evaluate model sensitivity. Table 9 presents the atmospheric component of the input - output calculations for PCBs, t-DDT, B[a]P, Mirex (Lake Ontario only) and lead fluxes for each of the Great Lakes. The input - output calculations indicate that the atmosphere is the dominant contributor of PCBs and DDT to the upper Great lakes of Superior, Michigan and Huron, and a relatively small contributor to the lower Great Lakes of Erie and Ontario.

TABLE 8. Great Lakes mass balance model calculations: PCBs.

LAKE SUPERIOR

Lake Parameters: Description	Symbol	Value	Chemical Parameters: Description	Symbol	Value
Tributary Inflow	Qtrib (m^3/yr)	5.4E+10	Tributary Conc.	Ctrib (ug/m^3)	1.0
Outflow from Lake	Qout (m^3/yr)	7.1E+10	Conc. in Connecting Channel to L.H.	Ccon (ug/m^3)	0.6
Surface Area	SA (m^2)	8.2E+10	Atm Vapor Conc.	Ca,v (ug/m^3)	4.00E-04
Lake Volume	V (m^3)	1.2E+13	Atm Particle Conc.	Ca,p (ug/m^3)	1.00E-04
Sedimentation Area	As (m^2)	4.1E+10	Total Rain Conc.	Cr (ug/m^3)	5.0
Resusp. Velocity	R (m)		Total Lake Conc.	Ct (ug/m^3)	0.6
Precipitation Rate	P (m/yr)	0.76	Dissolved Lake Conc.	Cd (ug/m^3)	0.4
Sedimentation Rate	Wsed (g/m^2-yr)	200	Lake Particle Conc.	Cp (ug/m^3)	0.2
Ice Cover Fraction	Icefrac	0.90	Surficial Sed. Conc.	Csed (ug/g)	0.03
Fraction of Year without Rain	f(1)	0.90	Atm Particle Dep Vel	Vd (m/yr)	3.2E+04
Fraction of Lake Accum. Sediments	f(sed)	0.5	Atm Part Washout Coef	Wo	0.00
			Atm/Water Mass Transfer Coef.	Kw (m/yr)	73.0
			Air/Water Distribution Coefficient	H/RT	0.0085

Flux into Lake Superior: Description	Symbol	Value	Flux out of Lake Superior: Description	Symbol	Value	% Atmospheric Contribution to Total Lake Input	Chemical Residence Time (yrs)
Tributary Flux	Ftrib (g/yr)	5.40E+04	Outflow from Lake	Fout (g/yr)	4.26E+04	90.4	3.3
Wet Deposition	Fa,w (g/yr)	3.12E+05	Sedimentation	Fsed (g/yr)	2.46E+05		
Dry Deposition	Fa,d (g/yr)	2.36E+05	Mass Transfer (Volatilization)	Fv (g/yr)	1.90E+06		
Other Loadings Unaccounted for Above (g/yr)							
Direct Wastewater Discharge		2.2E+03					
Direct Industrial Discharge		1.8E+03					
Total Flux In		6.06E+05	Total Flux Out		2.19E+06	Net Flux	-1.59E+06

TABLE 9 . Input-output calculations.

CHEMICAL/ Lake	INPUT kg yr^{-1}	% ATMOSPHERIC Direct	Total	OUTPUT kg yr^{-1}	% VOLATILIZATION of Total Output	NET FLUX kg yr^{-1}
PCBs						
Superior	606	90	90	2190	87	-1590
Michigan	685	58	58	7550	68	-6860
Huron	636	63	78	3400	75	-2760
Erie	2520	7	13	2390	46	130
Ontario	2540	6	7	1320	53	1220
t-DDT						
Superior	92.4	97	97	761	89	-669
Michigan	65.1	98	98	1070	44	-1000
Huron	91.7	72	97	794	62	-702
Erie	319	10	22	795	26	-476
Ontario	111	23	31	408	39	-296
Benzo[a]pyrene						
Superior	71.7	96	96	314	19	-242
Michigan	208	86	86	6250	6	-6050
Huron	290	63	80	1370	31	-1080
Erie	122	66	79	3720	2	-3600
Ontario	155	40	72	1290	3	-1130
Mirex						
Ontario	69.1	4.5	4.5	256	18	-187
Lead (10^3 kg yr^{-1})						
Superior	241	97	97	828	0	-587
Michigan	543	99.5	99.5	472	0	71
Huron	430	94	98	496	0	-66
Erie	567	39	46	2010	0	-1440
Ontario	426	50	73	490	0	-64

Figures 2-4 portray the direct and indirect atmospheric inputs as percentages of total inputs to the lakes. For example, the input calculations show that 90, 58 and 78% of the total PCB loading to Lakes Superior, Michigan and Huron, respectively, are derived from atmospheric deposition. In contrast, only 13 and 7% of total PCB inputs to Lakes Erie and Ontario are derived from atmospheric deposition. The primary reason for this difference between the upper and lower lakes is the large estimated loading of these chemicals to the connecting channels of the Detroit, St. Clair and Niagara Rivers

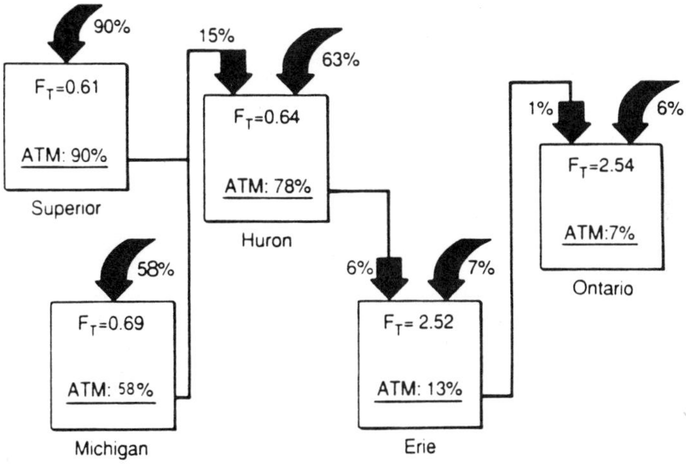

Units: 10^3 kg yr^{-1}

FIGURE 2. Atmospheric loading of PCBs to the Great Lakes.

by neighboring industrial dischargers, and leakage from chemical waste dump sites. Indirect atmospheric inputs to Lakes Huron, Erie and Ontario derived from "upstream" serve to increase the overall importance of total atmospheric inputs, typically up to 10%. For example, indirect atmospheric inputs increase the total contribution of PCB inputs from 68 to 78% for Lake Huron, but only from 6 to 7% for Lake Ontario. In contrast, an average of 81% of the B[a]P and 83% of the Pb loadings are derived from total atmospheric inputs to the lakes, most of which comes from direct atmospheric inputs. Only about 4.5% of the Mirex entering Lake Ontario is estimated to come from the atmosphere. Thus, the atmosphere represents an important if not dominant contributor of many chemicals to the Great Lakes when compared to inputs from all sources.

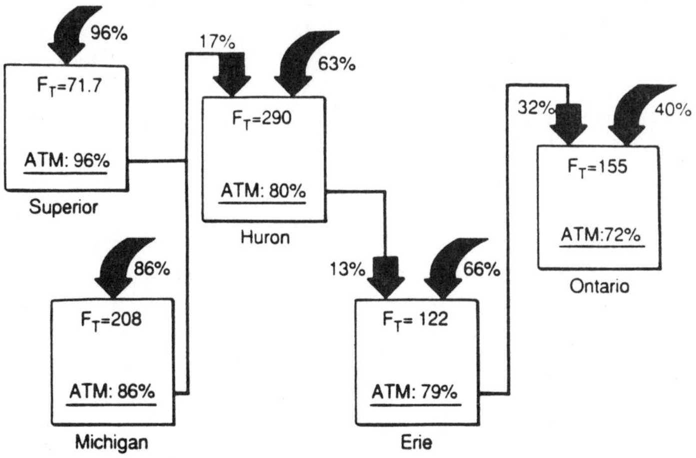

Units: kg yr^{-1}

FIGURE 3. Atmospheric loading of benzo[a]pyrene to the Great Lakes.

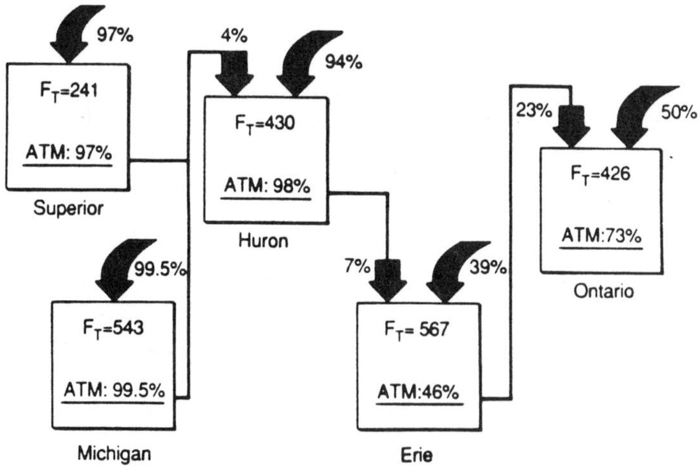

Units: 10^3kg yr^{-1}

FIGURE 4. Atmospheric loading of lead to the Great Lakes.

The ratio of wet:dry deposition for SOCs concentrated in the fine particles or gas phase is expected to be about 1.5-4.0:1. For the compounds of interest here, the following ratios were observed: PCBs, 1.3-1.5:1; Pb, 3.5-5.0:1; B[a]P, 0.3-0.5:1; t-DDT, 0.3-0.5:1. Except for PCBs and Pb, the wet:dry deposition ratios appear low. This suggests that either the estimates of wet deposition are too low, or more probably, that dry deposition values are too high.

Processes dominating chemical loss from the lakes are sedimentation and volatilization (Table 9). For PCBs and to a lesser extent, t-DDT, volatilization often exceeds sedimentation. For many chemicals and lakes, the net flux estimated from the mass budgets is <u>out</u> of the lake. That is, more chemical is leaving the lake than now entering, and should result in improved water and ecosystem quality. The observation that volatilization exceeds atmospheric and surface inputs for many chemicals, especially PCBs, supports the hypothesis that the Great Lakes are actively degassing organic contaminants deposited historically. The process of lake degassing is supported by modeling[11,37] and field studies[12]. Estimates of volatilization fluxes as shown in Table 9 are subject to large and unknown levels of uncertainty. The most appropriate strategy is to determine the atmospheric gas phase concentration of a specific SOC interacting with the aqueous dissolved, unassociated species. The difference between these concentrations should be linked to condition-specific mass transfer coefficients. Thus, field-measured chemical fugacities linked to mass transfer coefficients may be interpolated over various environmental conditions. In the mass balance model above, SOC concentrations were used as annual values in volatilization calculations. Although this approach provides uncertain results, it does serve to support the hypothesis that chemicals deposited in lakes over the last 40 years may be lost to the atmosphere by volatilization.

Eisenreich[11] and Mackay et al.[37] have described our present understanding of the relative importance of air-water transfer processes for SOCs in large lakes as follows. The dominant input pathway is the scavenging of particles from the atmosphere containing sorbed chemical species, and to a lesser extent, gas-phase compounds. Dry particle deposition appears to be much less important except near large sources. Chemicals on particles and dissolved in rain reaching the lake equilibrate with the aqueous environment and are distributed between the dissolved phase and biotic particles or detritus. SOCs in the dissolved phase seek to equilbrate with the atmospheric gas phase. Thus the major input is through precipitation scavenging of particles; the major losses are through volatilization and sedimentation. Conceivably measurements of SOC flux to the atmosphere may be more than offset by atmospheric inputs on particles. The overall process may be described as a dynamic, nonequilibrium process whereby the input is connected to loss via volatilization by the equilibrium partitioning of SOCs. This model suggests that volatilizing SOCs contribute to overlake concentrations, but to the extent of <10%. SOCs may cycle between water and air with intermittent periods of intense deposition followed by slower but prolonged volatilization. Sedimentary SOCs seasonally resuspended contribute to water column concentrations and losses to the atmosphere.[47,51,52]

GREAT LAKES ATMOSPHERIC DEPOSITION MONITORING AND RESEARCH PROGRAM

The Great Lakes National Program Office (GLNPO) is responsible for coordinating activities undertaken by the U.S. EPA to restore and maintain the water quality of the Great Lakes. In order to effectively address its coordination responsibilities and set priorities for Great Lakes remedial activities, GLNPO initiates and oversees numerous research and monitoring programs directed at identifying and quantifying pollutant inputs to the Great Lakes Basin. Since the atmosphere is recognized as an important contributor of anthropogenic pollutants to the Great Lakes, the GLNPO has modified its existing network to include many toxic chemicals. This portion of the paper will briefly describe the exisiting monitoring network and the evolution of the new research and monitoring network.

Great Lakes Atmospheric Deposition Network (GLAD)

Initiated in 1976, the purpose of the GLAD network was to provide atmospheric P loadings to Lake Erie. In 1981, this effort was expanded to monitor the deposition of a variety of nutrients and trace metals to the Great Lakes. The expansion was in part due to studies showing phosphorus[53,54] and metals[55] had significant atmopsheric pathways. The network grew to 36 monitoring sites on the U.S. side of the Great Lakes representing industrial, agricultural and urban source areas (Figure 5).

In 1985, GLNPO began evaluating the ability of the GLAD network to meet the evolving goals of the atmospheric deposition program. This evaluation brought to light several deficiencies, especially that GLAD as then constituted was not adequate to provide data on airborne loadings of trace pollutants to the Great Lakes. Of equal importance was the lack of coordination between the Canadian and U.S. efforts in this area, and the poor siting of many monitoring stations.. In light of these deficiencies, Environment Canada and the U.S. EPA decided to collaborate on the formation of a binational network having as its objectives: 1.) To determine the portion of total loadings of critical toxic pollutants contributed by atmospheric deposition; 2.) To recommend the extent to which additional remedial programs and international activities are needed to control atmospheric sources; and 3.) To provide source information for immediate regulatory action.

New GLAD Network

The proposed GLAD network will consist of atmospheric deposition research, monitoring and modeling activities designed to identify and quantify atmospheric inputs of toxic pollutants to the Great Lakes. The initial list of pollutants identified as important to include in initial monitoring activities include the metals As, Cd and Hg; the pesticides chlordane, DDT and metabolites, Dieldrin, mirex and toxaphene, other chlorinated compounds such as polychlorinated biphenyls (PCBs), and the combustion-related polycyclic aromatic hydrocarbons (PAHs). In addition, special research studies will be performed to assess the atmospheric pathway of PCDDs and PCDFs.

The proposed GLAD network will consist of two distinct types of atmospheric deposition monitoring sites: master sites and routine sites. The five master sites proposed for the Great Lakes will focus nontraditional, applied research activities aimed at supporting the scientific basis for assessing atmospheric loadings to the Great Lakes from the routine sites, as well as to provide quality assurance support. Instrumentation and sampling equipment at master sites will provide for collection and analysis of contaminants in rain, snow, air particulates and atmosphere. For those parameters being analyzed, a detailed annual analysis and interpretation of the data will be required. This involves calculation of annual loadings for each parameter via each precipitation type, evaluation of wet versus dry inputs, evaluation of the efficicy of various sampling equipment utilized, and determination of the relative importance of total atmospheric inputs to the region of the master site compared to all sources. Research activities focused at the site will investigate air-water exchange of toxic organic compounds, an evaluation of routine site location criteria, and analysis of new or previously undetected contaminants in precipitation.

Proposed instrumentation for each master site include three wet-only, integrating precipitation samplers capable of collecting SOCs, three directionally-operated high volume air samplers equipped with backup adsorbents, a Nipher snow gauge, one wet-only, precipitation sampler for nutrients, one high volume sampler for TSP and organic carbon measurements, two cascade impactors (metals and organics), and a 10 m meteorological tower to collect continuous data on wind speed, wind direction, humidity, temperature, precipitation intensity and solar radiation.

The modified GLAD network will also include 12 routine sites whose purpose is to conduct weekly/biweekly atmospheric deposition activities in support of deposition model calibration efforts and pollutant mass balance inventory development. The routine sites will be equipped with three wet-only, integrating precipitation samplers (nutrients, metals and toxic organics), a Nipher snow gauge, three high volume air samplers for organics and TSP/OC, one cascade impactor, and a meteorological tower equipped as above. In general, routine sites will be located in urban areas, on islands in lakes, and in areas removed from nearby sources. Data collected at urban sites will be used to define deposition gradients surrounding urban areas, and to quantify the contribution of urban emission sources to Great Lakes atmospheric deposition.

Participation by Environment Canada in the GLAD network will greatly enahnce the validity and utility of network data. The resulting joint US-Canada atmospheric deposition monitoring data bases will anable the derivation of more accurate estimates of lakewide pollutant loadings to each Great Lake. The IJC has recently recommmended the establishment of a joint, binational atmospheric deposition network under the amendments to the Water Quality Agreement. Canada is presently in the early stages of network implementation.

The cost of establishing and operating master and routine monitoring sites throughout the Great Lakes Basin is considerable. Costs (US Only) are projected at $406K, $562K, $772K, $813K and $752K for FY 1987 - FY 1991, respectively. Master sites will cost about $56K for one-time setup and about $110K per year for operating costs. The setup costs for a routine site are about $42K, with annual operating costs at $36K. Network implemetation schedule calls for establishing master sites in Lake Michigan in 1987, Lake Superior in 1989 and Lake Ontario in 1990 (US only). By 1991, there will be 12 routine sites established on the U.S. side of the Great Lakes; two on Lake Superior, four on Lake Michigan, and two each on Lakes Huron, Erie and Ontario. It is anticipated that the master site on Lake Ontario will be shared with Canada.

In 1987, a master site was established on lower Green Bay, Lake Michigan. In 1988, two routine sites were established on upper Green Bay. These sites are now generating atmospheric deposition data. In Canada, a master site was established at Point Petre on Lake Ontario in 1988. Discussions continue via the International Joint Commission on integrating binational activities of the evolving atmospheric deposition network.

SUMMARY

For more than two decades researchers have investigated the relative importance of the atmosphere as a long range carrier of persistent toxic chemicals. Large quantities of these compounds have been deposited into the Great Lakes. The processes responsible for input are wet and dry deposition and gas exchange at the air-water interface. Recent model calculations and field studies indicate that the atmosphere is an important if not dominant contributor of many trace organic and inorganic pollutants to the Great Lakes compared to all sources. Distant from emission sources such as urban and industrial centers, scavenging of atmospheric chemicals by rain is often the most important input pathway. Losses of organic chemicals from the lakes is dominated by sedimentation and volatilization with the latter often the most important. As a result of this phenomenon, a new Great Lakes atmospheric deposition network is evolving and includes both master (research) and routine (monitoring) sites. By 1991, this network will have 2.5 master sites and 12 routine sites deployed on the U.S. side of the Great Lakes, with a corresponding number in Canada. The GLAD network will provide the magnitude of atmospheric inputs of toxic chemicals to the Great Lakes, provide information on emission sources and direct management activities.

REFERENCES

1. Bidleman, T.F.; Olney, C.E. Science, **183**, 517-518 (1974).
2. Atlas, E.; Giam, C.S. Science, **211**, 163-165 (1981).
3. **ibid.**, In **The Role of Air-Sea Exchange in Geochemical Cycling**, P. Buat-Menard (eds.), NATO ASI Series No. 185, D. Reidel Publishing Co.: Dordrecht, 295-329 (1985).
4. Tanabe, S.; Hidaka, H.; Tatsukawa, R. Chemosphere, **12**, 277-288 (1983).
5. Murphy, T.J.; Rzeszutko, C.P. J. Great Lakes Res. 3£,305-312 (1977).
6. Doskey, P.V.; Andren, A.W. Environ. Sci. Tech., **15**, 705-711 (1981).
7. Eisenreich, S.J.; Looney, B.B.; Thornton, J.D. Environ. Sci. Tech., **15**, 30-38 (1981).
8. Murphy, T.J. In Toxic Contaminants in the Great Lakes, J.O. Nriagu; M.S. Simmons (Eds.), John Wiley and Sons, Inc.: New York, 53-79 (1984).
9. Strachan, W.M.J.; Huneault, H. J. Great Lakes Res., **5**, 61-68 (1979).
10. Strachan, W.M.J. Environ. Toxic. Chem. **4**, 677-683 (1985).
11. Eisenreich, S.J. In Sources and Fates of Aquatic Pollutants, R.A. Hites; S.J. Eisenreich (Eds.), Adv. Chem Series No. 216, American Chemical Society: Washington, D.C., 393-469 (1987).
12. Swackhamer, D.L.; Armstrong, D.E., Environ. Sci. tech., **20**, 879-883 (1986).
13. Rapaport, R.A.; Urban, N.R.; Baker, J.E.; Looney, B.B.; Gorham, E. Chemosphere, **14**, 1167-1173 (1985).
14. Rapaport, R.A.; Eisenreich, S.J. Atmos. Environ., **20**, 2367 (1986).
15. **ibid.**, Environ. Sci. Tech., **22**, 931-941 (1988).
16. Rice, C.P.; Sampson, P.J.; Noguchi, G. U.S. EPA Report, Atmospheric Transport of Toxaphene to Lake Michigan, EPA 600/S3-84-101.
17. Csuczwa, J.M.; Hites, R.A. Environ. Sci. Tech. **18**, 444-450 (1984).

18. **ibid.**, Environ. Sci. Tech., **20**, 195-200 (1986).
19. Strachan, W.M.J.; Eisenreich, S.J. Mass Balance Modeling of Toxic Chemicals in the Great Lakes: the Role of Atmospheric Deposition, Appendix I, International Joint Commission: Windsor, Ontario, 113 p., May (1988).
20. Eisenreich, S.J., Report on Atmospheric Deposition Workshop on Organic Contaminant Deposition to the Great Lakes Basin, US EPA No. R-005879-0 (1986).
21. Murphy, T.J. Design of a Great Lakes Atmospheric Inputs and Sources Network, EPA Report, EPA-905/4-87-001, 35 p. (1987).
22. Bidleman, T.F., Environ. Sci. Tech., **22**, 361-367 (1988).
23. Junge, C.E. In Fate of Pollutants in the Air and Water Environments, I.H. Suffet (Ed.), John Wiley and Sons New York, 7-25 (1978).
24. Pankow, J.F., Atmos. Environ., **21**, 2275-2283 (1987).
25. Yamasaki, H.; Kuwata, K.; Miyamoto, H., Environ. Sci. Tech., **16**, 189-194 (1982).
26. Bidleman, T.F.; Foreman, W.T. In Sources and Fates of Aquatic Pollutants, Adv. Chem. Ser. No. 216, Amer. Chem. Soc.:Washington, D.C., 27-56 (1987).
27. McVeety, B.D.; Hites, R.A., Atmos. Environ., **22**, 511-536 (1988).
28. Eitzer, B.; Hites, R.A., J. Environ. Anal. Qual., **27**, 215-230 (1987).
29. Ligocki, M.P.; Leuenberger, C.; Pankow, J.F., Atmos. Environ., **19**, 1609-1617 (1985a).
30. **ibid.**, Atmos. Environ., **19**, 1619-1616 (1985b).
31. Slinn W.G.N. et al., Atmos. Environ., **12**, 2055-2087 (1978).
32. Tateya, S.; Tanabe, S.; Tatsukawa, R. In Toxic Contamination in Large Lakes, N.W. Schmidtke (Ed.), Lewis Publishers: Ann Arbor, MI, Vol. III, 237-281 (1988).
33. Scott, B.C. In Atmospheric Pollutants in Natural Waters, S.J. Eisenreich (Ed.), Ann Arbor Science Publishers: Ann Arbor, MI, 3-22 (1981).
34. Slinn, W.G.N. In Air-Sea Exchange of Gases and Particles, P.S. Liss; W.G.N. Slinn (Eds.), NATO ASI Series, D. Reidel Publishing: Boston, MA, 299-406 (1983).
35. Slinn, S.A.; Slinn, W.G.N. Atmos. Environ., **14**, 1013-1016 (1980).
36. Giorgi, F. J. Geophys. Res., **91**, 9794-9806 (1986).
37. Mackay, D.; Paterson, S.; Schroeder, W.H. Environ. Sci. Tech., **20**, 810-816 (1986).
38. Liss, P.S.; Slater, P.G. Nature, **247**, 181-184 (1974).
39. Mackay, D.; Yuen, A.T.K. Environ. Sci. Tech., **17**, 211-217 (1983).
40. Liss, P.S. In Air-Sea Exchange of Gases and Particles, P.S. Liss; W.G.N. Slinn (Eds.), NATO ASI Ser., D. Reidel Publishing: Boston, MA (1983).
41. Gschwend, P.M.; Wu, S.C., Environ. Sci. Tech., **19**, 90-96 (1985).
42. Bierman, V.J.; Swain, W.R. Environ. Sci. Tech., **16**, 572-579 (1982).
43. Rogers, P.W.; Swain, W.R. J. Great Lakes Res., **9**, 548-558 (1983).
44. Thomann, R.V.; DiToro, D., J. Great Lakes Res., **8**, 695-699 (1983).
45. Imboden, D.M.; Schwarzenbach, R.P. In Chemical Processes in Lakes, W. Stumm (Ed.), Wiley-Interscience: New York, 1-28 (1985).
46. Capel, P.D.; Eisenreich, S.J., J. Great Lakes Res., **11**, 447-461 (1985).
47. Baker, J.E.; Eisenreich, S.J.; Johnson, T.C.; Halfman, B.M. Environ. Sci. tech., **17**, 854-861 (1985).
48. Baker, J.E.; Capel, P.D.; Eisenreich, S.J., Environ. Sci. Tech., **18**, 1136-1143 (1986).
49. Swackhamer, D.L.; Armstrong, D.E., J. Great Lakes Res., **13**, 24-36 (1987).
50. Karickhoff, S.W. J. Hydraul. Div. (ASCE), **110**, 707-735 (1984).

51. Eadie, B.J.; Chambers, R.L.; Gardner, W.S.; Bell, G.L., J. Great lakes Res., **10**, 307-321 (1984).
52. Baker, J.E.; Eisenreich, S.J. J. Great Lakes Res., in press (1989).
53. Murphy, T.J.; Doskey, P.V., J. Great Lakes res., **2**, 60-70 (1976).
54. Eisenreich S.J.; Emmling, P.J.; Beeton, A.M., J. Great Lakes Res., **3**, 291-304 (1977).
55. Eisenreich, S.J., Water, Air and Soil Poll., **13**, 287-301 (1980).
56. Kemp, A.L.W.; Dell, C.I.; Harper, N.S., J. Great Lakes Res., **4**, 276-287 (1978).
57. Bruland, K.W.; Koide, M.; Bowser, C.; Maher, L.J.; Goldberg, E.D., Quatern. Res., **5**, 89-98 (1975).
58. Evans, J.E.; Johnson, T.C.; Alexander, E.C., Jr.; Lively, R.S.; Eisenreich, S.J., J. Great Lakes Res., **7**, 299-310 (1981).
59. Edgington, D.N.; Robbins, J.A., Environ. Sci. Tech., **10**, 266-274 (1976).
60. Weininger, D.; Armstrong, D.E.; Swackhamer, D.L., In Physical Behavior of PCBs in the Great Lakes, D. Mackay; S. Paterson, S.J. Eisenreich; M.S. Simmons (Eds.), Ann Arbor Science: Ann Arbor, MI, 423-440 (1983).
61. Robbins, J.A. Sediments of Southern Lake Huron: Elemental Composition and Accumulation Rtaes, U.S. EPA Report EPA-600/3-80-080, 309 p. (1980).
62. Kemp, A.L.W.; Harper, N.S., J. Great Lakes Res., **3**, 215-220 (1977).
63. Kemp, A.L.W.; Anderson, T.W.; Thomas, R.L.; Mudrochova, A., J. Sed. Petrol., **44**, 207-218 (1974).
64. Nriagu, J.O.; Kemp, A.L.W.; Wong, H.K.T.; Harper, N., Geochim. Cosmochim. Acta, **43**, 247-258 (1979).
65. Thomas, R.L.; Jaquet, J.M.; Kemp, A.L.W.; Lewis, C.F.M., J. Fish. Res. Bd. Canada, **33**, 385-403 (1976).
66. Kemp, A.L.W.; Harper, N.S., J. Great lakes Res., **2**, 324-340 (1976).
67. Durham, R.W.; Oliver, B.G., **9**, 160-168 (1983).

IMPACT OF ACID RAIN ON LAKE WATER

QUALITY IN EASTERN CANADA

Dean S. Jeffries

National Water Research Institute
Canada Centre for Inland Waters
867 Lakeshore Road - P.O. Box 5050
Burlington, Ontario, Canada L7R 4A6

ABSTRACT

Forty-six percent of Canada is classified as having aquatic ecosystems that are highly sensitive to acidic deposition. The existing effect of acidic SO_4 deposition on surface waters has been evaluated for eastern Canada (east of the Ontario-Manitoba provincial border) by using a lake chemistry data base that contains chemical information for 7403 lakes.

Distribution statistics for alkalinity (ANC), sulphate, and base cation (Ca+Mg) concentrations are presented for 8 subregions which show that the Maritime provinces contain the highest proportion of acidic lakes even though the highest deposition (and lakewater sulphate concentrations) occur in Ontario and Quebec. Consideration of ion ratios (ANC:Ca+Mg, SO_4:Ca+Mg, and ANC:SO_4) show that the geographical pattern of acidification is a function of both deposition and terrain sensitivity (defined in terms of the terrain's ability to supply alkalinity and/or base cations to its water systems). The ratios provide evidence that a large area of south-central Ontario and southern Quebec has been acidified. The most dramatic evidence of acidification occurs in southern Nova Scotia and a small part of New Brunswick. Northwestern Ontario and Labrador show little evidence of the acidification effect.

Some of the data bases are now temporally long enough to demonstrate measured long term changes in lake chemistry. Depending on location, both continued lake acidification and "recovery" have been observed. Differences in chemical behavior are primarily related to whether or not the base cation export from a lake's terrestrial basin changed significantly in response to elevate SO_4 deposition.

INTRODUCTION

Canada contains approximately 4,000,000 km^2 (46% of the country) classified as having aquatic ecosystems that are highly sensitive to acidic deposition (Environment Canada, 1988). In central and eastern Canada, the area having this classification roughly coincides with the Precambrian geological formations collectively known as the Canadian Shield. It is primarily composed of non-carbonate bedrock overlain by glacially derived, coarse textured, shallow soils, and is characterized by many thousands of lakes. Wet sulphate deposition is maximum (30-40 kg $SO_4.ha^{-1}.yr^{-1}$)

in southern Ontario, southern Quebec, and southern New Brunswick and decreases northward, and eastward; areas receiving <10 $kg.ha^{-1}.yr^{-1}$ are not considered adversely affected by acid rain. For practical purposes, the location of the 10 $kg.ha^{-1}.yr^{-1}$ deposition isopleth is approximated by the 52° N latitude. Kelso et al. (1986) estimate that the area east of Manitoba and south of 52° (only a small portion of eastern Canada) contains 700,000 lakes corresponding to a surface area of 160,000 km^2 (excluding the Great Lakes).

The coincidence of a large area having little ability to neutralize acid inputs and existing elevated acidic deposition underlies the strong environmental concern now being expressed in Canada. This paper will present a brief overview of the chemical effect that acid rain has had on eastern Canadian surface waters. Several more comprehensive reviews of the impacts of acid rain on Canadian waters already exist (Harvey et al., 1981; Canada/US, 1983; RMCC, 1986; Cook et al., 1988).

REGIONAL CHEMISTRY OF LAKES

The data base used here to describe the water quality of eastern Canadian lakes is the same as employed by Jeffries (1986), an extended version of that presented by Jeffries et al. (1986). Schindler (1988) has also used the data in presenting an example of a geochemical acidification effect in his recent general review paper. Briefly, the data base is a compilation of several smaller federal and provincial government data sets and contains information for 7,403 lakes of which 4,895 are located in Ontario.

The data have been separated into 8 subregions as shown in Figure 1, most of which are defined by provincial boundaries. The Ontario subregions are the exception: northwest Ontario (NW ONT) was arbitrarily separated from northeastern Ontario (NE ONT) at 85° 20' W longitude, and NE ONT was separated from south-central Ontario (SC ONT) at 45° 50' N latitude.

It should be noted that all of the Labrador (LAB) subregion lies north of 52° N and so lakes in this area generally receive <10 $kg.ha^{-1}.yr^{-1}$ wet SO_4; this is also true of many lakes in NW ONT even though they are commonly south of 52° N. The Quebec (QUE) subregion is extremely large and experiences a wide gradient in deposition. Other subregions, particularly NE ONT, New Brunswick (NB), and Nova Scotia (NS) are geologically complex containing sensitive and insensitive terrain; NE ONT also contains the major point source emitter at Sudbury.

Lakewater ions most strongly influenced by acidic deposition are alkalinity (=ANC), SO_4, and base cations (approximated here as Ca+Mg). Table 1 gives the 5th, 50th (i.e. median), and 95th percentile concentrations ($\mu eq.L^{-1}$) for ANC, and Ca+Mg for the 8 Canadian subregions. ANC and Ca+Mg concentrations are generally greatest in Ontario due to the presence of small amounts of $CaCO_3$ in the glacial tills of the region. The quantity of $CaCO_3$ gradually increases northward towards the source area in the Hudson Bay Lowlands (Shilts, 1981) which accounts for the exceptionally high concentrations in the upper tail of the data distributions in Ontario. Negative ANC in the low tail of the NE ONT distribution reflects the significant number of acidic lakes occurring near Sudbury. Carbonate material is not present in the tills of most of Quebec and the Maritime provinces, and therefore, ANC and Ca+Mg are consistently lower than in Ontario. Nova Scotia (NS) has the lowest median ANC and Ca+Mg of all the subregions; over 30% of its lakes are currently acidic, i.e. ANC<0 (Jeffries et al., 1986). Labrador (LAB) is the region furthest downwind from the major source areas in the central part of the continent;

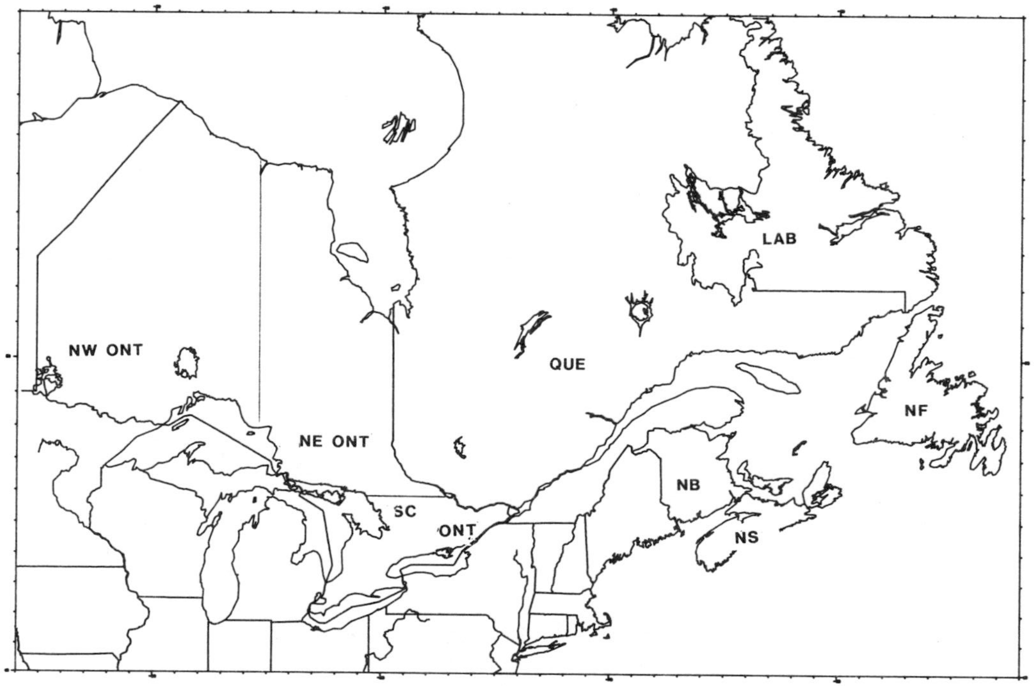

Figure 1. Subregions of eastern Canada used for analysis of lake survey data.

Table 1. Fifth, fiftieth, and ninety-fifth percentiles of lake water concentrations of ANC SO_4, and Ca+Mg ($\mu eq.L^{-1}$) for subregions of Eastern Canada (see Figure 1).

SUBREGION	n*	ALKALINITY			SULPHATE			CALCIUM + MAGNESIUM		
		5th	50th	95th	5th	50th	95th	5th	50th	95th
NW ONT	216-1078	46	254	1738	33	75	121	81	221	2093
NE ONT	793-1805	-11	165	2194	33	148	282	143	441	2623
SC ONT	705-1578	5	63	1331	105	158	201	126	199	447
QUE	1268-1452	4	52	536	22	93	186	65	163	713
LAB	180-203	14	46	166	9	24	43	31	69	245
NB	52-81	- 2	38	185	47	77	98	43	90	208
NS	275-413	-30	10	122	12	50	106	11	53	191
NF	176-268	- 0	28	244	10	31	55	21	79	461

*range in n depends on ion considered.

its median ANC concentration (46 $\mu eq.L^{-1}$) approximates a reasonable background value.

The distribution of SO_4 among subregions is also given in Table 1. The greatest median concentrations occur in SC ONT, NE ONT, and QUE where the deposition is also highest. In fact, the region to region pattern of median concentration closely follows the expected level of deposition across all eastern Canada. Hence, background SO_4 (i.e. in LAB) is approximately 24 $\mu eq.L^{-1}$ in agreement with the analysis of Wright (1983). High SO_4 in the upper tail for NE ONT once again reflects the influence of the major local source at Sudbury.

ION RATIOS

Wright (1983) used lake chemistry data from geographically diverse areas of North America to refine Henriksen's "acidification model." The model predicts that lake water SO_4, when elevated in concentration due to atmospheric deposition, is balanced by a decrease in ANC and/or an increase in Ca+Mg. Therefore consideration of ratios involving these 3 chemical parameters provides a simple tool for evaluating the extent of surface water acidification.

The dynamic interrelationship that exists among these ions means that, in response to elevated SO_4 deposition, the ratio ANC:Ca+Mg will deviate downwards from a "background" value of approximately 1. One definition of terrain sensitivity is its ability to supply base cations to surface waters through weathering or ion exchange reactions. Sensitive terrain has low Ca+Mg leading to a "sensitive" ratio (i.e. one that changes significantly in response to elevated acidic deposition) since Ca+Mg falls in the denominator. Conversely, if the terrain is geochemically insensitive due to a ready supply of ANC and Ca+Mg from carbonate bedrock or the like, the ratio will maintain a near background value independent of deposition. Naturally occurring organic acids may influence the ANC:Ca+Mg ratio, most likely by reducing the ANC concentration and thereby decreasing the ratio from the ideal unperturbed value of approximately one. The organic anion concentration (A^-) of waters is easily estimated if dissolved organic carbon (DOC) and pH are know (Oliver et al., 1983), and by assuming that the A^- content is independent of lake acidity and reflects an equivalent reduction in ANC, it is possible to "correct" the ratio. The importance of considering A^- is greatest when terrain (and ratio) sensitivity is high. Unfortunately, a paucity of DOC data precludes direct evaluation of the effect of A^- in this manner.

The SO_4:Ca+Mg ratio will ideally approach but not reach zero in areas of background deposition since background SO_4 concentrations are approximately 20 - 40 $\mu eq.L^{-1}$ (Wright, 1983; Jeffries et al., 1986); similarly, it will approach unity as SO_4 replaces ANC or A^- in high deposition areas. The ratio can in fact exceed unity when ANC becomes negative. Given that A^- more likely effects a reduction in background ANC rather than an increase in Ca+Mg, the SO_4:Ca+Mg ratio is relatively less affected by the presence of organic acids. Once again, since Ca+Mg is in the denominator of the ratio, the ratio's sensitivity will reflect the terrain's ability to supply base cations to its water systems.

The Alk:SO_4 ratio provides a direct indication of the SO_4 for ANC replacement that occurs during the acidification process. It will range from very large values (>10) when ANC is abundant (i.e. in relatively insensitive terrain) and SO_4 is low (i.e. low deposition areas) to values slightly <0 when ANC is negative. Important ratio values are 1.0 occurring when ANC = SO_4, 0.5 when SO_4 is 2-fold greater than ANC, and 0.2 when SO_4 is 5-fold greater than ANC.

The spatial variation for ratio values across eastern Canada is presented in Figures 2-4 (see also the ANC:Ca+Mg map presented by Schindler, 1988). The maps were prepared by segregating the data into a 20 X 20 minute latitude and longitude grid. The median ratio value for each grid square was determined and then the map manually contoured. The maps all show that the most dramatic indication of acidification occurs in southern NS and a small part of NB, principally due to terrain sensitivity rather than strongly elevated deposition. The fact that both the SO_4:Ca+Mg ratio and the ANC:Ca+Mg ratio exhibit much the same pattern gives credence to the conclusion of Jeffries et al. (1986) that A^- cannot explain on its own, the occurrence of many acidic lake systems in this area.

The maps also demonstrate that large areas of SC ONT, NE ONT and southwestern QUE exhibit strong evidence of lake acidification. The shape of the contour patterns are easily explained by the combined influences of geology (terrain sensitivity) and deposition intensity. The southern portion of the island of Newfoundland exhibits a gradient of acidification from highest in the south to lowest in the north, presumably reflecting the similar gradient in deposition. Finally, all the maps show that NW ONT and LAB have experienced minimal acidification to date.

AQUATIC RESOURCES AT RISK

By using steady state models to extrapolate from limited survey information to the aquatic resource present in eastern Canada south of 52°N, RMCC (1986) estimated that approximately 14,000 lakes are presently acidic ($pH < 5$), and that if current peak SO_4 deposition (36 $kg.ha^{-1}.yr^{-1}$) is maintained, 10,000 to 40,000 additional lakes will become acidic. Furthermore it also noted that the simulations were conservative, since the models did not account for a possible depletion of ANC within the terrestrial watersheds. While these estimates contain an unknown uncertainty, they do indicate the large magnitude of the freshwater resource that is at risk in eastern Canada.

RECOVERY OF AQUATIC SYSTEMS

An important test of acidification hypotheses and models is chemical recovery in response to decreased deposition. Emissions of SO_2 in North America have decreased from peak values in the 1970's. Furthermore, emission from the large point sources at Sudbury decreased >50% between 1973-78 and 1979-85. These reductions have resulted in several instances of water quality recovery, either increasing pH and ANC, or decreasing Ca+Mg. For example, Lazerte and Dillon (1984), Dillon et al. (1986), Hutchinson and Havas (1986), and Keller and Pitblado (1986) provide a detailed record of water quality improvements (generally increased pH and ANC) in response to deposition reductions. Other regions in Ontario not directly influenced by the Sudbury source also demonstrate decreased SO_4 and increased pH and ANC (e.g. the Algoma region west of Sudbury; Kelso and Jeffries, 1988). However, Dillon et al. (1987) showed that Plastic L. in SC ONT responded to decreased deposition by reducing base cation yield from the terrestrial catchment rather than increasing pH and ANC (also observed for certain cases by Kelso and Jeffries, 1988).

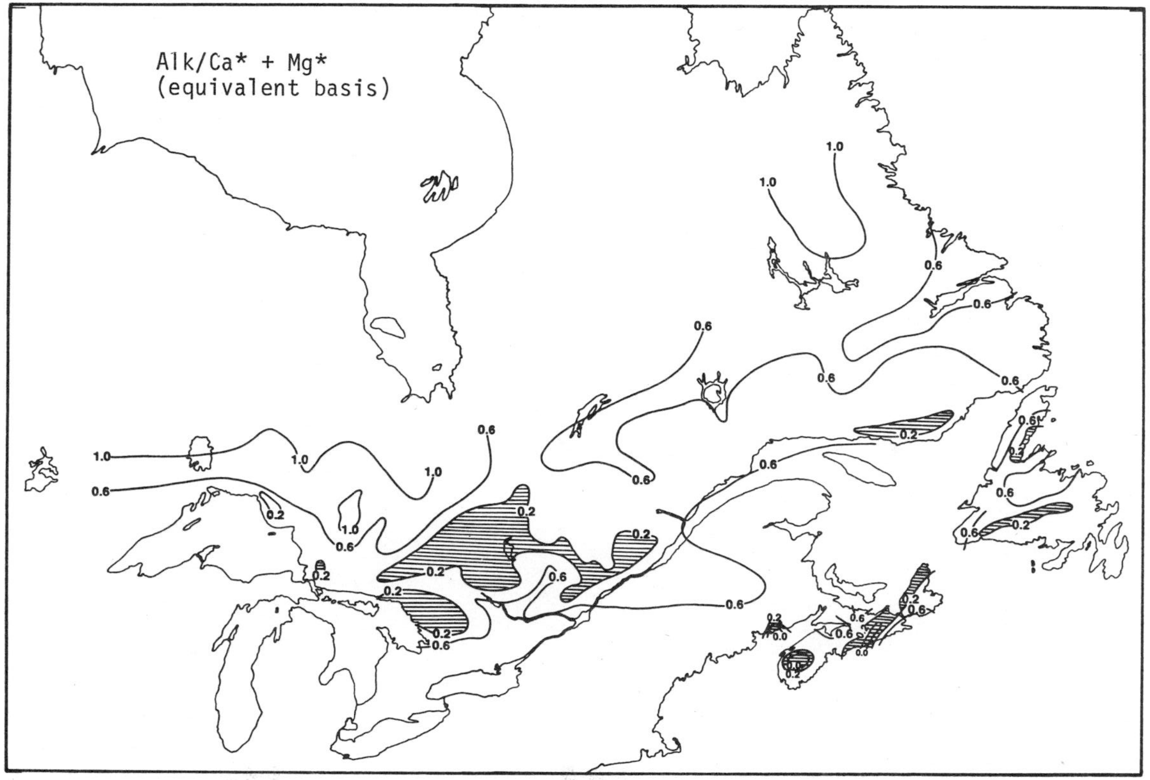

Figure 2. Geographical distribution of ANC:Ca+Mg ratio in lakewaters across eastern Canada (Ca+Mg was sea salt corrected). Crosshatched areas exhibit the most pronounced indication of surface water acidification. (after Jeffries, 1986).

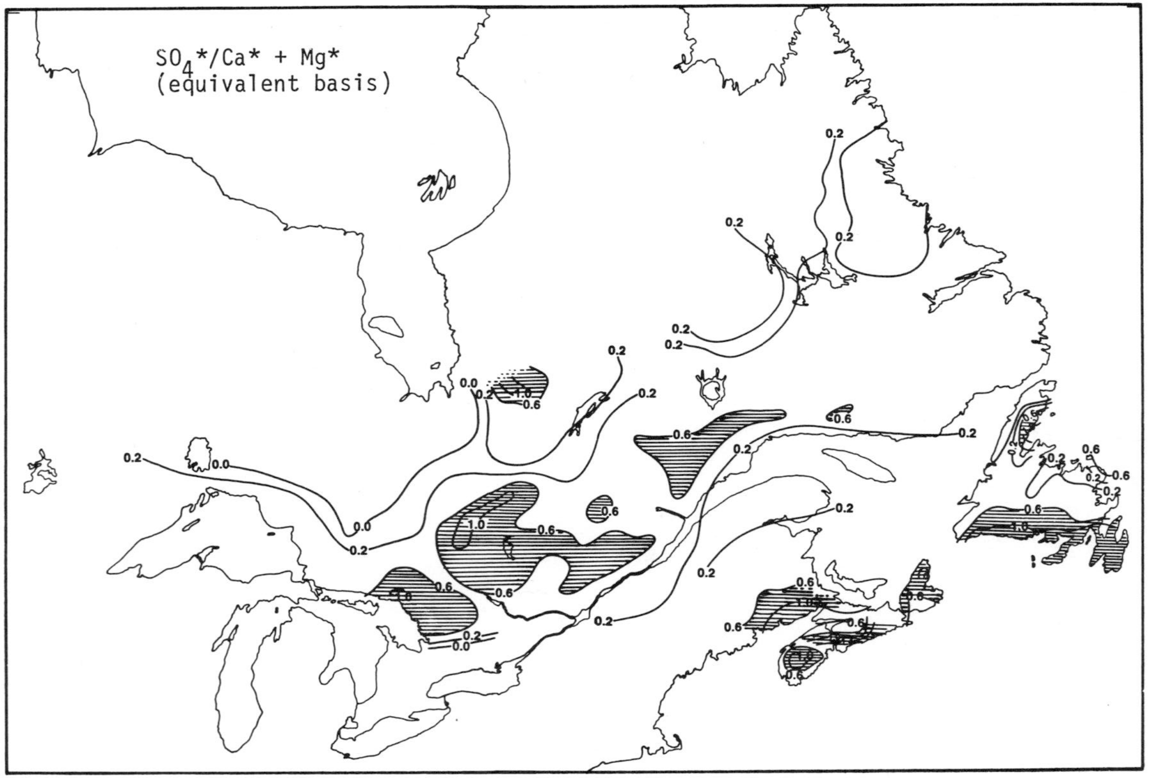

Figure 3. Geographical distribution of SO_4:Ca+Mg ratio in lakewaters across eastern Canada (all ions were sea salt corrected). Crosshatched areas exhibit the most pronounced indication of surface water acidification. (after Jeffries, 1986).

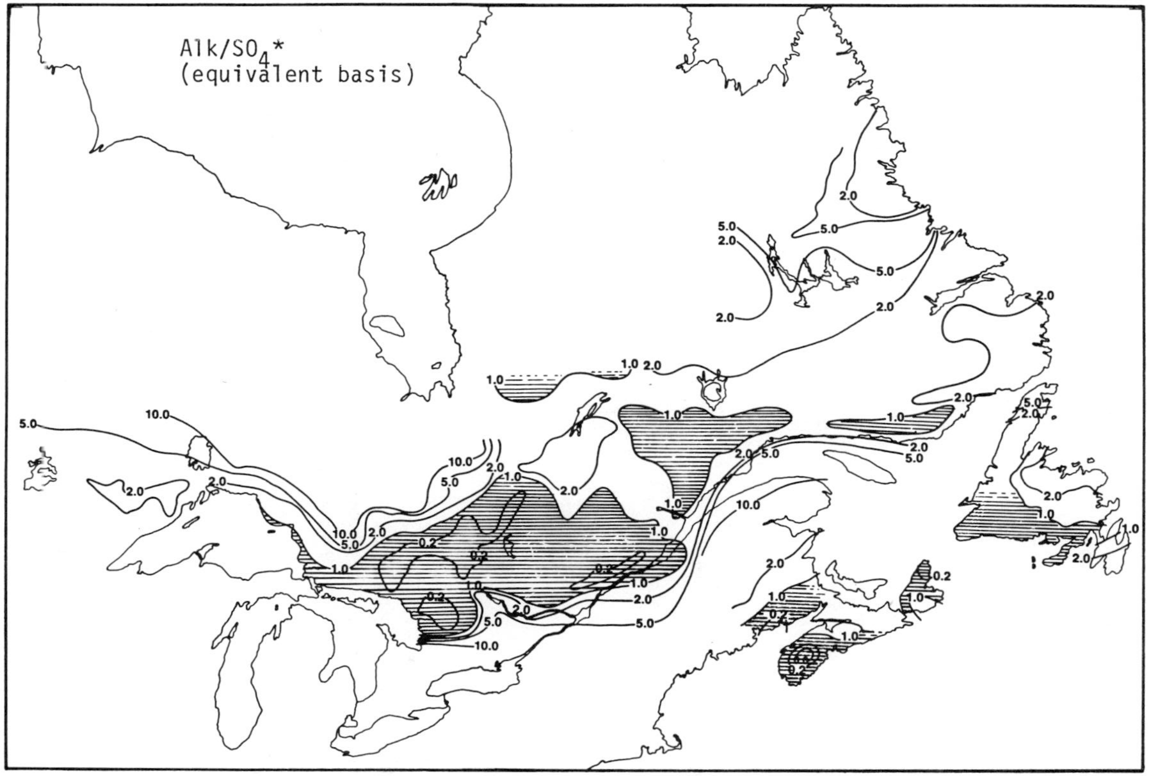

Figure 4. Geographical distribution of ANC:SO_4 ratio in lakewaters across eastern Canada (SO_4 was sea salt corrected). Crosshatched areas exhibit the most pronounced indication of surface water acidification. (after Jeffries, 1986)

REFERENCES

Canada/US. 1983 Memorandum of Intent on Transboundary Air Pollution, Impact Assessment, Working Group I.

Cook, R.B., M.L. Jones, D.R. Marmorek, J.W. Elwood, J.L. Malanchuk, R.S. Turner, and J.P. Smol. 1988. The effects of acidic deposition on aquatic resources in Canada: An analysis of past, present and future effects. Oak Ridge National Laboratory, Environ. Sci. Div. Publ. No. 3894, Oak Ridge, TN, 145pp.

Dillon, P.J., R.A. Reid, and R. Girard. 1986. Changes in the chemistry of lakes near Sudbury, Ontario, following reduction of SO_2 emission. Wat. Air Soil Pollut. 31: 59-65.

Dillon, P.J., R.A. Reid, and E. de Grosbois. 1987. The rate of acidification of aquatic ecosystems in Ontario, Canada. Nature 329: 45-48.

Environment Canada. 1988. Acid rain: A national sensitivity assessment. Inland Waters and Lands Directorate, Environmental Fact Sheet 88-1, Ottawa, Ontario, 5pp (+ map).

Harvey, H.H., R.C. Pierce, P.J. Dillon, J.R. Kramer, and D.M. Whelpdale. 1981. Acidification of the Canadian aquatic environment: Scientific criterion for assessment of the effects of acidic deposition on aquatic ecosystems. National Research Council of Canada Rep. No. 18475. National Research Council of Canada, Ottawa, Ontario, 369pp.

Hutchinson, T.C., and M. Havas. 1986. Recovery of previously acidified lakes near Coniston, Canada following reductions in atmospheric sulphur and metal emissions. Wat. Air Soil Pollut. 28: 319-333.

Jeffries, D.S. 1986. Evaluation of the regional acidification of lakes in eastern Canada using ion ratios. Proc. ECE Workshop on Acidification of Rivers and Lakes, Grafenau, FRG, April 28-30, 1986, 17-38.

Jeffries, D.S., D.L. Wales, J.R.M. Kelso, and R.A. Linthurst. 1986. Regional chemical characteristics of lakes in North America: Part I - Eastern Canada. Wat. Air Soil Pollut. 31: 55-567.

Keller, W. and J.R. Pitblado. 1986. Water quality changes in Sudbury area lakes: a comparison of synoptic surveys in 1974-1976 and 1981-1983. Wat. Air Soil Pollut. 29: 285-296.

Kelso, J.R.M., C.K. Minns, J.E. Gray, and M.L. Jones. 1986. Acidification of surface waters in eastern Canada and its relationship to aquatic biota. Can. Spec. Publ. Fish. Aquat. Sci. No. 87, 42pp.

Kelso, J.R.M. and D.S. Jeffries. 1988. Response of headwater lakes to varying atmospheric deposition in north-central Ontario, 1979-1985. Can. J. Fish. Aquat. Sci. 45: 1905-1911.

Oliver, B.G., R.L. Malcolm, and E.M. Thurman. 1983. The contribution of humic substances to the acidity of coloured natural waters. Geochim. Cosmochim. Acta, 47: 2031-2035.

RMCC. 1986. Assessment of the state of knowledge on the long-range transport of air pollutants and acid deposition. Federal-Provincial Research and Monitoring Coordinating Committee Report, Part 3 Aquatic Effects, Ottawa, Ontario.

Schindler, D.W. 1988. Effects of acid rain on freshwater ecosystems. Science 239: 149-157.

Shilts, W.W. 1981. Sensitivity of bedrock to acid precipitation: modification by glacial processes. Geol. Survey Can. Paper 81-14, 7pp.

Wright, R.F. 1983. Predicting acidification of North American lakes. Norwegian Institute for Water Research Tech. Rep. 4/1983, 165pp.

DISTRIBUTION PATTERN OF ATMOSPHERIC H_2O_2 IN LOS ANGELES AND ITS VICINITY AND ITS CONTROLLING ROLE AS AN OXIDANT OF SO_2

Hiroshi Sakugawa and Isaac R. Kaplan

Institute of Geophysics and Planetary Physics
University of California
Los Angeles, CA

ABSTRACT

Measurement of atmospheric H_2O_2 was carried out in Los Angeles (the concentration of gaseous H_2O_2, 0.03-1.35 ppb), the Mojave Desert (0.20-2.04 ppb) and the San Bernardino Mountains (0.43-1.72 ppb) during 1985-88. The H_2O_2 concentrations during midday (1200-1600) can be as high as 1 ppb in summer and ca. 0.2 ppb in winter in Los Angeles. Statistical analysis indicated that solar intensity is the most important factor controlling the seasonal variation of gaseous H_2O_2, although temperature, primary pollutants such as NO_x, and the height of the inversion layer are the determining factors in regulating the formation of O_3. Our results show that in summer, atmospheric H_2O_2 is the dominant oxidant of SO_2 in water droplets at low pH (<5.5), but the shortage of H_2O_2 in winter due to low photochemical generation of H_2O_2, limits the rate of S(IV) oxidation. Thus, the reduction of SO_2 emission from anthropogenic sources, which was carried out by local electric utilities during 1979-83, may not have resulted in an equivalent reduction of sulfuric acid levels in rainwater in Los Angeles, where rain events mainly occur in winter.

INTRODUCTION

Hydrogen peroxide in air is considered to be the most important oxidant of SO_2 in atmospheric water droplets (Penkett et al., 1979; Möller, 1980; Rodhe et al., 1981; Martin and Damschen, 1981; Hov, 1983; Kunen et al., 1983; National Research Council, 1983, 1986; Calvert et al., 1985; Kumar, 1986; NAPAP, 1987; Seigneur and Saxena, 1988). Moreover, H_2O_2 may participate in the aqueous phase oxidation of volatile organic compounds such as hydrocarbons and aldehydes via generating OH• free radicals (Graedel and Goldberg, 1983; Jacob, 1986; Graedel et al., 1986). H_2O_2 is also a toxic compound for plant cells and might be responsible for recent forest decline found in central Europe and the eastern North America (Masuch et al., 1986).

Atmospheric H_2O_2 in water droplets mostly comes from dissolution of gaseous H_2O_2, although formation in aqueous-phase and on solid surfaces has also been suggested (Chameides and Davis, 1982; Graedel and Goldberg, 1983; Chameides, 1984; Schwartz, 1984; Seigneur and Saxena, 1984; Calvert et al., 1985; McElroy, 1986; Sakugawa and Kaplan, 1987; Bahnemann et al., 1987). Until recently, measurement of gaseous H_2O_2 was very limited due to analytical problems (Kok et al., 1978 a and b; Groblicki and Ang, 1985; Jacob et al., 1986; Lazrus et al., 1986; Slemr et al., 1986; Tanner et al., 1986; Hartkamp and Buchhausen, 1987; Sakugawa and Kaplan, 1987; Kleindienst et al., 1988). Thus, the distribution pattern and generation mechanisms of gaseous H_2O_2 are not well understood. Theoretical studies indicate that several atmospheric pollutants such as NO_x, NMHC (non-methane hydrocarbon), CO, O_3 and aldehydes, and meteorological factors may affect the generation of gaseous H_2O_2 (Rodhe et al., 1981; Calvert and Stockwell, 1983, 1984; National Research Council, 1983; Hov, 1983; Kleinman, 1984, 1986; Calvert et al., 1985; Stockwell, 1986).

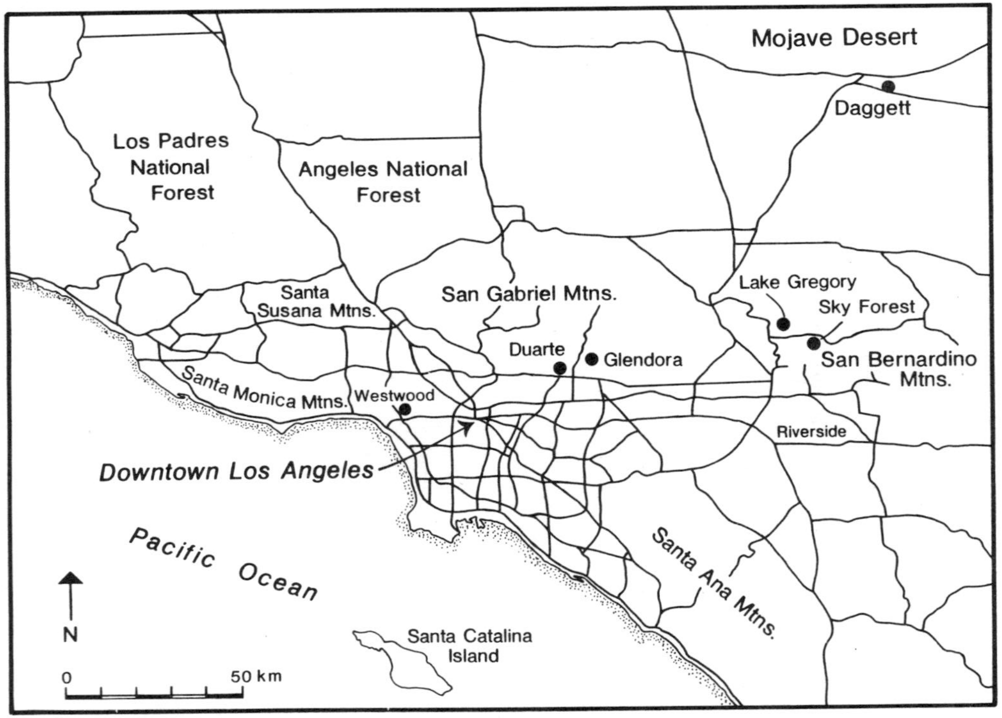

Fig. 1. Map of Los Angeles and its adjacent areas. Gaseous H_2O_2 was collected at Westwood, Duarte and Glendora in the Los Angeles Basin, at Daggett in Mojave Desert, and at Sky Forest and Lake Gregory in San Bernardino Mountain. Solid lines indicate the major highway systems in the region.

Sulfuric acid generation in Los Angeles air has been extensively investigated during the last decade (Appel et al., 1978, 1982; Cass, 1979, 1981; Liljestrand and Morgan, 1981; Waldman et al., 1982, 1985; Brewer et al., 1983; Munger et al., 1983; Richards et al., 1983; Zeldin et al., 1983, 1985; South Coast AQMD, 1984; Zeldin and Ellis, 1984; Jacob et al., 1985). The observations indicate that the concentration ranges of sulfuric acid in rain, cloud and fogwater are 11-51, 100-7310 and 62-20,000 μeq l^{-1}, respectively. In all the processes leading to the formation of sulfuric acid, H_2O_2 is expected to be a major oxidant for SO_2 in water droplets (Seigneur and Saxena, 1984), although the oxidation by O_3 and O_2 catalyzed with trace metals, Mn and Fe may represent other important processes (Jacob and Hoffman, 1983; Hoffmann and Jacob, 1984; Seigneur and Saxena, 1984).

In the Los Angeles Basin, atmospheric SO_2 concentration has largely decreased during the last decade, because electrical utilities in the Basin reduced their SO_2 emissions by more than 90% during 1979-83 due to fuel switching from oil to natural gas (South Coast AQMD, 1984). The resultant decrease in total SO_2 emissions in the Basin was estimated to be 42% during 1979-84 (Young et al., 1986). Consequently, ambient SO_2 concentrations in the coastal and central parts of the Basin, where major SO_2 point sources are located, have decreased from 25 to 30% during 1979-84 (CARB, 1979-84; Young et al., 1986). However, a statistically significant trend in the decrease of volume-weighted sulfate levels in rainwater has not been uniformly detected in the basin (Zeldin and Ellis, 1984; Young et al., 1986; CARB, 1988). Because the amount of precipitation in Los Angeles is small (377 mm per year, average) and nearly all rain events occur during the cold months, annual trends of rainwater sulfate chemistry may depend on the availability of oxidants of SO_2.

The UCLA group studied atmospheric H_2O_2 in Los Angeles and its vicinity during the last three years (1985-88) to evaluate its role as an oxidant of SO_2 and organic compounds in rain and fogwaters. We first determined the temporal distribution of gaseous H_2O_2 and then evaluated the possible factors controlling its concentrations using statistical analyses. We also measured the concentration of H_2O_2 in rain water. Finally, we calculated the oxidation rates of SO_2 by H_2O_2 and compared them with those of O_3 and O_2-catalyzed Fe and Mn, in order to evaluate the role of H_2O_2 as an oxidant of SO_2 in the Los Angeles atmosphere.

Here we report the results of field measurements of H_2O_2 and discuss the effect of H_2O_2, as a powerful atmospheric oxidant, on rain and fogwater chemistry in Los Angeles.

METHODOLOGY

Sampling Sites

Gaseous H_2O_2 was collected on the roof of the Geology Building (20 m height above ground surface) at the University of California, Los Angeles (UCLA), Westwood, which is located 8 km east of Santa Monica Bay, North Pacific and 20 km west of downtown Los Angeles (Fig. 1). Gaseous H_2O_2 was also collected on the roof of the City of Hope Hospital (10 m height), at Duarte, which is located 25 km northeast of downtown Los Angeles. The collection of gaseous H_2O_2 in the Mojave Desert was carried out at the Southern California Edison Co. Air Monitoring Station at Daggett, 15 km east of Barstow, California. A collector for gaseous H_2O_2 was set up at 1.5 m height from the ground surface. The sampling sites in the San Bernardino Mountains was at Sky Forest (1790 m height above sea level), 5 km southeast of Lake Arrowhead and at the Crescent Air Monitoring Station of the South Coast AQMD, located at Lake Gregory (1378 m height). A collector for gaseous H_2O_2 was set up at 1.5 m height from ground surface at both sites.

Westwood is located in an area in the Los Angeles Basin where there are no primary industrial emissions. Seabreezes (south-west) dominate during midday in all the seasons except for occasional flows of hot-dry landbreezes (north to east winds) so-called Santa Ana winds. Two major intersecting highways, which lie 1.5 km west of the collection site, as well as several major local streets, may be large contributors of primary pollutants from automobile emissions. Power plants and refineries, as stationary sources of pollutants, are located, 10-40 km south of Westwood (South Coast AQMD, 1984). Hazy weather (visibility, < 5 km), which was frequently observed in the coastal area of the Basin throughout the year, resulted from intrusions of marine layers with low stratus clouds from the North Pacific. Hazy weather conditions during daytime were associated with fog events at nighttime.

Duarte is located near the foothills of the San Gabriel Mountains, and is impacted by pollutants carried by seabreezes from coastal and central areas of the Basin. Gaseous H_2O_2 was simultaneously collected at Duarte, Westwood and Glendora, 10 km east of Duarte, during the Carbonaceous Species Method Comparison Study (sponsored by the California Air Resources Board), on August 11-21, 1986.

Daggett is located in the Mojave Desert. An east wind dominates during midday and is later replaced by west winds from Barstow, Los Angeles and/or San Joaquin Valley at night during the summer. Major sources of primary pollutants may be from two highways, which are located within 3 km of the monitoring station.

Sky Forest is located on a south slope of the San Bernardino Mountains. South to southwest winds are dominant in summer. However, during Santa Ana conditions, north to east winds are dominant. Lake Gregory is located west side of the San Bernardino Mountains. South to west winds are dominant at daytime in summer. Major sources of pollutants at both sites are thought to result from the transport of pollutants from the coastal areas of Southern California (South Coast Air Basin), carried by daily seabreezes (south to west winds). A nearby highway may also be a local source of pollutants in the mountain.

Measurement of Gaseous H_2O_2

Gaseous H_2O_2 in air was collected by the method of Sakugawa and Kaplan (1987). After air collection, the frozen water vapor, which was trapped from air, was defrosted. To the liquid samples were added fluorescence reagents, p-hydroxyphenyl acetic acid and peroxidase. The fluorescence of H_2O_2 in the fixed solution was determined by a fluorometer within 12 hr after the samples were collected and brought back to the UCLA laboratory. Blank and standard solutions of H_2O_2 were also prepared in the field to calibrate the concentration of H_2O_2 in the air samples. No significant decomposition of the fluorescence was found during transportation of the samples. The concentration of gaseous H_2O_2 (ppb) in air was calculated from the fluorescence value, total volume of water vapor trapped from the air and total air mass pulled through the trap.

Atmospheric Pollutants and Meteorological Data

The data on atmospheric pollutants at Westwood were supplied by the South Coast AQMD which monitors these pollutants at the Veterans Administration Hospital, 2 km west of UCLA. The detection limits of O_3, NO_x, NMHC and CO_2 were 10 ppb and were 2.5 ppb for NO and SO_2. For values lower than 10 ppb of NO and SO_2, the concentration was roughly estimated at 2.5 ppb intervals. The meteorological data (temperature, solar radiation, relative humidity, wind direction and speed) at Westwood was provided by the Department of Atmospheric Sciences, UCLA. The visibility was determined by observation of distant buildings from the UCLA campus.

The height of the inversion layer in Los Angeles was measured at 12 pm by South Coast AQMD using radiosones in warm months.

Ozone data at Daggett and Lake Gregory were supplied by Southern California Edison Co. and South Coast AQMD, respectively.

Rainwater Sampling

Rainwater was collected on a teflon-coated steel funnel (70 cm diameter) in a glass bottle (4 L) reservoir. The lid of the collector (teflon) was opened immediately after the start of a rain event by an automatic trigger system. After collection of rainwater, the sample was preserved with $HgCl_2$ and subjected to H_2O_2 analysis within 15 min by a fluorescence technique (Sakugawa and Kaplan, 1987).

Statistical Analysis

All the statistical analysis for H_2O_2 and O_3 data studied were performed by a "SAS" program (SAS Institute, Inc., North Carolina, USA) using an IBM computer. Principal component analysis was first applied to our field data in order to select best fit regressors for the H_2O_2 and O_3 data from many environmental factors and selecting for minimum collinearity among the regressors.

Single or multiple linear regression analysis was conducted to determine which factors most strongly affect the concentration of gaseous H_2O_2 and O_3 at a statistically significant level.

Determination of Oxidation Rate of SO_2 by H_2O_2, O_3 and O_2

A calculation was made on the basis of the observed H_2O_2, O_3 and SO_2 concentrations in ambient air, their Henry's law constants and aqueous phase concentrations of Fe(III) and Mn(II). Oxidation rates of SO_2 by H_2O_2, O_3 and O_2 in the air were determined in summer and winter, using known temperature, pH of water droplets and liquid water content (LWC). Here we assume that all the dissolved H_2O_2 in water droplets come from the dissolution of gaseous H_2O_2.

H_2O_2 concentration (24 hr mean) was assumed as 1.0 and 0.2 ppb in the air during the summer (June) and winter (December) respectively from our three year data base at Westwood. O_3 concentration was taken as 50 and 20 ppb of O_3 for summer and winter 24 hr mean concentrations, respectively, from the South Coast AQMD data. Assumed temperature and SO_2 concentration were 25°C and 2.5 ppb, and 15°C and 5 ppb in summer and winter, respectively. A value of 4.6 was used for the pH of water droplets, which is the average rainwater pH in Los Angeles (South Coast AQMD, 1984). The concentrations of Fe(III) and Mn(II) were assumed to be $1x10^{-5}$ and $1x10^{-6}$ M, respectively, which are common concentrations in fog and cloud droplets (Waldman et al., 1982; Brewer et al., 1983; Munger et al., 1983; Hoffmann and Jacob, 1984; Jacob et al., 1985).

RESULTS

Diurnal Variation of Gaseous H_2O_2

Study of diurnal variation of gaseous H_2O_2 indicated that the concentration of gaseous H_2O_2, as well as O_3, is at maximum in early afternoon at all the sites studied when the sky is clear or partially cloudy. Figure 2 shows the diurnal variation of gaseous H_2O_2 at Westwood, Duarte and Glendora in the Los Angeles Basin on August 19, 1986 during the Carbonaceous Species Method Comparison Study. The relative concentration of gaseous H_2O_2 were Glendora > Duarte > Westwood in order of highest concentrations. The transport of pollutants from coastal area to inland area and the resulting accumulation of the pollutants at the foothills of the San Gabriel Mountain under a low inversion layer (< 300 m height of inversion layer, monitored by South Coast AQMD) may be the reason why gaseous H_2O_2 was higher at Glendora and Duarte than at Westwood. Gaseous H_2O_2 was also high in the early afternoon at rural areas studied, such as in the Mojave Desert and the San Bernardino Mountains. On the other hand, H_2O_2 concentration largely decreased at night at all the sites studied. These results strongly suggest that gaseous H_2O_2 is photochemically generated in the atmosphere.

Seasonal Variation of Gaseous H_2O_2

We measured gaseous H_2O_2 in Los Angeles and its vicinity during August 1985-September 1988 to study the seasonal variation of H_2O_2. H_2O_2 was collected at midday (12 pm - 4 pm) at all sites, when photochemical generation of H_2O_2 is at a maximum. The H_2O_2 sampling was carried out only under clear or partially cloudy skies. Seasonal variation of H_2O_2 was high (~ 1.0 ppb) in summer, but low (0.1-0.3 ppb) in winter at Westwood, Los Angeles (Fig. 3a). Insufficient data were available to evaluate seasonal variation of H_2O_2 at the other sites studied because no determination of H_2O_2 was conducted during the cold months at the other sites. However, the concentrations of H_2O_2 were relatively high in warm months at Duarte (0.12-0.78 ppb, n=16), Daggett (0.20-2.04 ppb, n=9), and Sky Forest (0.43-1.45 ppb, n=10) and Lake Gregory (0.82-1.72 ppb, n=3). Seasonal trends of O_3 were also studied during the period to compare with the H_2O_2 data (Fig.3b). No clear seasonal trend of O_3 was found at Westwood, although the O_3 concentration was generally higher in warm months than in cold months. The correlation between the data of H_2O_2 and O_3 was poor at all the sites studied. The correlation coefficients were 0.21, 0.44, 0.00 and 0.19 at Westwood, Duarte, Daggett and Lake Gregory, respectively. These results suggest that either the generation or decomposition mechanisms of H_2O_2 and O_3 are different from each other.

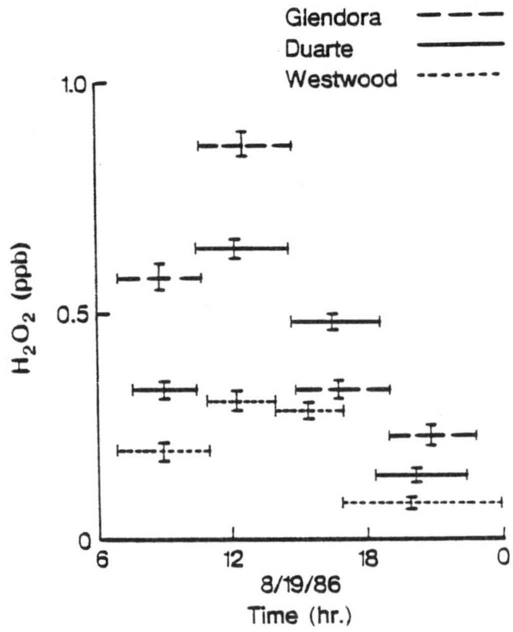

Fig. 2. Simultaneous measurement of gaseous H_2O_2 at Westwood, Duarte and Glendora in the Los Angeles Basin on August 19, 1986. Gaseous H_2O_2 was collected every 3 or 4 hr periods, except for 8 hr (1700-0100 hr) at Westwood.

Table 1. Simultaneous measurement of ambient gaseous H_2O_2 and O_3 (average concentration $\pm$SD, $\mu g\ m^{-3}$) at Westwood (W_{1-4}) and Duarte (D) in the Los Angeles Basin, Daggett (Da) in Mojave Desert and Sky Forest (SF) and Lake Gregory (LG) in the San Bernardino Mountain.

Locations simultaneously measured		W	D	Da	SF	LG
W_1-D	H_2O_2	0.47$\pm$0.21	0.59$\pm$0.26	-	-	-
(n=11)	O_3	165$\pm$90	223$\pm$135	-	-	-
W_2-Da	H_2O_2	0.64$\pm$0.35	-	1.65$\pm$0.94	-	-
(n=8)	O_3	141$\pm$36	-	156$\pm$36	-	-
W_3-SF	H_2O_2	0.71$\pm$0.38	-	-	1.44$\pm$0.55	-
(n=9)	O_3	126$\pm$49	-	-	ND	-
W_4-LG	H_2O_2	0.96$\pm$0.20	-	-	-	2.11$\pm$0.61
(n=3)	O_3	167$\pm$54	-	-	-	294$\pm$36

ND: not determined

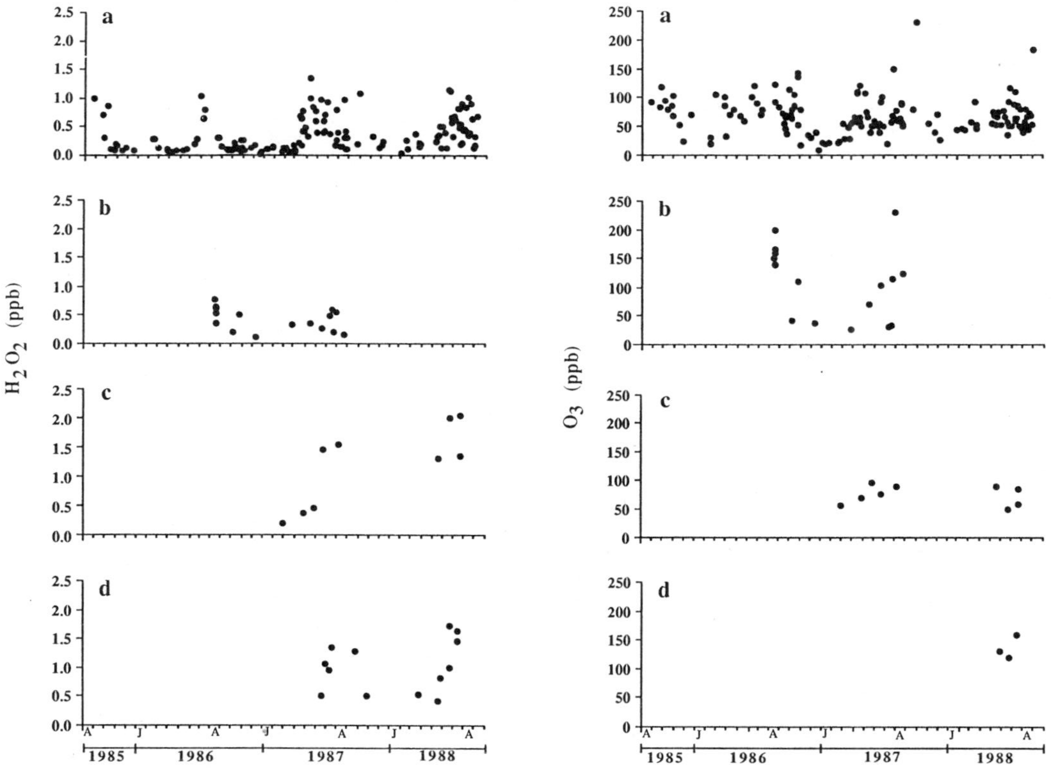

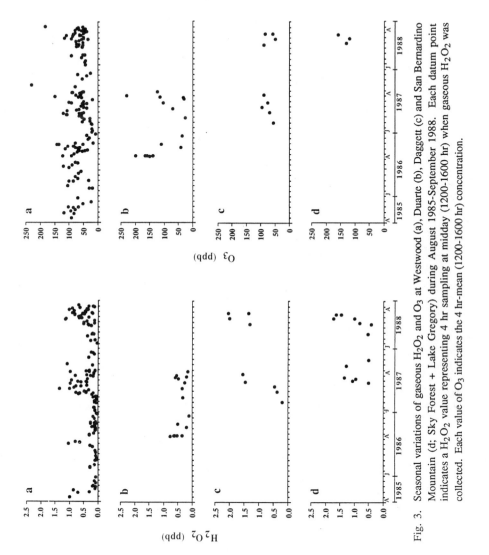

Fig. 3. Seasonal variations of gaseous H_2O_2 and O_3 at Westwood (a), Duarte (b), Daggett (c) and San Bernardino Mountain (d; Sky Forest + Lake Gregory) during August 1985-September 1988. Each datum point indicates a H_2O_2 value representing 4 hr sampling at midday (1200-1600 hr) when gaseous H_2O_2 was collected. Each value of O_3 indicates the 4 hr-mean (1200-1600 hr) concentration.

Simultaneous measurements of H_2O_2 at different sites were carried out to find the regional variation of H_2O_2. The measurement of H_2O_2 was carried out from 12 to 4 pm between March-October 1986-88 in multiple locations. The results indicate that the concentration of gaseous H_2O_2 is higher in the rural areas than in Los Angeles - about twice higher at Mojave Desert and San Bernardino Mountains than at Westwood and Duarte (Table 1). Substantially higher concentrations of primary pollutants such as NO_x and SO_2 are present in Los Angeles than at the rural areas, which may result in the lower concentration of gaseous H_2O_2. Ambient NO_x and SO_2 concentrations at Daggett were usually less than the detection limit (< 1 ppb) at midday, whereas in Los Angeles NO_x and SO_2 contents in air were 20-60 and 2.5-10 ppb. O_3 concentration (4 hr mean) was higher at Duarte than at Westwood due to the accumulation of pollutants at Duarte, at the foothills of the San Gabriel Mountains. O_3 concentration (4 hr mean) was also higher at Lake Gregory than at Westwood, whereas no significant difference of O_3 concentration between Westwood and Daggett was found. At Lake Gregory, the height of inversion layer is near surface at midday due to the high altitude of the site. Transport of O_3 from the Los Angeles area to high elevations of the mountain by an upslope flow of marine air is significant in the afternoon (Miller et al., 1986). Natural hydrocarbons emitted from higher plants may be available for the formation of O_3 in the San Bernardino Mountains, where mixed conifer trees are the dominant flora at an elevation >1000 m. On the other hand, high NO_x concentration in Los Angeles where the ratio of NMHC/NO_x is at a range of 4-10 at midday in summer could suppress the formation of O_3. O_3 concentration rapidly decreases at night in Los Angeles due to the decomposition of O_3 by NO, whereas O_3 is preserved at night in the rural areas due to limited anthropogenic sources of NO_x. Thus, annual average O_3 in the concentration is higher in the San Bernardino Mountains than in Los Angeles (South Coast AQMD, 1983).

Controlling Factors of Gaseous H_2O_2 and O_3

To determine the generation and decomposition mechanisms for gaseous H_2O_2 and O_3, each of 150 measured datum points obtained at Westwood has been compared with atmospheric pollutants and meteorological data. Six atmospheric pollutants (NO, NO_2, NO_x, SO_2, CO, NMHC) and eight meteorological parameters (temperature, RH (relative humidity), SR (solar radiation), wind direction and speed, the IH (inversion height), cloudiness and visibility) were obtained during the study.

The data on wind direction indicate that seabreezes from the North Pacific (south to west wind, mostly southwest; n=147) were dominant at Westwood when air sampling was conducted. Thus, we have only considered the local pollution in the coastal area of the Basin, where seabreezes originating from the North Pacific via Santa Monica Bay coast passed over Westwood. The landbreezes observed during midday on November 8, 1985, November 5, 1986 and December 7, 1987 were the northeasterly Santa Ana winds and carried a very dry air mass. However, pollutant concentrations in the air mass did not show much difference with those carried by seabreezes during other days studied. Thus, the effect of the landbreezes on the formation of H_2O_2 and O_3, was not decisively resolved in this study.

Case Study 1

Ten factors including H_2O_2, O_3, NO, NO_2, NMHC, NMHC/NO_x, SO_2, CO, Temp and SR were chosen as variables using 139 of the total data set from 1985-88. The data sets (n = 11) which have one or more than one missing parameters were omitted for the statistical analysis. The correlation matrices (Table 2) indicate that there are strong correlations among primary pollutants, particularly NO_2, NMHC and CO, which are major emittants from automobiles. The correlation coefficients among these three compounds were higher than 0.80. The correlations among other parameters were, however, not strong (<0.70) in the matrices. Principal component analysis was carried out to seek the intercorrelations among the parameters. Four components, were found to account for 87% of the total variance of the original data set (Table 3). Factor 1 group consists of NO_2, CO, NMHC and NO, whereas Factor 2 group is composed of O_3 and Temp. Factor 3 is composed of H_2O_2 and SR; Factors 4 and 5 are represented by NMHC/NO_x and SO_2 respectively. The results suggest that H_2O_2 correlates most strongly with SR among all the factors studied, whereas O_3 correlates best with temperature.

SR was chosen as an independent variable for a linear regression analysis for H_2O_2, because (1) SR has no interference from other parameters studied and (2) it was most correlated with H_2O_2 in this data set. SO_2 was also used as an independent variable, because SO_2 is considered to be a factor controlling gaseous H_2O_2

Table 2. Matrix of correlation coefficients (r) among environmental parameters in Case Study 1.

	H_2O_2	O_3	NO	NO_2	NMHC	CO	SO_2	NMHC/NO_x	Temp	SR
H_2O_2	1.00									
O_3	0.20	1.00								
NO	-0.27	-0.36	1.00							
NO_2	-0.21	0.15	0.57	1.00						
NMHC	-0.31	0.28	0.45	0.82	1.00					
CO	-0.23	0.24	0.46	0.81	0.81	1.00				
SO_2	-0.20	0.20	0.20	0.44	0.37	0.33	1.00			
NMHC/NO_x	-0.19	0.11	-0.22	-0.18	0.28	0.00	-0.07	1.00		
Temp	0.20	0.65	-0.14	0.19	0.26	0.31	0.24	0.01	1.00	
SR	0.53	0.23	-0.40	-0.38	-0.46	-0.46	0.04	-0.23	0.21	1.00

Number of variables: 10
Number of data set: 139

Table 3. Principal component analysis for three case studies.

Case study Seasons in data set n	1 All seasons 139	2 Warm months 73	3 April-October 1987-88 72
Factor 1	NO_2, CO, NMHC, NO	O_3, Temp, WS(-), IH(-)	O_3, Temp, NO_2
Factor 2	O_3, Temp	H_2O_2, SR	H_2O_2, SO_2(-)
Factor 3	H_2O_2, SR	O_3, NO_2	NO
Factor 4	NMHC/NO_x	NO, WS(-)	SR
Factor 5	SO_2	SO_2	
Total variance (%)	87	86	86

*1 Four or five components (factors), which account for >80% of total variance of original data sets, were extracted in all the case studies.

*2 Environmental parameters, which are composed of each factor group in the principal component analysis, are listed here only when the regression coefficient of each parameter was more than 0.5 or less than -0.5.

*3 (-) indicates that the regression coefficient of the environmental parameter is negative.

concentration by heterogeneous decomposition in water droplets. NO, NO_2, NMHC and CO were not used as independent variables for H_2O_2 in the regression analysis, due to their collinearity with SR. On the other hand, temperature and NO were chosen as independent variables for the regression analysis of O_3, because their correlation with O_3 was high in the correlation matrix and they may be important factors for regulating O_3 concentration. The results indicate that the H_2O_2 concentration should increase with increasing SR, but will decrease with increasing SO_2 content, at statistically significant levels ($p<0.05$), whereas O_3 concentration increases with increasing temperature, but decreases with increasing NO content in air (Table 4). The single regression analysis for H_2O_2 by SR indicates that SR accounts for 29% (R^2=0.29) of total variance of H_2O_2. Inclusion of SO_2 improved the variance a little (R^2=0.33). On the other hand, Temp

Table 4. Single or multiple linear regression analysis for H_2O_2 and O_3 data in this study.

Case study	n	Dependent variable	Number of independent variables	Independent variable	Sign of regression coefficient	Variance (r^2)*1
1 All seasons	139	H_2O_2	1	SR	+	0.29
		H_2O_2	2	SR SO_2	+ -	0.33
		O_3	1	Temp	+	0.42
		O_3	2	Temp NO	+ -	0.49
2 Warm months	73	H_2O_2	1	SR	+	0.22
		H_2O_2	2	SR SO_2	+ -	0.31
		O_3	1	Temp	+	0.40
		O_3	1	NO_2	+	0.32
		O_3	1	IH	-	0.28
		O_3	4	Temp NO_2 IH NO	+ + - -	0.62
3 April-October, 1987-88	72	H_2O_2	1	SO_2	-	0.23
		H_2O_2	2	SO_2 NO	- -	0.29
		O_3	1	Temp	+	0.60
		O_3	1	NO_2	+	0.44
		O_3	2	Temp NO_2	+ +	0.66

*1: All listed here are statistically significant ($p < 0.05$).

largely explains the variance (R^2=0.42) of O_3. Inclusion of NO raises the total variance to 49%. It should be noted that all the variables used as independent variables in the regression analysis, except SR, are not strictly independent variables because temperature depends on SR and primary pollutants are also affected by meteorological factors. However, the correlation matrix indicates that such interferences among the independent variables appear to be insignificant. Thus the results of the regression analyses are not significantly affected by the "multicollinearity effects", which means that in this model a regressor is nearly a linear combination of other regressors.

Case Study 2

The effect of IH on H_2O_2 and O_3 chemistry was investigated using only data from the warm seasons, when a strong and low inversion layer is frequently developed in Los Angeles. The nine parameters H_2O_2, O_3, NO, NO_2, SO_2, Temp, SR, IH and WS were chosen as variables in the statistical analysis using 73 data sets. The results of the factor analysis are given in Table 3.

The results indicate that (1) Temp, O_3, WS and IH can be classified into one group (Factor 1). (2) Factor 2 group is composed of H_2O_2 and SR, (3) Factor 3 consists of O_3 and NO_2, (4) Factor 4 is composed of NO and WS. (5) Factor 5 is composed of SO_2 only. Regression analysis for H_2O_2 indicates that SR only or both SR and SO_2 account for 22 and 31%, respectively, of total variation of H_2O_2 (Table 4). On the other hand, Temp, IH, and NO_2 are best fit regressors for O_3 in the data set. In fact, these variables modestly correlate with each other when a low inversion layer develops in the Los Angeles Basin. When IH is low, wind speed is low and temperature increases because of a relatively small penetration of moist and relatively cool marine air masses from the North Pacific into the Los Angeles Basin, passing under a low inversion layer. It should be noted that temperature has an inverse relation with RH (r= -0.78) in this case study.

Case Study 3

The effect of haziness in air on the concentration of H_2O_2 was studied during April-October 1987 and April-September 1988, when (1) SR was high (> 1.0 calcm^{-2}min^{-1}) and (2) humid air and haze events were frequently measured at Westwood. During this period, RH was relatively high (ave. 68$\pm$18%). Thirty-three (33) haze events were recorded and they were strongly associated with high RH (ave. 75$\pm$11%). All the haze events, except one case (RH=48%), was found when RH was >50%.

The environmental factors used as variables for the statistical analysis were H_2O_2, O_3, NO, NO_2, SO_2, Temp and SR using 72 of the data sets. Four principal components were extracted from the original data set in the factor analysis for the case study (Table 3). Factor 1 group consists of O_3, Temp and NO_2, which indicates that O_3 is largely associated with Temp and primary pollutants. Factor 2 consists of H_2O_2 and negatively SO_2, whereas NO and SR produced independent factor groups (Factors 3 and 4). The relationship between H_2O_2 and SR was very poor (r= -0.04) in this data set. Regression analysis of the data set indicates that SO_2 only or both SO_2 and NO accounted for 23 and 29% of total variance of H_2O_2, respectively (Table 4). Regression analysis for O_3 indicates that Temp and NO_2 explain a large part of total variation of O_3 (R^2=0.60 and 0.44).

Availability of H_2O_2 and O_3 for oxidizing SO_2

From the seasonal variation of gaseous H_2O_2 and O_3, we have estimated below the availability of H_2O_2 and O_3 as oxidants of dissolved SO_2 (S(IV)) in water droplets in the Los Angeles atmosphere. The oxidation rate of O_2 catalyzed by Fe(III) and Mn(II), which may be additionally important oxidants of SO_2, was also estimated. Because a large portion of Fe(III) is in the form of $Fe(OH)_3$ (solid) at high pH (>4.0), the influence of Fe(III) on the oxidation of SO_2 is less important. We therefore, only considered Mn(II) as a catalyst of the O_2-S(IV) reaction at the assumed pH (4.6) in the calculation. The results indicate that the initial rates of sulfuric acid production by H_2O_2 in the atmospheric water droplet were 661 and 542 μMhr^{-1} in summer and winter, respectively (Table 5). In contrast, the oxidation rates by O_3 were 15 and 16 μMhr^{-1}. Similarly, the oxidation rates by O_2 were 28 and 22 μMhr^{-1} in summer and winter, respectively. Thus, H_2O_2 would be the dominant oxidant for SO_2 in all the seasons. Although the pH of rainwater is mostly 3.5-6.0, the pH range of fog and cloudwater in Los Angeles is 2.0-6.0. Our calculations indicate that H_2O_2 would be the dominant oxidant of SO_2 in water droplets when pH is <5.5, whereas O_3 would be dominant at pH of >5.5. However, it should be noted that the effect of O_2 catalyzed by Mn(II), as well as O_3 oxidation will become more significant at high pH (>5.0) due to increase in the solubility of SO_2 with increasing pH.

The characteristic time for depletion of SO_2 and oxidants (H_2O_2, O_3 and O_2) by S(IV)-oxidant reaction in water droplet has been calculated to determine which reagent is limiting in the reaction (Kelly et al., 1985). The results of such a calculation indicate that SO_2 would be a limiting reagent for the reaction of S(IV) and H_2O_2 in summer and H_2O_2 would be limiting in winter, when LWC was assumed to be 1.0cm^{3}m^{-3}. On the other hand, when LWC is 0.1cm^{3}m^{-3}, which is the common water content for fog water, H_2O_2 would be a limiting reagent in both seasons for the S(IV)•H_2O_2 reaction. Thus, except for the case when LWC = 1.0cm^{3}m^{-3} in summer, only a relatively small proportion of the available S(IV) can be oxidized in water droplets during all other times due to the lack of H_2O_2.

After H_2O_2 depletion, O_3 and O_2 would be dominant oxidants of S(IV), despite their low oxidation rates. The characteristic times for depletion of SO_2 as the limiting reagent for the S(IV)-O_3 reaction would be 6.9

Table 5. Calculated oxidation rate of S(IV) with H_2O_2, O_3 and O_2 in water droplet in Los Angeles air.

Season	Summer			Winter		
Oxidant (24hr-mean gaseous concentration)	H_2O_2 (1 ppb)	O_3 (50 ppb)	O_2 catalyzed by Mn(II)	H_2O_2 (0.2 ppb)	O_3 (20 ppb)	O_2 catalyzed by Mn(II)
Initial rate of sulfuric acid production in the droplet (μM hr^{-1})	661	15	28	542	26	22
Limiting reagent						
LWC $1.0cm^3m^{-3}$	SO_2	SO_2	SO_2	H_2O_2	SO_2	SO_2
LWC $0.1cm^3m^{-3}$	H_2O_2	SO_2	SO_2	H_2O_2	SO_2	SO_2
Characteristic time* for depletion of limiting reagent						
LWC $1.0cm^3m^{-3}$	0.16	6.9	3.7	0.07	8.4	10
LWC $0.1cm^3m^{-3}$	0.73	68	36	0.22	82	97

T = 25°C and SO_2 = 2.5 ppb, and T = 15°C and SO_2 = 5.0 ppb as 24 hr-mean temperature and contents of SO_2 in summer and winter, respectively. The pH of water droplet is assumed to be 4.6, which is the average pH of Los Angeles rainwater. 1.0 or $0.1cm^3m^{-3}$ as liquid water content (LWC) in air. Henry's law constants for H_2O_2, SO_2 and O_3 are $7.4x10^4$ (25°C) and $1.6x10^5$ (15°C) $Matm^{-1}$ (Lind and Kok, 1986), 1.24 (25°C) and 1.81 (15°C) $Matm^{-1}$ (Maahs, 1983), and $1.15x10^{-2}$ (25°C) and $1.55x10^{-2}$ (15°C) $Matm^{-1}$ (Chameides, 1984), respectively. First and second dissociation constants for $(SO_2)_{aq}$ are $1.32x10^{-2}$ (25°C) and $1.67x10^{-2}$ (15°C), and $6.42x 10^{-8}$ (25°C) and $7.61x10^{-8}$ (15°C) M, respectively (Maahs, 1983), where $[S(IV)] = [(SO_2)_{aq}] + [(HSO_3^-)_{aq}] + [(SO_3^{2-})_{aq}]$. The oxidation rate constants of $(SO_2)_{aq}$ with H_2O_2 and the rate constants of S(IV) with O_3 are $8x10^5$ (25°C) and $5.2x10^5$ (15°C) $M^{-1}sec^{-1}$ (Chameides, 1984; Martin, 1984) and $4.4x10^6$ (25°C) and $3.8x10^6$ (15°C) $M^{-1}sec^{-1}$ (Maahs, 1983), respectively. Concentration of Mn(II) in water droplet is $1x10^{-6}$M and the oxidation constants of HSO_3^- with Mn (II) are $5x10^3$ (25°C) and $1x10^3$ $M^{-1}sec^{-1}$ (15°C) (Martin, 1984).

* Characteristic time is defined as the total moles of gaseous and aqueous phase S(IV), H_2O_2 or O_3 contained in $1m^3$ of air divided by the initial reaction rate (moles $m^{-3}hr^{-1}$).

and 8.4hr in the summer and winter, respectively, when LWC is $1.0cm^3m^{-3}$, which is much longer than the residence time (1hr) of atmospheric water droplets during precipitation. The characteristic times for the S(IV)-O_2 reaction would be 3.7 and 10hr in the summer and winter, respectively, when LWC is $1.0cm^3m^{-3}$. The depletion time would be longer with decreasing LWC. O_2 catalyzed Fe(III) and Mn(II) may be more important than H_2O_2 as an oxidant of SO_2 in fog chemistry (Jacob and Hoffmann, 1983; Hoffmann and Jacob, 1984; Seigneur and Saxena, 1984) because no photochemical generation of gaseous H_2O_2 is expected at night. For Los Angeles, our data suggest that after 1hr reaction, 106 and $86nMm^{-3}$ of sulfuric acid would be generated in water droplets by the oxidation of SO_2 with H_2O_2 and O_3 in summer and winter, respectively, when LWC is $1.0cm^3m^{-3}$; or that the formation rate of sulfuric acid in winter is 81% of that in summer. On the other hand, 53 and $17nMm^{-3}$ of the sulfuric acid would be generated when LWC is $0.1cm^3m^{-3}$; or 32% of the summer value of SO_4^{2-} would be formed in winter. These results indicate that significantly less SO_2 would be oxidized in winter due to lack of sufficient amounts of oxidants.

Gaseous H_2O_2 During Fog Event

We found that the concentration of gaseous H_2O_2 was extremely low (< 0.03 ppb, n=5) during fog events observed on October 1, 2, 27-28, 1986 and April 16-27, 1987, compared with 0.01-0.48 ppb (ave. 0.19, n=57) of H_2O_2 in all the nighttime air samplings (each 2-4 hr) conducted during 1985-87 at Westwood (Sakugawa and Kaplan, unpublished data).

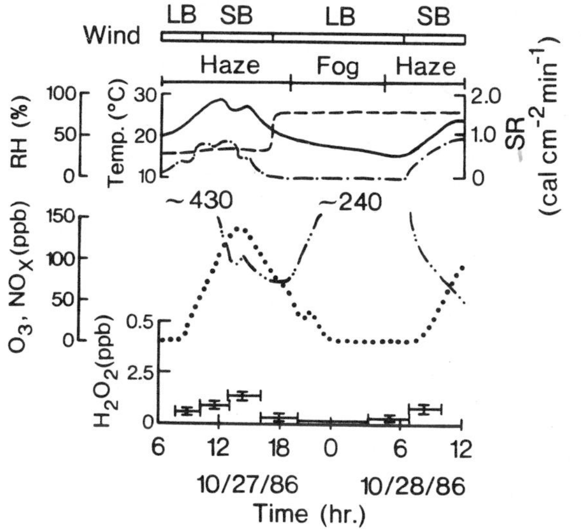

Fig. 4. Diurnal variations of gaseous H_2O_2 (⊢I⊣), O_3 (.....), NO_x (-..-), temperature (—), relative humidity (RH, - - -) and solar radiation (SR, -.-) at Westwood on October 27-28, 1986. Gaseous H_2O_2 was collected every 3 hr during the period. The dominant wind was seabreeze (SB) or landbreeze (LB). A dense fog (visibility <100m) occurred during the late evening through early morning.

Dense fog and haze was observed on October 26-28, 1986 at Westwood, both during day and nighttime. The weather pattern during these days was described as low stratus clouds under a low (IH = 150-240m) and strong (ΔT = 1.4-7.4°C) inversion layer. The wind speed of the landbreeze at 0000-0600 was 1.0-2.5ms^{-1} at Westwood. The highest H_2O_2 concentration was observed in the early afternoon on October 27 when O_3 was at a maximum, but NO_x was at a minimum (Fig. 4). No early evening peak of H_2O_2 was observed on October 27. Visibility largely decreased in the evening and then a fog event occurred. During the fog event (visibility <100m), the concentration of gaseous H_2O_2 was less than the detection limit (<0.01 ppb) while SO_2 content was about 10 ppb. These results suggest that gaseous H_2O_2 is utilized for chemical reactions in fog droplets.

H_2O_2 in Rain and Fogwaters

The concentration of H_2O_2 in rainwater was measured at Westwood during 1985-88. Thirty-four events of rain were collected and subjected to the chemical analysis of inorganic and organic compounds. Twenty-eight of a total 34 events were collected during cold months (October-April). The results show that the concentrations of organic and ionic components in rain largely depend on the amount of precipitation. Short rain events or drizzles, which was usually observed in warm months, showed high concentrations of the chemical constituents, and their concentrations substantially decreased with increasing amount of precipitation. Precipitation-weighted mean concentration of H_2O_2 was 4.8μM with a range of 0.1-95μM in rainwater, which is within a range of the predicted value of H_2O_2 (in the order of magnitude) in summer (~74μM) and in winter (~32μM), calculated from the Henry's law constants of H_2O_2 (assuming 1.0 and 0.2 ppb concentrations of gaseous H_2O_2 in summer and winter, respectively; Table 5). Time-series collections of rainwater indicated that the concentration of H_2O_2 is high at the afternoon, but low at night through early morning, a pattern which is in good agreement with that of gaseous H_2O_2 as mentioned earlier. These results support the conclusion that a major source of rainwater H_2O_2 is gaseous H_2O_2, which is photochemically generated in the atmosphere.

The UCLA group also measured H_2O_2 concentration in fogwater collected at Hanninger Flats (770m height), in the San Gabriel Mountains in the city of Pasadena on June 23-24, 1983 (Personal communication with K. Kawamura, 1988). The measured concentration at that time was in the range of 1.0-5.5μM (ave. 2.7, n=5).

DISCUSSION

The results of our multivariate analysis for the data sets including H_2O_2 data suggest that solar radiation is the most important environmental factor for controlling gaseous H_2O_2 in the air among the factors studied. High solar intensity would result in intense photochemical reaction of chemical species in the air, i.e. photochemical oxidation of primary pollutants and increase in the generation rate of radical species such as OH•, RO•, HO_2• and RO_2• (the precursors for H_2O_2) from volatile organic compounds (Calvert and Stockwell, 1983; Kleinman, 1986; Stockwell, 1986). Kleinman's theoretical study (1986) on the formation of gaseous H_2O_2 in the boundary layer indicated that high SR results in higher concentration of gaseous H_2O_2 when other environmental factors are assumed to be constant. Aircraft observations on the distribution of gaseous H_2O_2 over the eastern and central North America (Heikes et al., 1987; Van Valin et al., 1987) indicate a latitude dependency of the H_2O_2 concentration with higher concentrations in the south, probably resulting from higher average SR.

The amount of volatile organic carbon (VOC) in the air may also be important for the generation of gaseous H_2O_2. In urban areas such as Los Angeles, the concentration of VOC is very high due to their emissions by industrial activity. The greatest uncertainty in this study, is the amount of "reactive VOCs" such as olefins, aldehydes and ketones in the air during the measurement period, which may be largely responsible for the generation of radical species and the resulting gaseous H_2O_2 formation (Kleinman, 1986; Stockwell, 1986). The data available from AQMD indicates that the concentration of NMHC, which gives no information regarding individual organic compounds, is low during the warm months. However the concentration of reactive moieties could be high due to photochemical generation of secondary species, such as aldehydes, which are important sources of free radicals in urban air (Calvert and Stockwell, 1983; Stockwell, 1986). The photochemical generation of gaseous aldehydes is high during the middle of the day, when O_3 is at a maximum in Los Angeles (Tuazon et al., 1981; Grosjean, 1982). Emission of isoprenoid compounds from higher plants in summer may also contribute to the generation of H_2O_2, although the amount of isoprenes in the air at Westwood is unknown.

The relationship between NO_x and H_2O_2 was not statistically decisive in this study, although high H_2O_2 concentration was associated with relatively low NO_x (r= -0.27 for NO, in Case study 1). This result implies that NO_x may not be a major factor controlling the seasonal variation of gaseous H_2O_2. However, the following reasons suggest other possible interpretations. (1) There is an inverse relation between NO_x and SR, and a strong collinearity of NO_x with NMHC, CO and SO_2. This collinearity makes it difficult to estimate the sole effect of NO_x on the H_2O_2 formation in statistical analysis. A multiple regression analysis for Case study 3 indicates an inverse relation of H_2O_2 with NO. However, multicollinearity effects may again lead to a misinterpretation. (2) Because NO_x indirectly affects the formation of gaseous H_2O_2 by scavenging free radical species from the air (Calvert and Stockwell, 1983, 1984; National Research Council, 1983; Kleinman, 1986; Stockwell, 1986), the amounts of VOC and resulting free radical species, have to be known to determine the NO_x effect correctly. Thus, our field observations do not yet exclude possible effects of NO_x on the seasonal trend of gaseous H_2O_2 in the Los Angeles air.

During the period April through October 1987 and April through September 1988, although SR was mostly high, significant fluctuation in the concentration of gaseous H_2O_2 was found. A statistical analysis indicates that there is an inverse relation between H_2O_2 and SO_2. Heterogeneous reaction of H_2O_2 with SO_2 is fast and is believed to be a very important process for the formation of sulfuric acid in water droplets in the air. During the periods when gaseous H_2O_2 was collected, the air was mostly humid and haze events were frequently observed in the period. Humid air, together with high SO_2, may be the primary condition for the decomposition of H_2O_2 in the air. Field observations, performed by other workers, showed that in cloud-atmospheres, H_2O_2 concentrations were low (Römer et al., 1985; Daum et al., 1984; Kelly et al., 1985; Heikes et al., 1987), whereas gaseous SO_2 concentration was at a modest level. These field studies imply that a significant portion of gaseous H_2O_2 is utilized for the oxidation of SO_2 in humid air where large amounts of water droplets exist.

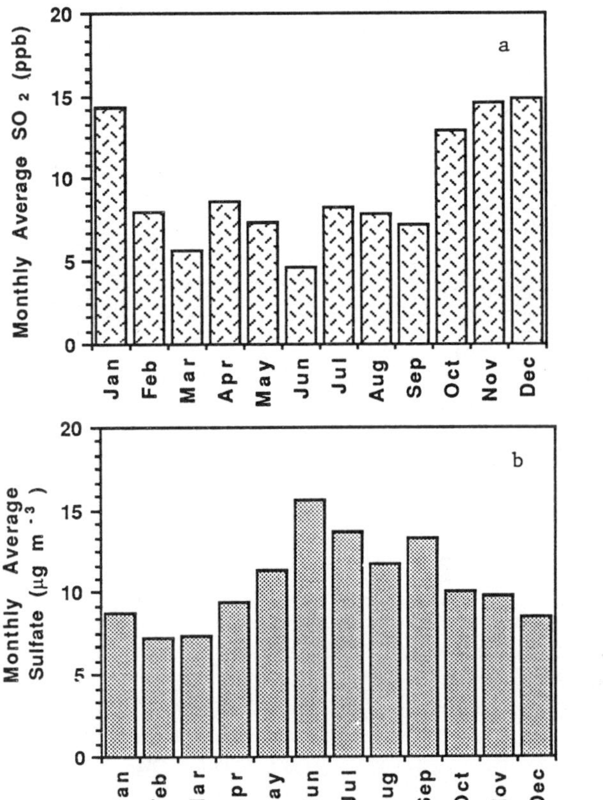

Fig. 5. Seasonal variation of ambient SO_2 at Lennox in Los Angeles in 1983 (a) and seasonal variation of aerosol sulfate in downtown Los Angeles during 1980-83 (b), according to South Coast AQMD (1983).

Heterogeneous reactions of SO_2 in humid air is now generally recognized as being an important process for the formation of aerosol sulfate (Cass, 1981; Middleton et.al., 1982; McMurry and Wilson, 1983; Bizjak et al., 1986; Lamb et al., 1987; Saxena and Seigneur, 1987). A good correlation between RH and the concentration of aerosol sulfate was observed in Los Angeles (Hidy et al., 1978; Henry and Hidy, 1979; Cass, 1979; Zeldin et al., 1983). Seasonal variation of ambient aerosol sulfate is high in summer when low stratus clouds are often found in Los Angeles, whereas the SO_2 content in air is highest in winter (Fig. 5), suggesting that, in addition to the gas phase oxidation by OH• (Cass, 1981), the heterogeneous oxidations of SO_2 in air may determine the seasonal trend of aerosol sulfate. Our study indicates that H_2O_2 decomposition by SO_2 may occur in humid air from spring through early autumn (Case study 3), whereas the formation rate of H_2O_2 is highest in warm months. These statistical relationships are supported by our calculations on the production rate of sulfate in Los Angeles air which indicates that gaseous H_2O_2 is the dominant oxidant for SO_2 in atmospheric water droplets, especially in summer.

Case study 2 provides information on why temperature, which is a good indicator for air mass stability in Los Angeles, is influential in the formation of O_3. A low IH causes stagnant air and an accumulation of primary pollutants in the atmosphere at near ground level with a resultant increase in temperature. Such conditions clearly favor the generation of O_3 due to the availability of NO_x, NMHC and CO which are source materials for photochemical reactions of O_3. Temperature dependency of ambient O_3 concentration in Los Angeles air has been previously recognized by several workers (Henry and Hidy, 1979; Zeldin et al., 1983; South Coast AQMD, 1985; Kuntasal and Chang, 1987). In contrast, the relationship between H_2O_2 and the local meteorology was not decisive in this study.

Whereas fog water chemistry in Los Angeles air clearly indicates that gaseous H_2O_2 is a dominant oxidant for SO_2 during the early stages of fog events, O_2 catalyzed by Fe(III) and Mn(II) could be dominant after the depletion of H_2O_2 (Jacob and Hoffmann, 1983; Hoffmann and Jacob, 1984; Seigneur and Saxena, 1984). According to these studies, gaseous H_2O_2 will be depleted within several minutes (<30 min.) at night when limited availability of H_2O_2 is present and when no photochemical generation of gaseous H_2O_2 is expected.

Clearly, further studies on the availability of oxidants and the production rate of sulfate in the air is necessary to estimate future trends of sulfate deposition in Los Angeles and elsewhere. The effect of NO_x on the formation of gaseous H_2O_2 also needs to be further investigated. NO_x content in the air has gradually decreased in recent years in Los Angeles due to the reduction in NO_x emission from mobile sources and the electric utility industry (8% reduction during 1979-83; CARB, 1985). If NO significantly reduces the H_2O_2 atmospheric content, the current decreasing trend of NO_x emission could increase the gaseous H_2O_2 content in the atmosphere and consequently may lead to an increase in sulfate concentration in precipitation. Finally, the distribution of individual VOCs and their emission and/or photochemical generation process must be understood in order to estimate the extent of VOCs involved in the generation of gaseous H_2O_2 in the Los Angeles atmosphere.

CONCLUSION

(1) The ambient concentrations of gaseous H_2O_2 measured were 0.36 (0.03-1.35, n=150), 0.42 (0.12-0.78, n=16), 1.19 (0.20-2.04, n=9) and 1.02 ppb (0.43-1.72 ppb, n=13) at Westwood, Duarte, Daggett and the San Bernardino Mountains, respectively during August 1985-September 1988. The diurnal variation of H_2O_2 was highest in the early afternoon at all the sites studied when the photochemical activity was at a maximum, suggesting that H_2O_2 is photochemically generated. A clear seasonal variation of gaseous H_2O_2 was found at Westwood during 1985-88. Simultaneous measurement of gaseous H_2O_2 suggests that it can reach highest concentrations in rural areas such as the Mojave Desert and the San Bernardino Mountains due to lower concentrations of primary pollutants, which could react with H_2O_2 or inhibit the formation of H_2O_2.

The seasonal trend of O_3 was less obvious at Westwood, although the O_3 concentration was generally higher in summer than in winter. There was no decisive relationship between the data of H_2O_2 and O_3 in the statistical analysis performed in this study, suggesting that the generation or decomposition mechanisms of H_2O_2 and O_3 are different from each other.

(2) Statistical analysis indicated that the concentration of gaseous H_2O_2 is best correlated with SR among the environmental parameters studied. A regression analysis indicates that the concentration of gaseous H_2O_2 increases with increasing SR at a statistically significant level ($P<0.05$). R^2 of SR for the H_2O_2 variance was 0.29 in the regression analysis. SR is considered to be of primary importance for the generation of free radical species, the precursors of H_2O_2, in the air.

SO_2 was found to be inversely correlated with gaseous H_2O_2 when haze events were observed at Westwood. The concentration of gaseous H_2O_2 was low when SO_2 content was high under hazy air conditions. These results strongly suggest that SO_2 is a major factor controlling the concentration of H_2O_2, due to heterogeneous reaction of H_2O_2 with SO_2 in water droplets. These results (which confirm studies performed by other workers elsewhere), suggest that gaseous H_2O_2 may largely determine the formation of aerosol sulfate in humid air during warm months.

On the other hand, the total variance of O_3 can be largely explained by temperature ($R^2=0.42$). Our results suggest that temperature is a good indicator of air mass stability in Los Angeles because high temperature is largely associated with a low inversion layer and stagnant air, which result in an optimal condition for the accumulation of primary pollutants, such as NO_x, NMHC and CO, and the resulting O_3 formation.

(3) Estimation of production rates of sulfuric acid by H_2O_2, O_3 and O_2 (catalyzed by Fe and Mn) oxidation of SO_2 suggests that (a) H_2O_2 is the dominant oxidant for SO_2 at low pH (<5.5) in water droplets, and (b) the production rate of sulfuric acid is lower in winter than in summer due to limited availability of oxidants. Thus, the reduction in SO_2 emissions from anthropogenic sources, which has been accomplished in Los Angeles by use of natural gas for power generation, may not lead to an equivalent reduction in particulate sulfate or in dissolved sulfate, where rain events principally occur in the winter season.

(4) Concentration of gaseous H_2O_2 during fog events was very low (< 0.05 ppb, n=5) at Westwood, when SO_2 content in the air was high (about 10ppb), suggesting the heterogeneous oxidations of SO_2 by H_2O_2 in fogwater. The actual concentration of H_2O_2 in rain (0.1-95μM) was within the range of predicted H_2O_2 values from its Henry's law constant. Diurnal patterns of rainwater H_2O_2 correspond to those of gaseous H_2O_2. These results suggest that H_2O_2 in rainwater mostly comes from gaseous H_2O_2.

REFERENCES

Appel, B.R., Kothny, E.L., Hoffer, E.M., Hidy, G.M and Wesolowski, J.J., 1978, Sulfate and nitrate data from the California Aerosol Characterization Experiment (ACHEX), Environ. Sci. Technol., 12: 418.

Appel, B.R., Hoffer, E.M., Tokiwa, Y. and Kothny, E.L., 1982, Measurement of sulfuric acid and particulate strong acidity in the Los Angeles Basin, Atmos. Environ., 16: 589.

Bahnenmann, D.W., Hoffman, M.R., Hong, A.P. and Kormann, C., 1987, Photocatalytic formation of hydrogen peroxide, in: "The Chemistry of Acid Rain," R.W. Johnson and G.E. Gordon, eds., American Chemical Society, Washington, D.C..

Bizjak, M., Hudnik, V., Hanse, A.D.A. and Novakov, T., 1986, Evidence for heterogeneous SO_2 oxidation in Ljubljana, Yugoslavia, Atmos. Environ., 20: 2199.

Brewer, R.L., Gordon, R.J. and Shepard, L.S., 1983, Chemistry of mist and fog from the Los Angeles urban area, Atmos. Environ., 17: 2267.

California Air Resources Board, 1979-84, California Air Quality Data.

California Air Resources Board, 1985, The Effect of Oxides of Nitrogen on California Air Quality.

California Air Resources Board, 1988, Fifth Annual Report to the Governor and Legislature on the Air Resources Board's Acid Deposition Research and Monitoring Program.

Calvert, J.G., and Stockwell, W.R., 1983, Acid generation in the troposphere by gas-phase chemistry, Environ. Sci. Technol., 17: 428A.

Calvert, J.G., and Stockwell, W.R., 1984, Mechanism and rates of the gas-phase oxidations of sulfur dioxide and nitrogen oxides in the atmosphere, in: "SO_2, NO and NO_2 oxidation Mechanisms: Atmospheric Considerations," J.G. Calvert, ed., Butterworth Publishers, Boston.

Calvert, J.G., Lazrus, A., Kok, G.L., Heikes, B.G., Walega, J.G., Lind, J. and Cantrell, C.A., 1985, Chemical mechanisms of acid genertion in the troposphere, Nature, 317: 27.

Cass, G.R., 1979, On the relationship between sulfate air quality and visibility with examples in Los Angeles, Atmos. Environ., 13: 1069.

Cass, G.R., 1981, Sulfate air quality control strategy design, Atmos. Environ., 15: 1227.

Chameides, W.L., 1984, The photochemistry of a remote marine stratiform cloud, J. Geophys. Res., 89: 4739.

Chameides, W.L., and Davis, D.D., 1982, The free radical chemistry of cloud droplets and its impact upon the composition of rain, J. Geophys. Res., 87: 4862.

Daum, P.H., Kelly, T.J., Schwartz, S.E. and Newman, L., 1984, Measurement of the chemical composition of stratiform clouds, Atmos. Environ., 18. 2671.

Graedel, T.E., and Goldberg, K.I., 1983, Kinetic studies of raindrop chemistry - I. Inorganic and organic processes, J. Geophys. Res., 88: 10865.

Graedel, T.E., Mandich, M.L., Weschler, C.J., 1986, Kinetic model studies of atmospheric droplet chemistry. 2. Homogeneous transition metal chemistry in raindrops, J. Geophys. Res., 91: 5205.

Groblicki, P.J., and Ang, C.C., 1985, Measurement of H_2O_2 without ozone interference, in Symposium on Heterogeneous Processes in Source-dominated Atmospheres, October 8-11, New York, Sponsored by NSF and U.S. Department of Energy, LBL-20261 Abstract, Report #36.

Grosjean, D., 1982, Formaldehyde and other carbonyls in Los Angeles ambient air, Environ. Sci. Technol., 16: 254.

Hartkamp, H., and Bachhausen, P., 1987, A method for the determination of hydrogen peroxide in air, Atmos. Environ., 21: 2207.

Heikes, B.G., Kok, G.L., Walega, J.G. and Lazrus, A.L., 1987, H_2O_2, O_3 and SO_2 measurements in the lower troposphere over the Eastern United States during fall, J. Geophys. Res., 92: 915.
Henry, R.C., and Hidy, G.M., 1979, Multivariate analysis of particulate sulfate and other air quality variables by principal components - Part I. Annual data from Los Angeles and New York, Atmos. Environ., 13: 1581.
Hidy, G.M., Mueller, P.K. and Tong, E.Y., 1978, Spatial and temporal distributions of airborne sulfate in parts of the United States, Atmos. Environ., 12: 735.
Hoffmann, M.R., and Jacob, D.J., 1984, Kinetics and mechanisms of the catalytic oxidation of dissolved sulfur dioxide in aqueous solution: an application to nighttime fog water chemistry, in: "SO2, NO and NO_2 Oxidation Mechanisms: Atmospheric Considerations," J.G. Calvert, ed., Butterworth Publishers, Boston.
Hov, Ø., 1983, One-dimensional vertical model for ozone and other gases in the atmosphere boundary layer, Atmos. Environ., 17: 535.
Jacob, D.J., 1986, Chemistry of OH in remote clouds and its role in the production of formic acid and peroximonosulfate, J. Geophys. Res., 91: 9807.
Jacob, D.J., and Hoffmann, M.R., 1983, A dynamic model for the production of H^+, NO_3^-, and SO_4^{2-} in urban fog, J. Geophys. Res., 88: 6611.
Jacob, D.J., Waldman, J.M., Munger, J.W., and Hoffmann, M.R., 1985, Chemical composition of fogwater collected along the California coast, Environ. Sci. Technol., 19: 730.
Jacob, P., Tavares, T.M. and Klockow, D., 1986, Methodology for the determination of gaseous hydrogen peroxide in ambient air, Fresenius Z Anal. Chem., 325: 359.
Kelly, T.J., Daum, P.H. and Schwartz, S.E., 1985, Measurements of peroxides in cloudwater and rain, J. Geophys. Res., 90, 7861.
Kleindienst, T.E., Shepson, P.B., Hodges, D.N., Nero, C.M., Arnts, R.R., Dasgupta, P.K., H. Hwang, H., Kok, G.L., Lind, J.A., Lazrus, A.L., Mackay, G.I., Mayne, L.K. and Schiff, H.I., 1988, Comparison of techniques for measurement of ambient levels of hydrogen peroxide, Environ. Sci. Technol., 22: 53.
Kleinman, L.I., 1984, Oxidant requirements for the acidification of precipitation, Atmos. Environ., 18: 1453.
Kleinman, L.I., 1986, Photochemical formation of peroxides in the boundary layer, J. Geophys. Res., 91: 10889.
Kok, G.L., Darnall, K.R., Winer, A.M., Pitts Jr., J.N. and Gay, B.W., 1978a, Ambient air measurements of hydrogen peroxide in the California south coast air basin, Environ. Sci. Technol., 12: 1077.
Kok, G.L., Holler, T.P., Lopez, M.B., Nachtrieb, H.A. and Yuan, M., 1978b, Chemiluminescent method for determination of hydrogen peroxide in the ambient atmosphere, Environ. Sci. Technol., 12: 1072.
Kumar, S., 1986, Reactive scavenging of pollutants by rain: a modeling approach, Atmos. Environ., 20: 1015.
Kunen, S.M., Lazrus, A.L., Kok, G.L. and Heikes, B.G., 1983, Aqueous oxidation of SO_2 by hydrogen peroxide, J. Geophys. Res., 88: 3671.
Kuntasal, G., and Chang, T.Y., 1987, Trends and relationships of O_3, NO_x and HC in the South Coast Air Basin of California, J. Air Pollution Control Association, 37: 1158.
Lamb, D., Miller, D.F., Robinson, N.F. and Gertler, A.W., 1987, The importance of liquid water concentration in the atmospheric oxidation of SO_2, Atmos. Environ., 21: 2333.
Lazrus, A.L., Kok, G.L., Lind, J.A., Gitlin, S.N., Heikes, B.G. and Shetter, R.E., 1986, Automated fluorometric method for hydrogen peroxide in air, Anal. Chem., 58: 594.
Liljestrand, H.M., and Morgan, J.J., 1981, Spatial variations of acid precipitation in Southern California, Environ. Sci. Technol., 15: 333.
Lind, J.A., and Kok, G.L., 1986, Henry's law determinations for aqueous solutions of hydrogen peroxide, methylhydroperoxide, and peroxyacetic acid, J. Geophys. Res., 91, 7889.
Maahs, H.G., 1983, Kinetics and mechanism of the oxidation of S(IV) by ozone in aqueous solution with particular reference to SO_2 conversion in nonurban tropospheric cloud, J. Geophys. Res., 88: 10721.
Martin, L.R., 1984, Kinetic studies of sulfate oxidation in aqueous solution, in: "SO_2, NO and NO_2 Oxidation Mechanisms: Atmospheric Considerations," J.G. Calvert, ed., Butterworth Publishers, Boston.
Martin, L.R., and Damschen, D.E., 1981, Aqueous oxidation of sulfur dioxide by hydrogen peroxide at low pH, Atmos. Environ., 15: 1615.
Masuch, G., Kettrup, A., Mallant, R.K.A.M., Slanina, J., 1986, Effects of H_2O_2-containing acidic fog on young trees, Inter. J. Environ. Anal. Chem., 27: 183.
McElroy, W.J., 1986, Sources of hydrogen peroxide in cloudwater, Atmos. Environ., 20: 427.
McMurry, P.H., and Wilson, J.C., 1983, Droplet phase (heterogeneous) and gas phase (homogeneous) contributions to secondary ambient aerosol formation as functions of relative humidity, J. Geophys. Res., 88: 5101.

Middleton, P., Kiang, C.S. and Mohnen, V.A., 1982, The relative importance of various urban sulfate aerosol production mechanisms - a theoretical comparison, in: "Heterogeneous Atmospheric Chemistry," D.R. Schryer, ed., American Geophysical Union, Washington, D.C..

Miller, P.R., Taylor, O.C. and Poe, M.P., 1986, Spatial variation of summer ozone concentrations in the San Bernardino Mountains, Presented at the 79th Annual Meeting of the Air Pollution Control Association, Minneapolis, Minnesota, June 22-27.

Möller, D., 1980, Kinetic model of atmospheric SO_2 oxidation based on published data, Atmos. Environ., 14: 1067.

Munger, J.W., Jacob, D.J., Waldman, J.M. and Hoffmann, M.R., 1983, Fogwater chemistry in an urban atmosphere, J. Geophys. Res., 88: 5109.

National Acid Precipitation Assessment Program, 1987, Vol. I-IV, Interim Assessment, The Causes and Effects of Acidic Deposition.

National Research Council, 1983, Acid Deposition: Atmospheric Processes in Eastern North America, National Academy Press, Washington, D.C..

National Research Council, 1986, Acid Deposition: Long-Term Trends, National Academy Press, Washington, D.C..

Penkett, S.A., Jones, B.M.R., Brice, K.A. and Eggleton, A.E.J., 1979, The importance of atmospheric ozone and hydrogen peroxide in oxidizing sulphur dioxide in cloud and rainwater, Atmos. Environ., 13: 123.

Richards, L.W., Anderson, J.A., Blumenthal, D.L., Mcdonald, J.A., Kok, G.L. and Lazrus, A.L., 1983, Hydrogen peroxide and sulfur(IV) in Los Angeles cloud water, Atmos. Environ., 17: 911.

Rodhe, H.P., Crutzen, P. and Vanderpool, A., 1981, Formation of sulfuric and nitric acid in the atmosphere during long-range transport, Tellus, 33: 132.

Römer, F.G., Viljeer, J.W., van den Beld, L., Slangewal, H.J., Veldkamp, A.A. and Reijnders, H.F.R., 1985, The chemical composition of cloud and rainwater. Results of preliminary measurements from an aircraft, Atmos. Environ., 19, 1847.

Sakugawa, H., and Kaplan, I.R., 1987, Collection of atmospheric H_2O_2: Comparison of cold trap method with impinger bubbling method, Atmos. Environ., 21, 1791.

Saxena, P., and Seigneur, C., 1987, On the oxidation of SO_2 to sulfate in atmospheric aerosols, Atmos. Environ., 21: 807.

Schwartz, S.E., 1984, Gas- and aqueous-phase chemistry of H_2O_2 in liquid-water clouds, J. Geophys. Res., 89: 11589.

Seigneur, C., and Saxena, P., 1984, A study of atmospheric acid formation in different environments, Atmos. Environ., 18: 2109.

Seigneur, C., and Saxena, P., 1988, A theoretical investigation of sulfate formation in clouds, Atmos. Environ., 22: 101.

Slemr, F., Harris, G.W., Hastie, D.R., Mackay, G.I. and Schiff, H.I., 1986, Measurement of gas phase hydrogen peroxide in air by tunable diode laser absorption spectroscopy, J. Geophys. Res., 91: 5371.

South Coast AQMD, 1983, Summary of Air Quality in California's South Coast Air Basin.

South Coast AQMD, 1984, Acid Deposition in the South Coast Air Basin: An Assessment.

South Coast AQMD, 1985, Air Quality Trends in the South Coast Air Basin 1975-1984.

Stockwell, W.R., 1986, A homogeneous gas phase mechanism for use in a regional acid deposition model, Atmos. Environ., 20, 1615.

Tanner, R.L., Markovits, G.Y., Ferren, E.M. and Kelly, T.J., 1986, Sampling and determination of gas-phase hydrogen peroxide following removal of ozone by gas-phase reaction with nitric oxide, Anal. Chem., 58: 1857.

Tuazon, E.C., Winer, A.M. and Pitts Jr., J.N., 1981, Trace pollutant concentrations in a multiday smog episode in the California South Coast Air Basin by long path length fourier transform infrared spectroscopy, Environ. Sci. Technol., 15: 1232.

Van Valin, C.C., Ray, J.D., Boatman, J.F. and Gunter, R.L., 1987, Hydrogen peroxide in air during winter over the south-central United States, Geophys. Res. Letters, 14: 1146.

Waldman, J.M., Munger, J.W., Jacob, D.J., Flagan, R.C., Morgan, J.J. and Hoffman, M.R., 1982, Chemical composition of acid fog, Science, 218: 677.

Waldman, J.M., Munger, J.W., Munger, D.J. and Hoffmann, M.R., 1985, Chemical characterization of stratus cloudwater and its role as a vector for pollutant deposition in a Los Angeles pine forest, Tellus, 37B: 91.

Young, J.R., Collins, J.F. and Coyner, L.C., 1986, Analysis of the Southern California Edison Precipitation Chemistry Data Base for Southern California, Environmental Research & Technology, Document No. P-D578-300, ERT Inc., Newberry Park, California, 91320.

Zeldin, M.D., Farber, R.J. and Keith, R.W., 1983, Statistical and case study meteorological relationships for sulfate formation in the Los Angeles Basin, in The Proceedings of the 76th Annual Meeting of the

Air Pollution Control Association, June 19-24, Atlanta.
Zeldin, M.D., and Ellis, E.C., 1984, Trends in precipitation chemistry in Southern California, in The Proceedings of the 77th Annual Meeting of the Air Pollution Control Association, June 24-29, San Francisco.
Zeldin, M.D., Ellis, E.C. and Huang, A.A., 1985, Empirical evidence for rapid heterogeneous sulfate formation in the Los Angeles urban airshed, in Symposium on Heterogeneous Processes in Source-Dominated Atmospheres, U.S. Department of Energy, LBL-20261 Abstracts, Report #19.

DETERMINATION OF DRY DEPOSITION OF ATMOSPHERIC AEROSOLS FOR POINT AND AREA SOURCES BY DUAL TRACER METHOD

Hilary Hafner Main and Sheldon K. Friedlander

Department of Chemical Engineering and National Center
for Intermedia Transport Research
University of California
Los Angeles, CA 90024, U.S.A.

INTRODUCTION

Dry deposition is the removal of gases and particles from the atmosphere to the earth's surface by processes other than precipitation. Rates of atmospheric dry deposition are usually characterized by the deposition velocity, v_d. Estimates of v_d are used in models which predict the fate of atmospheric pollutants. Measurements and predictions of v_d can vary among investigators by several orders of magnitude (Sehmel, 1980). This variation occurs because v_d is a complex function of particle size, shape, and composition; the calculation or measurement method; meteorology; and deposition surface properties.

In a previous study, Friedlander et al. (1986), showed that an average v_d for atmospheric aerosols can be estimated by measuring the concentration ratio of a depositing species (such as lead, ZnS) to a non-depositing, conserved species (such as carbon monoxide, SF_6), provided both species originate from the same source (dual tracer method). We have successfully applied this method to both area (urban vehicular emissions) and point sources.

During the South Coast Air Quality Study (SCAQS), summer 1987, we measured the size distribution of Pb in ambient and vehicle source (tunnel) aerosols with an 8-stage low pressure impactor (LPI) (Hering et al., 1978). We originally planned to use the Friedlander et al. (1986) method for calculating v_d as a function of particle size. When viewed as an average over a large size range, such as 0-1 or 1-10 μm, this method accounted for the diffusion or interception processes which remove some particles during transport. However, when we reviewed the size-segregated data, of those particles which remain, many may move from smaller to larger sizes by growth (condensation) or coagulation. We then modified the calculation of v_d to estimate the contribution of particle growth.

To calculate v_d for particles from point sources, we used literature data from field studies of simultaneous releases of depositing and nondepositing tracers. The dependence of v_d on wind speed was then determined.

EXPERIMENTAL

Area Source

Ambient concentration ratios of Pb and CO ([Pb]/[CO]) were measured at the SCAQS site in Claremont, CA, located approximately 30 miles east of Los Angeles. Samples were collected on June 19, 24, 25, July 13-15, August 27-29, and September 2-3. Both Pb and CO originate primarily from vehicular exhaust in the L. A. basin. Samples from the Claremont area had been shown to represent the average ambient air in the basin (Huntzicker, 1975). To characterize vehicular source emissions, measurements of [Pb]/[CO] were obtained in a separate study of the Caldecott tunnel, a freeway commuter tunnel located near Berkeley, CA (Turner, 1988).

The Hering LPI has 50% efficiency cutoffs of 4.0 (stage 1), 2.0, 1.0, 0.5, 0.26, 0.12, 0.075, and 0.05 (stage 8) μm aerodynamic diameter. Impaction surfaces in the LPI were teflon filters oiled with oleic acid. The oil-coated filter maintains a solid particle sticking efficiency >90% independent of loading (Turner and Hering, 1987). Samples were analyzed for Pb by atomic absorption. CO was measured continuously with a Dasibi Model 3003 CO monitor (Wolff and Ruthkowsky, 1987) in Claremont and with an Ecolyzer 2106 CO detector in Berkeley. Further details of sampling and analysis are presented elsewhere (Main, 1988.)

Point Source

The Pacific Northwest Laboratory measured airborne concentrations during simultaneous releases of ZnS particles, mass mean diameter (mmd) = 4.8 μm, and SF_6 tracer gas (Doran and Horst, 1985). The test site was located in a semi-arid region of flat terrain covered with desert grasses and 1- to 2- m high sagebrush. The two tracers were released from a height of 2 m and sampled at 100, 200, 800, 1600, and 3200 m downwind. Wind speeds ranged from 1.2 to 3.5 m s^{-1} at the surface (2 m). All releases were made during moderately stable to near-neutral atmospheric conditions.

SIMPLE ESTIMATE OF DEPOSITION VELOCITY

Area Source

Friedlander et al. (1986) applied a continuously stirred atmosphere model (ie. box model, Hanna et al. (1982)) to the Los Angeles basin to estimate v_d for submicron lead-containing particles. Data from the field study indicated that particle growth (by secondary pollutant gas-to-particle conversion) may be important in the 0.26 to 1 μm size range (Main and Friedlander, 1988). For example, this size range corresponds to the peak normally observed in Los Angeles sulfate size distributions (Hering and Friedlander, 1982). (Particle growth by coagulation could also occur, but aerosols in the size range 0.1-1.0 μm grow principally by gas-to-particle conversion (Heisler and Friedlander, 1977).) Thus, when applying the stirred-atmosphere model on a size-segregated basis, the balance on the depositing species should incorporate a term which accounts for particle growth.

The general equation, where i represents the particle size range (impactor stage):

$$Q_{2i} = c_{2i}q + f_{di}a + c_{gi}q \tag{1}$$

Dividing eqn. (1) by the balance on the conserved species ($Q_1 = c_1q$) and rearranging gives β_i, the fraction of Pb emitted which deposits over a and δ_i, the net fraction of particle growth:

$$\frac{f_{di}a}{Q_{2i}} = \beta_i - \delta_i \tag{2}$$

The deposition velocity is given by:

$$v_{di} = \frac{f_{di}}{c_{2i}} \tag{3}$$

and after some rearranging the resulting general expression for v_{di}:

$$v_{di} = \frac{(\beta_i - \delta_i)\; Q_{10}}{1-\beta_i \quad c_{10}a} \qquad \text{(for LPI, i = 2 - 8)} \tag{4}$$

For no net particle growth, $\delta_i = 0$, eqn. (4) reduces to the relationship used by Friedlander et al. (1986). (Stage 1, 4 μm cutoff diameter, does not have a well-defined upper cut point.)

Point Source

Several methods have been proposed to incorporate deposition losses into the dispersion model for pollutant releases from point sources. The simplest method of Chamberlain (1953) and Hanna et al. (1982), referred to as the "source depletion" model, assumes that the source strength decreases with distance from the source. Friedlander et al. (1986) used this method to calculate v_d from dual tracer data (subscripts 1 and 2 refer to the conserved and depositing species, respectively):

$$v_d = u\,(\pi/2)^{\frac{1}{2}}\,(1/I)\,\ln\left[\frac{C_1(x)}{Q_{10}}\;\frac{Q_{20}}{C_2(x)}\right] \qquad (5)$$

RESULTS AND DISCUSSION

Area Source

Figure 1 shows the [Pb]/[CO] ratios as a function of size for the ambient and source (Caldecott tunnel). The source [Pb]/[CO] ratios were greater than the ambient ratios in all size ranges except 0.26-1 μm. In this size range, the losses and gains of particulate Pb approximately balance.

Applying eqn. (4) as we originally intended, ie. $\delta_i = 0$, we calculated $v_d = 0$ for 0.26 - 1 μm particles; the range where source and ambient [Pb]/[CO] ratios were equal. Thus, to extract v_d for particles in that size range, we added the growth terms, δ_6, corresponding to growth out of size range 6 (0.12 - .26 μm), δ_5, net growth in size range 5 (.26 - .5 μm), and δ_4, growth into size range 4 (0.5 - 1 μm). It follows that $\Sigma\, \delta_i = 0$. By adding these new variables, we now need new equations to obtain a solution.

For particles in the diffusion-control range, $v_d \sim D^{2/3}$. (Fernandez de la Mora and Friedlander, 1982). This proportionality provides two additional equations for the diffusion range, stages 5 and 6 (0.12 - 0.5 μm). Rates of particle removal efficiency from the atmosphere pass through a minimum in the size range around 0.3 μm (see for example, Sehmel, 1980). In this study, we did not consider additional growth terms because we assumed diffusion or interception were the most significant removal mechanisms of particles in the size ranges < 0.12 to > 0.5 μm, respectively (Schack et al., 1985). In summary: $v_{di} = 0.24\beta_i/(1 - \beta_i)$ for i = 2, 3, 7, 8; $v_{di} = 0.24(\beta_i - \delta_i)/(1 - \beta_i)$ for i = 4-6; $v_{di}/v_{dj} = (D_i/D_j)^{2/3}$ for i = 5, 6, j = 7 or 8; and $\Sigma\, \beta_i = 0$.

The resulting v_d's are shown in Figure 2. The curve has the characteristic v-shape with a minimum in the 0.1 to 0.4 μm range. We have plotted v_d for particle densities of 1 and 6 g cm^{-3}. The density of the Pb-containing particles is unknown. We have chosen a range of densities for Figure 2 corresponding to lead halide particles (density = 6 g cm^{-3}) and exhaust soot particles (density = 1 g cm^{-3}). Auto exhaust Pb is present primarily as lead halide (Habibi, 1973). Soot is abundant in submicron particles emitted from vehicles (Kittelson et al., 1978; Habibi, 1973) and some lead may be associated with these particles. The curves shown in Figure 2 compare well to curves based on published models for v_d (Milford and Davidson, 1987).

Repeating the calculation of v_{di} without size resolution (ie. for one size range between 0.05 -1 μm), we obtain $v_d = 0.13$ cm s^{-1} for particles < 1 μm. (Note that the net growth in this case = 0.) Friedlander et al. (1986) calculated $v_d = 0.26$ cm s^{-1} for the same particle size range. Different measurement and analysis techniques were used to determine Pb concentrations in the earlier estimate of v_d. For example, 60% of the ambient total particle sample was assumed to be < 1 μm. Considering the uncertainty in this assumption, the v_d values agree well.

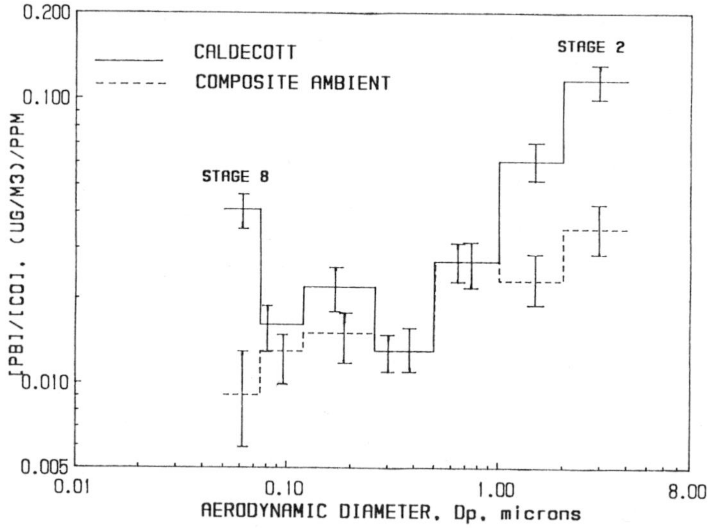

Figure 1. [Pb]/[CO] ratios for the Caldecott tunnel (vehicular source, Berkeley, CA) and for a composite of five ambient aerosol samples (Claremont, CA). The ambient data represent a total sampling time of 260 hours.

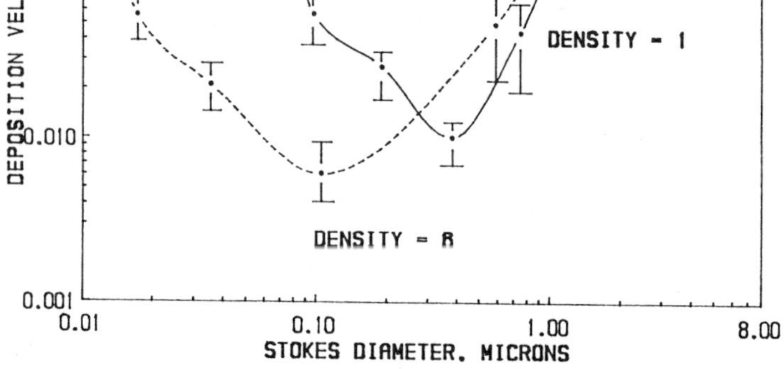

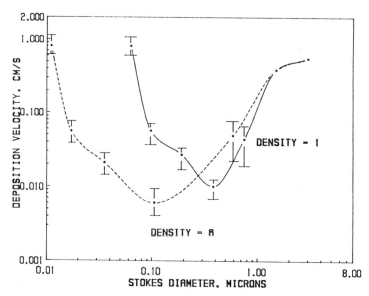

Figure 2. Deposition velocity as a function of particle size for particle densities of 1 and 6. Only significant error bars are shown. The mass average deposition velocity over the entire size range (0.05 - 4 μm) = 0.28 cm s^{-1}.

Fernandez de la Mora and Friedlander (1982) showed theoretically that in the interception regime for well-defined collectors, $v_d - v_s \sim dp^2$. As a check on our calculations for the larger particles, we plotted $v_d - v_s$ vs dp in Figure 3. The slope for the composite ambient data was 1.8 for a particle density = 6 g cm^{-3} and 1.9 for density = 1 g cm^{-3}. The expected relationship based on theory is a slope of 2. The calculated slopes agree with the theoretical value within experimental error. More information on the density of Pb- containing particles would be needed to determine v_d (and v_s) with more certainty.

Point Source

Fernandez de la Mora and Friedlander (1982) have shown theoretically that for particle deposition by interception $v_d - v_s \sim u^{3/2}$. We calculated v_s = 0.28 cm s^{-1} (Friedlander, 1977) based on a particle density of 4 (ZnS) and the estimated mmd = 4.8 μm. Figure 4 shows v_d-v_s plotted versus the average wind speed at 2 m. Agreement of our calculated deposition velocities with the theoretical dependence of v_d in the interception regime is very good.

Table I shows the dependence of $v_d - v_s$ on wind speed as a function of downwind distance. The calculated dependence on wind speed, exponent b, is closer to the expected value for interception of 1.5 at the most distant receptor. We have assumed an average v_s which remains constant with distance. However, since the ZnS particle size distribution changes with distance, we would expect the average v_s to change as well. In fact, by using an arbitrary value for v_s of 0.5 cm s^{-1} at 800 m, the source depletion model gives b = 1.48. Thus, the average v_s based on the mmd may be too small for samples collected close to the source.

CONCLUSIONS

We applied the stirred-atmosphere model to aerosol samples taken in the L. A. basin to calculate an average v_d for aerosol Pb from urban vehicular emissions as a function of particle size. We found that particle growth by gas-to- particle conversion may be an important mechanism for particles in the 0.12 - 1 μm size range; we added a term in the v_d calculation to estimate this contribution. Average deposition velocities calculated in this way varied over three orders of magnitude depending on the particle size. Results from the area source field study were encouraging; we recommend that additional measurements be made in one geographical area for source and ambient aerosols using the same measurement methods at all sites.

Deposition velocities were calculated from experimental data using the source depletion model (Friedlander et al., 1986) in a simple and straightforward manner. Deposition velocities calculated in this way were approximately proportional to $u^{3/2}$ as shown theoretically for the interception mechanism of deposition by Fernandez de la Mora and Friedlander (1982).

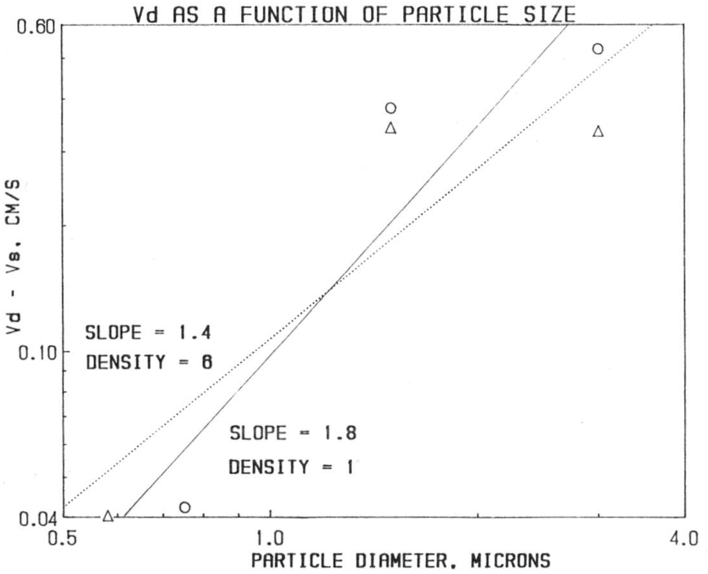

Figure 3. Deposition velocity as a function of aerodynamic diameter for particles in the interception regime. In this regime, the theoretical dependence of v_d on particle diameter is $v_d - v_s \sim dp^2$.

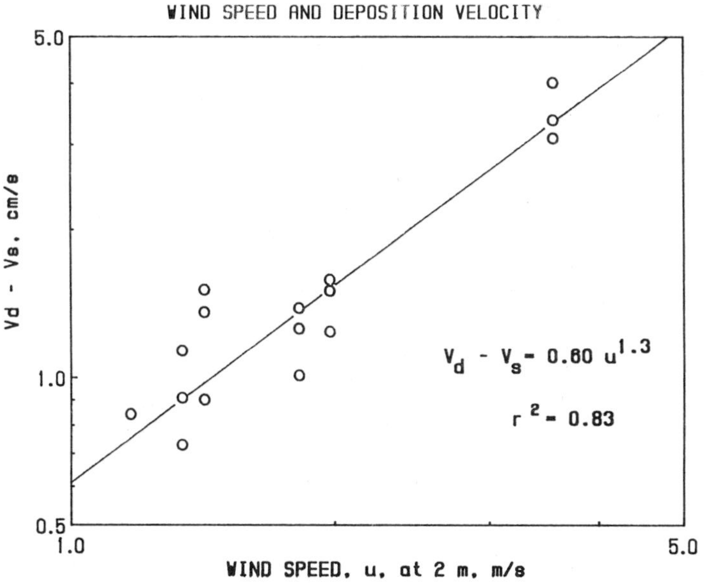

Figure 4. Deposition velocities calculated by the source depletion model using field data from Doran and Horst, (1985). Theory for deposition by interception predicts $v_d - v_s \sim u^{3/2}$. Agreement of the model estimates of v_d with theory is good.

Table 1. Dependence of Deposition Velocity on Wind Speed. Fit of Experimental Data: $V_d - V_s = a\ u^b$

Source Depletion Model			
x(m)*	a	b	r^2
800	.670	1.25	.840
1600	.726	1.30	.951
3200	.465	1.49	.926

Average wind speed, u, at the source measured at 2 m high.

v_s = 0.28 cm s^{-1} based on particle density of 4 and mass mean diameter of 4.8 μm.

* x = distance downwind.

Acknowledgements This work was supported in part by the U.S. Environmental Protection Agency assistance agreement with the National Center for Intermedia Transport Research at UCLA (Grant CR-812771). The contents do not necessarily reflect the views and policies of the Environmental Protection Agency. S.K. Friedlander is Parsons Professor of Chemical Engineering. We extend our appreciation to Mr. J. Turner for the Caldecott tunnel samples, Mr. H. Pourmand and Dr. S. Hering for assistance in the field study, Dr. E. Steinberger for her dispersion modeling expertise, and Mr. M. Ruthkowsky and Dr. G. Wolff at General Motors for unpublished data.

Nomenclature

dp = aerodynamic diameter

D = diffusion coefficient

c_1 = atmospheric concentration of conserved, nondepositing species (CO)

c_2 = atmospheric concentration of depositing species (Pb)

c_{10}, c_{20} = concentration measured at the source

c_{gi} = net growth into the size range i

Q_1 = rate of emission of nondepositing species (CO)

Q_{10}, Q_{20} = rate of emission from source

β_i = fraction of Pb emitted which deposits, $1 - \dfrac{c_{2i}c_{10}}{c_1 c_{20i}}$

δ_i = net fraction of particle growth, $\dfrac{c_{gi}c_{10}}{c_1 c_{20i}}$

q = steady volumetric flow rate of air through the L.A. basin

f_{di} = deposition flux

a = surface area over which deposition occurs

$\dfrac{Q_1}{c_1 a}$ = 0.24 cm s^{-1} for CO in the Los Angeles Basin (Friedlander et al., 1986) u = wind speed at the sampling height

$$I = \int_0^x \frac{dx}{\sigma_z \exp(h^2/\sigma_z^2)}$$

x = downwind distance

σ_z = vertical dispersion coefficient as defined by Hanna et al., 1982

h = release height

v_s = settling velocity (Friedlander, 1977)

$$= \rho \frac{gdp^2 C}{18\mu} \left(1 - \frac{\rho}{\rho_p} \right)$$

ρ_p = particle density

ρ = gas density

g = acceleration due to gravity

C = Cunningham slip factor

μ = gas viscosity

References

Chamberlain, A.C., (1953), Aspects of travel and deposition of aerosol and vapour clouds. UKAEA Report No. AEREHP/R- 1261, Harwell, Berkshire, England.

Doran, J.C., and Horst, T.W., (1985), An evaluation of gaussian plume depletion models with dual-tracer field measurements, Atm. Environ., 19: 939-951.

Fernandez de la Mora, J., and Friedlander, S.K. (1982), Aerosol and gas deposition to fully rough surfaces: filtration model for blade-shaped elements, Int. J. Heat Mass Transfer, 25: 1725-1735.

Friedlander, S.K., (1977), Smoke, Dust and Haze, J. Wiley and Sons.

Friedlander, S. K., Turner, J.R., and Hering, S.V. (1986), A new method for estimating dry deposition velocity for atmospheric aerosols," J. Aerosol Sci., 17: 240-244.

Habibi, K., (1973), Characterization of particulate matter in vehicle exhaust, Environ. Sci. Technol., 7: 223-234.

Hanna, S. R., Briggs, G.A., and Hosker, Jr., R.P., (1982), "Handbook on Atmospheric Diffusion", NTIS U. S. Dept. Commerce, Springfield, VA 22161.

Heisler, S.L., Friedlander, S.K. (1977), Gas to particle conversion in photochemical smog: Aerosol growth laws and mechanisms for organics, Atmos. Environ., 11, pp. 157-168.

Hering, S.V., Flagan, R. C., and Friedlander, S. K. (1978), Design and evaluation of new low-pressure impactor. I., Environ. Sci. Technol., 12: 667-673.

Hering, S.V., and Friedlander, S.K., (1982), Origins of aerosol sulfur size distributions in the Los Angeles basin, Atmos. Environ., 16: 2647-2656.

Huntzicker, J.J., Friedlander, S.K., Davidson, C.I., (1975), Material balance for auto-emitted lead in Los Angeles basin, Environ Sci. Technol., 9, pp. 448-456.

Main, H.H. (1988), M.S. Thesis in Chemical Engineering, University of California, Los Angeles.

Main, H.H., and Friedlander, S.K., (1988), Dry deposition of atmospheric aerosols by dual tracer method. Part I: Area source, Atmos. Environ., submitted for publication.

Milford, J.B., and Davidson, C.I., (1987), The sizes of particulate sulfate and nitrate in the atmosphere - A review, JAPCA, 37: 125-134.

Schack, C.J., Pratsinis, S.E., Friedlander, S.K. (1985), A general correlation for deposition of suspended particles from turbulent gases to completely rough surfaces, Atmos. Environ., 19: 953-960.

Sehmel, G. (1980), Particle and gas dry deposition: A review, Atmos. Environ., 14: 983-1011.

Turner, J.R., and Hering, S.V., (1987), Greased and oiled substrates as bounce-free impaction surfaces, J. Aerosol Sci., 18: 215-224.

Turner, J.R. (1988), The dynamics of carbon thermograms, Chemical Engineering Department, Washington U., St. Louis.

Wolff, G., and Ruthkowsky, M. (1987), Personal communication.

TRANSPORT OF TRACE ELEMENTS TO MAN BY ATMOSPHERIC AEROSOL

Walter John

Air and Industrial Hygiene Laboratory
California Department of Health Services
Berkeley, CA 94704

INTRODUCTION

Pollutants undergo complicated exchanges between the air, land and water media, following multiple pathways to a receptor (Figure 1). In considering the transport of material through the media, special attention is warranted for those pathways leading to man. One of the pathways which has received insufficient attention is via atmospheric aerosol. Bowen (1979) stated that, "They constitute a neglected, but sometimes important, source of elements for organisms and must be considered in any quantitative

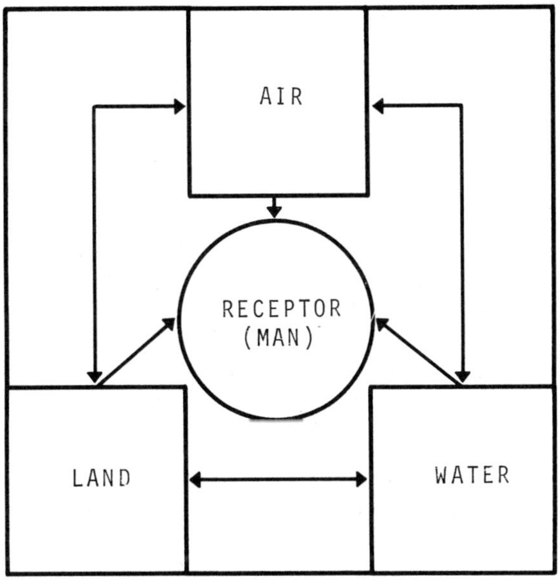

Figure 1. Trace elements are interchanged between the media and are deposited on receptors, including man.

studies of elemental cycling." Some of the trace elements transported by aerosol are toxic; anthropogenic sources are contributing an important and increasing fraction of these elements (Nriagu and Pacyna, 1988).

Trace element concentrations in human blood are strongly correlated to those in atmospheric aerosol (John, 1983). The probable reason for this correlation is that aerosol carries the trace elements on the major pathways leading to man. These pathways are discussed in the present paper.

TRACE ELEMENTS IN HUMAN BLOOD

Hamilton, et al. (1972/1973) published measurements of trace elements in the blood from 2,500 persons in the United Kingdom (U.K.), the largest single data set on multielemental concentrations in blood. Moreover, the sampling and analysis were done under stringent conditions to prevent contamination. This is very important since the concentrations in blood range down to about 1 ppb. Hamilton, et al. (1972/1973) mentioned that the trace elemental composition of blood bore some similarity to that in crustal rock and in sea water. However, John (1983) found that R^2 (R = correlation coefficient) was only 0.29 for concentrations in blood vs. those in crustal rock and the slope of the regression line was only 0.45. For sea water, R^2 was 0.41 and the slope was 0.33.

A much stronger correlation is found with trace elements in atmospheric aerosol. In Figure 2, the trace element concentrations in U.K. blood from Hamilton et al. (1972/1973) are plotted vs. those in "global"

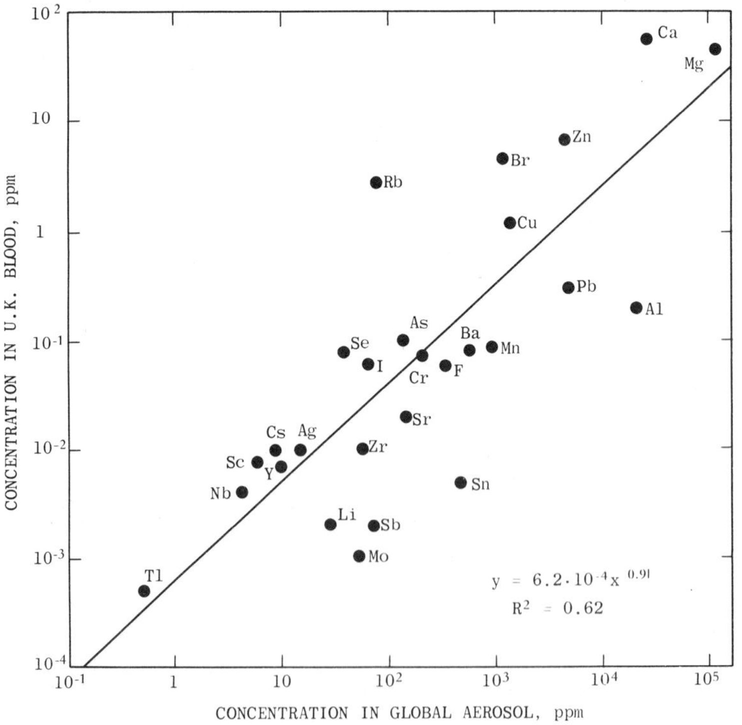

Figure 2. Trace element concentrations in U.K. Blood vs. that in global aerosol, with regression line (from John, 1983)

aerosol. "Global" aerosol was taken from the average of 104 aerosol data sets complied by Rahn (1976). These data were taken worldwide and should be broadly representative of aerosol composition. Use of more limited data from the United Kingdom does not alter the results qualitatively. The regression in Figure 2 has $R^2 = 0.62$, and the slope is 0.91, a strong correlation.

A further examination (John, 1983) of the data of Hamilton, et al., showed that the trace element concentrations in lung, brain and lymph node tissues were very strongly correlated with the concentrations in blood. Thus blood concentrations were found to be representative of those in body tissues.

PATHWAYS TO MAN

The present hypothesis is that trace element concentrations in blood are correlated to those in atmospheric aerosol because aerosol transports trace elements along the major pathways to man. In Figure 3, the pathways trace elements may take, beginning with atmospheric aerosol, are shown schematically. The simplest route is direct deposition in the lungs via respiration. In a few cases this is important, but, in general, it is not a major pathway. For example, even in the case of lead, diet is more important. Another relatively direct pathway is particle deposition into drinking water. Again, this is not the major pathway, but one which cannot be neglected.

Undoubtedly, on the basis of mass balance, the most important pathway is through diet. This is a complicated pathway. Particles are deposited on surfaces by the process of wet or dry deposition. Deposition onto plant surfaces can lead to direct ingestion by man through consumption of vegetables. This is particularly obvious in the case of leafy salad vegetables. Livestock consume plants containing particles and then man eats

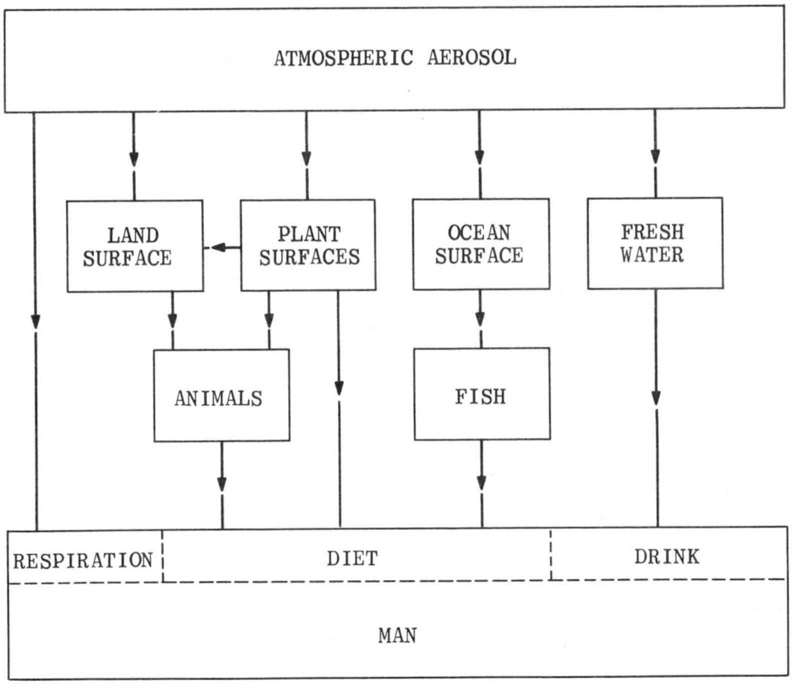

Figure 3. Schema of trace element pathways to man.

the meat. Grazing animals also ingest a considerable amount of soil and with it the particles which have deposited on the land surface. Some trace elements in particles depositing on land surfaces can be taken up by plants. The finely divided particulate matter, much of which is delivered with rain, should be readily available to plants.

Particles also deposit on the surfaces of the oceans and enter the marine food chain which begins with the phytoplankton on the water surface. To generalize, aerosol deposits on the interface between the atmosphere and land and the interface between the atmosphere and water, precisely where the food chain begins.

The above arguments are corroborated by the strong correlation between the trace element concentrations in the U.K. blood and the U.K. diet found by John (1983) using data from Hamilton, et al. (1972/1973). In the case of organic chemicals, Travis (1988) believes diet to be the primary pathway for human exposure, with most of the organic pollutants in plants coming through the air rather than via the roots. Thus, similar conclusions are reached concerning the major pathways for organic and for elemental pollutants. In both cases, quantitative assessments of the pathways are needed.

SOURCES OF TRACE ELEMENTS IN AEROSOL

Nriagu and Pacyna (1988) have published a new inventory of worldwide emissions of trace metals. Their data are mainly for anthropogenic sources, but in a few cases they list estimates of natural emissions. For these latter cases, it is possible to compute the total emissions and to plot them vs. aerosol concentrations, where the latter are also available. For the plot in Figure 4, the regression results are $R^2 = 0.66$, and the slope is 0.63. The uncertainties are relatively large and the data are few, but

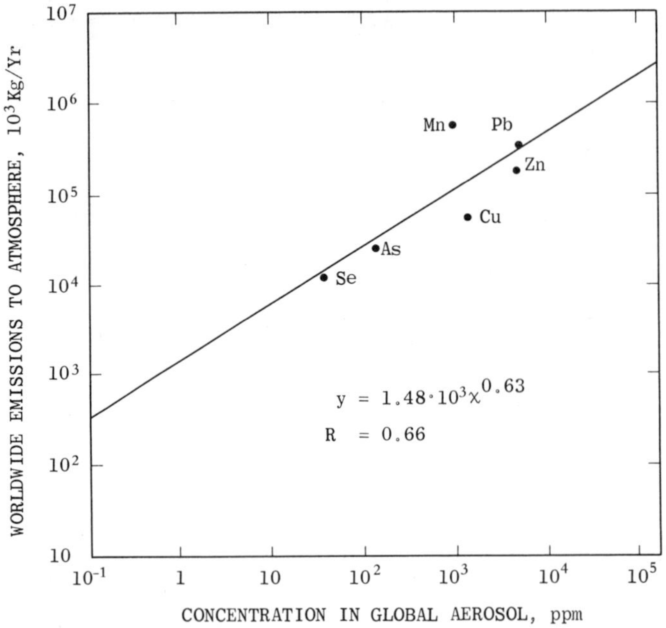

Figure 4. Worldwide emissions of trace metals to the atmosphere (Nriagu and Pacyna, 1988) vs. concentrations in global aerosol.

Table 1. Percentage of worldwide emissions of trace elements in aerosol which are estimated to have anthropogenic sources, based on data from Nriagu and Pacyna (1988)

Element in Aerosol	% Anthropogenic Emissions
Pb	95
Cd	88
Zn	74
As	71
Ni	68
Cu	65
V	57

there is a fair correspondence between the estimated emissions and airborne concentrations.

Also, from the data of Nriagu and Pacyna (1988) it is possible to calculate the percentage of worldwide emissions of a given element which is anthropogenic. The cases where anthropogenic emissions are more than 50% of the total are listed in Table 1. It should be noted that some of the listed elements including Pb, Cd, As, Ni and possibly V are toxic. Nriagu and Pacyna (1988) concluded that human activities now have a major impact on the global and regional cycles of most of the trace elements. There is significant contamination of fresh water resources and an accelerating accumulation of toxic metals in the food chain.

Trace elements which are not toxic can serve as tracers for the transport of toxic elements and for some toxic compounds. It is well known that atmospheric aerosols have a large organic chemical content.

Enrichment factors provide a means of comparing the composition of a material such as aerosol to a reference material such as crustal rock. The definition of the enrichment factor (EF) of element Z is:

$$(EF,\ Z)_{Blood} = \frac{(C_Z/C_{AL})_{Blood}}{(C_Z/C_{AL})_{Rock}}$$

where C_Z is the concentration of element Z, and C_{AL} is the concentration of reference element AL.

Figure 5 shows that elements which are enriched in U.K. aerosol are also enriched in U.K. blood. The underlined elements with mainly anthropogenic emission sources have very high enrichment factors.

SUMMARY AND CONCLUSIONS

The observed correlation between trace element concentrations in human blood and atmospheric aerosol is interpreted to mean that aerosol affords a major pathway to man. Aerosol transports trace elements efficiently on a global scale, depositing on land and water surfaces, the interfaces with the atmosphere where the food chain begins. Pb, As, Cu and Zn, whose atmospheric sources are mainly anthropogenic emissions, are highly enriched in both aerosol and blood.

There is very little data on trace element concentrations in blood. Systematic multielemental measurements are needed in order to establish baseline values and to observe future trends. Blood is a convenient human substance to monitor since it can be obtained from healthy living individuals and is more readily obtained and analyzed than organs. Also, elemental analysis is simpler than for organic compounds which may be metabolized. There is need for coordinated multielemental measurements on aerosol. Aerosol sampling should be conducted in agricultural areas where aerosol enters the food chain. The study of trace elements in blood and aerosol affords a major opportunity for research with important implications for human health.

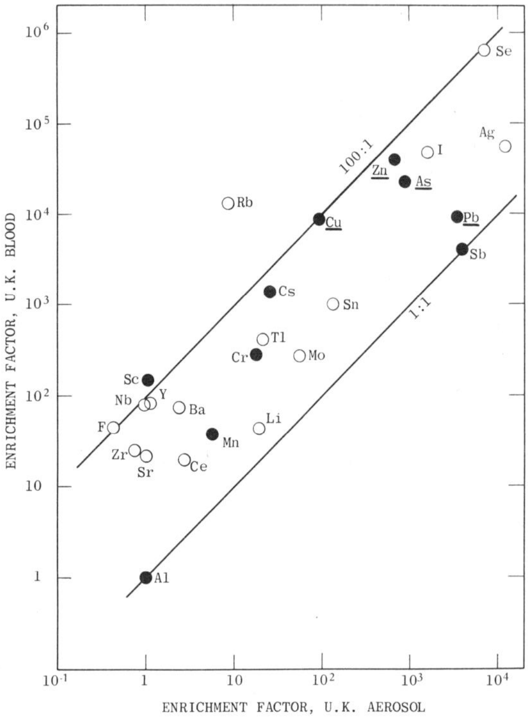

Figure 5. Enrichment factors for trace elements in U.K. blood vs. those for U.K. aerosol

REFERENCES

Bowen, H.J.M., 1979, "Environmental Chemistry of the Elements," Academic Press, London, p. 9.

Hamilton, E.I., Minsky, M.J., and Cleary, J.J., 1972/1973, The concentration and distribution of some stable elements in healthy human tissues from the United Kingdom -- An environmental study, Sci. Total Environ., 1:341-373.

John, W., 1983, Relationship between trace element concentrations in human blood and atmospheric aerosol, Sci. Total Environ., 27:21-32.

Nriagu, J.O., and Pacyna, J.M., 1988, Quantitative assessment of worldwide contamination of air, water and soils by trace metals, Nature, 333:134-139.

Rahn, K., 1976, "The chemical composition of the atmospheric aerosol", Technical Report, Graduate School of Oceanography, University of Rhode Island, July 1, 1976.

Travis, C.C., 1988, Multimedia partitioning of dioxin, this volume.

TRANSPORT OF POLLUTANTS FROM SOILS, LAKES AND OCEANS TO THE ATMOSPHERE

MULTIPHASE CHEMICAL TRANSPORT IN POROUS MEDIA

Patrick A. Ryan and Yoram Cohen

University of California, Los Angeles
Department of Chemical Engineering
Los Angeles, California 90024

INTRODUCTION

Chemical transport from soil sources can result in potentially harmful pollutant emissions into the air we breath. The monitoring of pollutant fluxes from the soil can be used to assess the current risk to humans from soil sources at a field site. Monitoring, by itself, however, is inadequate for preventing or predicting potential toxic situations that may arise in the future. In order to a priori ascertain the potential volatilization risk associated with future chemical sources in soils, a theoretical model that describes chemical transport from the soil into the atmosphere is needed. The formulations of an accurate chemical transport model should mimic features of the soil environment that significantly affect the transport of pollutants.

Previous theoretical studies of chemical volatilization from the soil have considered chemical volatilization from both nearly dry and moist soils [Farmer et. al., 1972; Hamaker, 1972; Mayer et. al., 1974; Bonazountas, 1983; Jury et. al., 1984 a,b,c]. In general, previous studies have considered the soil as an isothermal system, so variations in transport coefficients, in the model equations, with temperature were not considered. The first non-isothermal model was utilized in the recent study by Freeman and Schroy (1986) for describing the volatilization of dioxin from a dry soil. The model accounted for variations in the air phase diffusion coefficient with temperature, but the adsorption coefficient for dioxin between the air and solid phase was set to a constant. It is noted that a comparison of non-isothermal and isothermal model results were not made. Hence, the importance of including variations in the air phase diffusion coefficient with temperature was not demonstrated. Finally, none of the above models have considered the effect that moisture variations in the soil can have on chemical transport.

A non-isothermal moisture varying model for chemical transport requires a detailed knowledge of the moisture and equations. It is noted that there exists little or no data on the variation of these coefficients with temperature and/or moisture for field conditions. This is a major stumbling block to providing an adequate model for chemical transport in unsaturated soils. Despite the difficulties enumerated above the theoretical study of chemical transport from the soil under non-isothermal field conditions is warranted. The use of an a priori description for chemical transport is needed to discover the potential for hazardous emissions before they arise, so that steps can be taken to prevent the problem from occurring rather then address the issue after the fact.

Model results can provide a systematic evaluation of non-isothermal effects on chemical volatilization and can be used to demonstrate differences between non-isothermal and isothermal models. Also, simulations can be used to identify those terms in the model equations most important for describing chemical transport. Moreover, a comparison of predicted and monitored values for chemical transport can be employed in a recursive technique to improve subsequent model formulations.

MULTIPHASE CHEMICAL TRANSPORT MODEL

The transport of a contaminant in the multiphase soil matrix consisting of soil-air, soil-water, and soil-solids phases can be described by the following set of unsteady convection-diffusion mass balance equations,

$$\frac{\partial(\theta_i C_i)}{\partial t} = \frac{\partial}{\partial z}\left[\frac{\theta_i}{\tau_i} D_i \frac{\partial C_i}{\partial z}\right] - \frac{\partial(\theta_i v_i C_i)}{\partial z} + \theta_i D_{vi} \frac{\partial^2 C_i}{\partial z^2} + \sum_{j=1}^{N} (Ka)_{ij} (C_j H_{ij} - C_i) \quad [1]$$

in which C_i is the chemical concentration in phase i moving at an interstitial velocity [cm/s] v_i. The volume content [cm^3/cm^3] of phase i is designated by θ_i, and H_{ij} is the chemical i-j partition coefficient ($H_{ij} = C_j/C_i$ at equilibrium). The chemical molecular diffusion and convective diffusion coefficients [cm^2/s] in phase i are D_i and D_{vi}, and the phase i tortuosity is given by τ_i. Finally, $(Ka)_{ij}$ is the overall i phase volumetric mass transfer coefficient between phases i and j, and N is the number of phases in contact with phase i.

In this paper, the effect of measured temperature variations in a nearly dry field soil and of diurnally measured temperature and moisture variations for a moist soil on chemical transport are demonstrated. In order to establish a comprehensive analysis on the effect that temperature and moisture have on volatile and slightly volatile chemicals the transport of benzene, dieldrin, dioxin, and lindane are illustrated. First, an illustration of the current accuracy of diffusion model predictions is provided, assuming chemical equilibrium exists in the soil phases, to assess potential biases in model simulations.

In the absence of convection, and when chemical equilibrium is assumed to exist between soil phases, the transport of pollutants in the soil matrix can be described by simplifying eqn. 1 into the following one-dimensional diffusion equation:

$$\frac{\partial C_{soil}}{\partial t} = D \frac{\partial^2 C_{soil}}{\partial z^2} \qquad [2]$$

subject to the appropriate boundary conditions, where t is time and z is depth into the soil. The soil equilibrium diffusion coefficient, D, is perhaps the simplest coefficient to describe in the soil transport equations. Also, the technique of its measurement under laboratory conditions is well established. It is defined by [Jury at. el., 1984c; Cohen and Ryan, 1985; Enfield et. al., 1986],

$$D = \left[\frac{\theta_a D_a/\tau_a + H_{aw} \theta_w D_w/\tau_w}{\theta_a + \theta_w H_{aw} + \theta_s H_{ws}} \right] \qquad [3]$$

where the subscripts a, w, and s denote the air, water and solid phases of the porous media, respectively. The phase i molecular diffusion coefficient is denoted by D_i, and the volume fraction by θ_i. The tortuosity of the diffusion path in phase i is given by τ_i. Finally, H_{ij} is the ij partition coefficient.

Experimental measurements for the soil diffusion coefficient when two mobile phases (e.g. water and air) are present in the soil indicate that the measurement for the diffusion coefficient is not reproducible [Ehlers et. al., 1969; Shearer et. al., 1973], see Figure 1. van Brakel and Heertjes (1974) point out that differences in measured diffusion values can be due to packing differences between experiments, which result in a change in the value for the tortuosity. In order to demonstrate the sensitivity of model results to the tortuosity model employed, the value for the chemical equilibrium diffusion coefficient is predicted based on three tortuosity models. Model results are compared to diffusion data for lindane to determine the tortuosity model that yields the best fit to experimental data.

Finally, present predictions for the diffusion coefficient based on the generally accepted model (using a constant H_{ws}) are incapable of describing the value of the diffusion coefficient in nearly dry soils, below a moisture content by weight of 4%. For example, the study of Jury et. al. (1983) demonstrates that, the present model for the diffusion coefficient can overpredict the soil diffusion coefficient for lindane in a (nearly dry) soil of moisture content 1% (by weight) by two orders of magnitude. Also, the strict application of eqn. 1 with moisture independent partition coefficients cannot describe the moisture dependence of the diffusion coefficient data for lindane [Ehlers et. al., 1969; Shearer et. al., 1973] and dieldrin [Farmer and Jensen, 1970]. The diffusion coefficient data for lindane (see Figure 1) depict an increase in the soil diffusion coefficient by at least an order of magnitude as the moisture content increases from 1% to 3% (corresponding

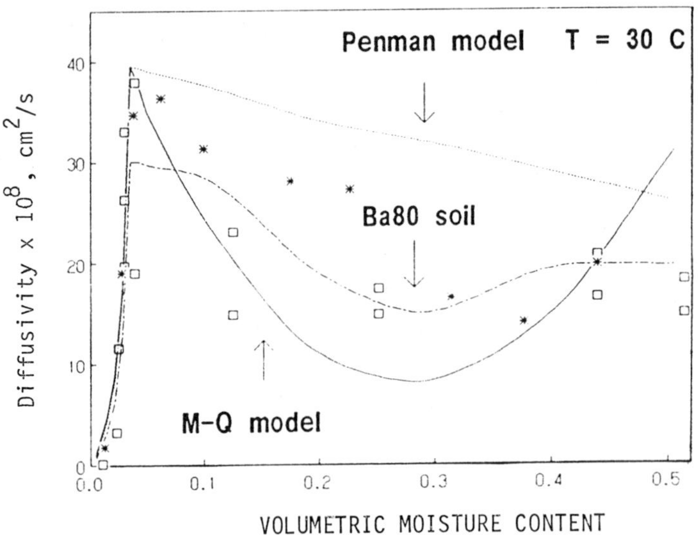

Figure 1 Predicted and Experimental diffusion coefficient data for Lindane at 30°C. Volumetric moisture content is defined as volume of water present in the soil per unit volume of soil.
Legend: (*) Data of Ehlers et. al., 1969
(□) Data of Shearer et. al., 1973

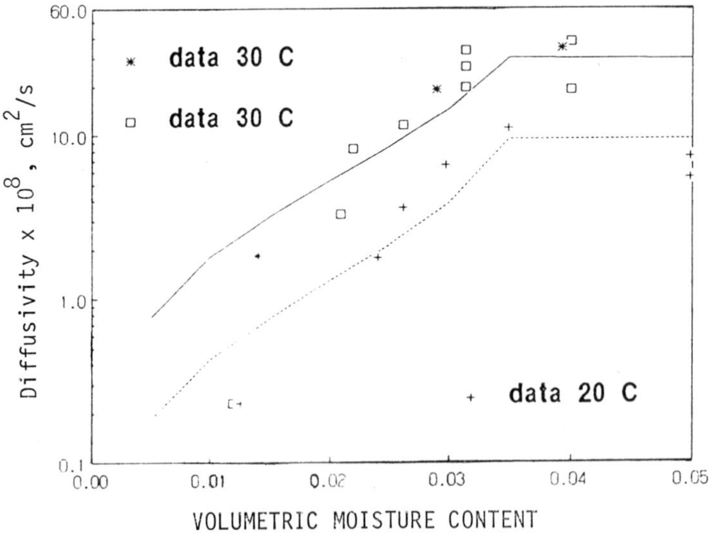

Figure 2 Predicted and Experimental Diffusion data for Lindane at low moisture content using the Ba80 tortuosity model. Volumetric moisture content is as defined in Figure 1.
Legend: (+,*) Data of Ehlers et. al., 1969
(□) Data of Shearer et. al., 1973

respectively), a local maximum in the diffusion coefficient at a moisture content of 3-4%, and a decrease in the diffusion coefficient with increasing moisture content above a 4% moisture level.

Recently, Ryan and Cohen (1988) developed a model for the adsorption coefficient that was moisture dependent in nearly dry soils, below a critical moisture content. The model is applicable for organic chemicals with a melting temperature greater then the soil temperature. The resulting soil diffusion model is expressed by:

$$D = \frac{H_{wa}\ \theta_a\ D_a/\tau_a\ +\ \theta_w\ D_w/\tau_w}{H_{wa}\ \theta_a + \theta_w + \theta_s\ K_{oc}f_{oc}\rho_s} \qquad : \theta_w \geq \theta_{wc} \qquad [4a]$$

and

$$D = \frac{H_{wa}\ \theta_a\ D_a/\tau_a\ +\ \theta_w\ D_w/\tau_w}{H_{wa}\ \theta_a + \theta_w + \theta_s\{\theta_w/\theta_{wc} + (1-\theta_w/\theta_{wc})p_l^{(s)}/p_s^{(s)}k'\}\ K_{oc}f_{oc}\rho_s}$$

$$: \theta_{wr} \leq \theta_w \leq \theta_{wc} \qquad [4b]$$

in which θ_{wr} is the residual soil water content. Eqn. 4a is applicable for moisture contents greater then the critical value, θ_{wc}. Eqn. 4b is valid for locations in the soil were θ_w is below the critical moisture content but above the residual moisture level. $p_l^{(s)}$, $p_s^{(s)}$ are the liquid and solid phase vapor pressures for the chemical. The term $K_{oc}f_{oc}\rho_s$ is the water/solid adsorption coefficient accurate for moisture levels above the critical moisture content. Finally, k' is a thermodynamic parameter that accounts for the non-ideality of wet-solid and dry-solid adsorption, for dilute concentrations k' may be assumed to be unity.

In order to demonstrate the accuracy of the present model formulation the model predictions are compared to experimental data for lindane. The predicted diffusion coefficient values for lindane (assuming k' = 1) are plotted for the Millington and Quirk, Penman, and Ba80 tortuosity models (see Figures 1 and 2). It is clear that the analysis of Ryan and Cohen (1988) explains, for the first time, the variation of the diffusion data over the entire range of possible moisture contents encountered in soils. The predictions of all three models are essentially identical below the critical moisture content. The three models, however, deviate significantly above the critical moisture content. Finally, no one model appears to provide the best fit to the diffusion data, but the Ba80 and Millington and Quirk models are the most accurate.

Simulation results for chemical volatilization from the soil for both nearly dry and moist soil conditions are described below. It is noted that for the nearly dry soil case water was assumed to exist as a discontinuous phase (no water movement) and the moisture content of the soil was assumed constant, invariant with depth. Finally, for the moist soil case simulation results included the effect of water movement and liquid phase dispersion.

An example of model results for chemical transport in nearly dry soils is presented below for the volatilization of lindane. The results are for a soil uniformly contaminated with lindane in the top 10 cm (a more in-depth presentation of results for a variety of cases are presented by Cohen et. al., 1988). Results show that by accounting for non-isothermal effects in the model coefficients only a slightly larger amount of lindane volatilizes, see Figure 3. It is noted, however, that the lindane air phase concentration varies significantly between day and night and that this effect cannot be predicted based on an isothermal model, see Figure 4. After 10 days, about 97.8% of the initial mass of lindane volatilized when non-isothermal coefficients were used, and only 98.0% volatilized when time-averaged constant values for the coefficients were utilized (see Figure 3).

Also, Cohen et. al. (1988) have shown that natural convection, induced by temperature gradients in the soil, is insignificant relative to molecular diffusive effects for almost all environmental soil situations. Furthermore, the authors showed that an isothermal diffusion model, using the diffusion coefficient as an adjustable parameter, duplicated field concentration data for dioxin twelve years after the initial contamination incident. Thus, the authors concluded that it may be possible to utilize isothermal model predictions at sufficiently long time periods past the initial contamination. Such an approach is valid when the isothermal diffusion model parameters are determined from field data at a time period that is well past the time of ininial contamination.

CHEMICAL TRANSPORT IN MOIST SOILS

It is noted that there is a dilemma in investigating chemical transport in soils with non-uniform moisture distributions under non-isothermal conditions. Simultaneous temperature and moisture records are needed along with chemical transport measurements (e.g. chemical volatilization fluxes) as a function of time. The authors are unaware of such a data source. In fact, to the knowledge of the authors, the study of Jackson (1987) for the Adelanto loam soil is the only source for accurate data of simultaneous temperature and moisture profiles. Thus, we have employed the data from the Adelanto loam soil to other similar soils, acknowledging that this may lead to some differences between the "true" and model employed values for the chemical and soil dependent transport and partition coefficients. It is our contention, however, that the quantitative effects of moisture and temperature variations are of sufficient accuracy to assess the approximate magnitude of the reported volatilization fluxes.

A major difference between moist and dry soils is that liquid water movement is possible in moist soils. Also, the water phase velocity induces a dispersion effect for chemical transport in the liquid phase. The combined effect of convection and dispersion can result in a large difference in the chemical volatilization rate compared to dry soil conditions.

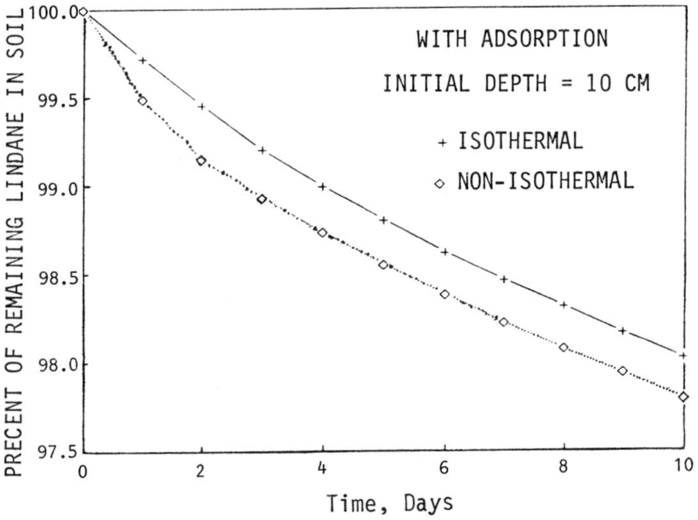

Figure 3 Comparison of nonisothermal and isothermal calculations of lindane volatilization. Initial contamination depth of 10 cm at an initial soil-air concentration of 1.2 μg/L, in equilibrium with the adsorbed phase.

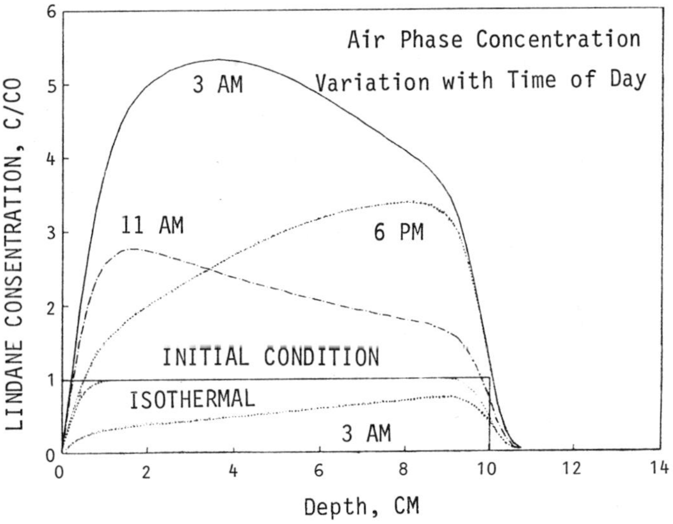

Figure 4 Simulation of concentration profile for nonisothermal lindane diffusion. (Initial conditions as in Figure 3).

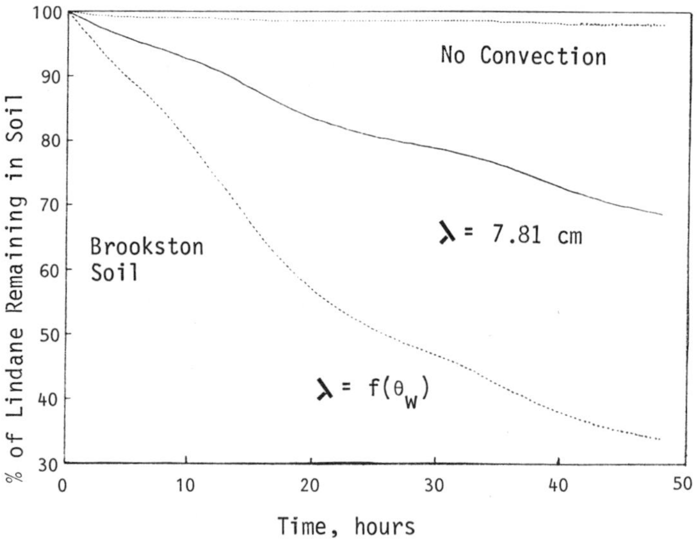

Figure 5 Simulation of lindane volatilization from Brookston soil. Initial contamination depth top 10 cm.

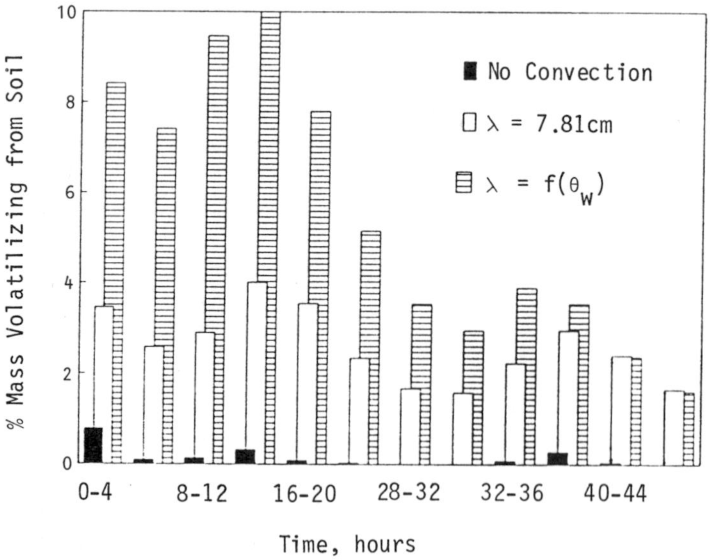

Figure 6 Percentage of lindane mass volatilizing from Brookston soil for consecutive four hour periods.

magnitude increase in the volatilization flux for moist compared to dry soil conditions. Also, the results indicate that two reasonable (but different) expressions for the dispersion coefficient yield significant differences in the predicted volatilization flux for lindane. Clearly, model results are sensitive to the dispersion relationship employed. The volatilization flux for lindane is also shown to vary diurnally (Figure 6). The largest effect is apparent when liquid phase convection is not considered; an increase by a factor of about 60 in the volatilization flux from night to day is predicted.

Results were also obtained for benzene using the same soil conditions as above. Volatilization fluxes were shown to be only slightly affected by the diurnal temperature and moisture cycle (see Figure 7). At most a factor of three increase in the volatilization flux is predicted for the day time compared to night time soil conditions. Also, the effect of liquid water movement on benzene transport is not as important as for lindane. As demonstrated in Figure 7 the effect of adsorption/desorption kinetics for benzene, however, is shown to be as important in affecting benzene volatilization as accounting for liquid water movement.

The above test cases indicate that the chemical volatilization flux may change over a diurnal cycle by a significant amount. The accuracy of these model predictions, however, needs confirmation. Thus, predicted volatilization fluxes for dieldrin are compared with measured volatilization fluxes from a field study in Ohio [Parmele et. al., 1972]. Unfortunately, the lack of temperature and moisture data for the Ohio soil site allow us only a qualitative comparison of model predictions to field data. Model predictions, in the absence of liquid convection, are in reasonable agreement with the experimental data, as indicated in Figure 8. The maximum predicted volatilization flux is at most 40% higher than the measured value, and the minimum predicted volatilization flux is at most 75% lower than the measured data. The qualitatively good results obtained with dieldrin and for dioxin in nearly dry soils were both obtained using a simple diffusion model. Hence, it seems that chemical movement in the upper soil zone is primarily by molecular diffusive transport, and that liquid convection is important only for moist field sites.

Finally, it is worth noting that the current results suggest that diurnal volatilization flux variations may lead to substantially different predictions of chemical exposures (previously based on an isothermal constant moisture content soil system) associated with landfills or other soil sources.

SUMMARY

A simple model for the soil diffusion coefficient was shown to provide the first quantitative prediction of the diffusion data for dieldrin and lindane at low moisture levels. This new model predicts as much as an order of magnitude smaller value for the chemical equilibrium diffusion coefficient compared to previous models. Hence, in soils of low moisture content using the current model should result in significantly improved predictions of chemical

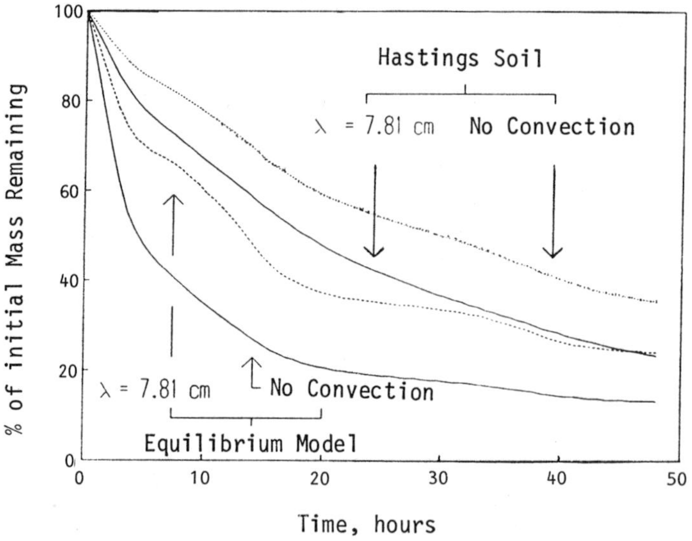

Figure 7 Simulation of benzene volatilization from the top soil. Initial contamination depth top 10 cm.

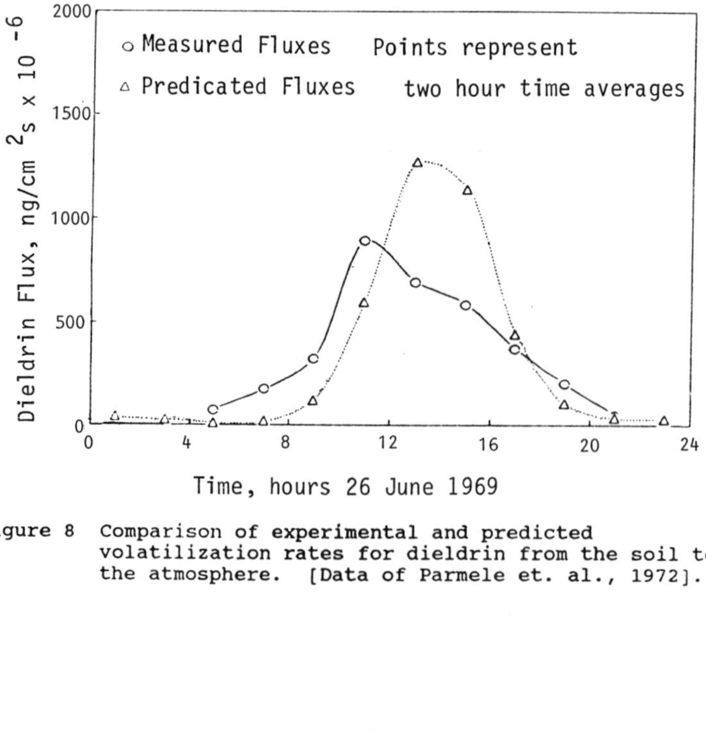

Figure 8 Comparison of experimental and predicted volatilization rates for dieldrin from the soil to the atmosphere. [Data of Parmele et. al., 1972].

The use of a simple formulation for multiphase chemical transport in the soil was shown to provide a reasonable description of chemical volatilization at field conditions. The results from this study demonstrate the importance of temperature and moisture on the volatilization flux. Also, the study has shown that model results are sensitive to the dispersion model employed. Thus, field data are needed to establish the proper variation of the dispersion coefficient with moisture. The results have also indicated that the diurnal swing in the volatilization flux is most apparent for slightly volatile chemicals as compared to volatile species (e.g. benzene). There is a need for future monitoring field studies that will collect data on chemical volatilization fluxes, temperature and moisture profiles as a function of time. Data from these studies should then allow a more definitive evaluation of the present model, and also aid in establishing the temperature and/or moisture dependence of model coefficients.

REFERENCES

Bonazountas, M., In Fate of Chemicals in the Environment, eds. R.L. Swann and A. Eshenroeder, ACS Symposium Series No. 225, Americal Chemical Society, Washington D.C.

Cohen, Y. and P.A. Ryan, Env. Sci. & Tech., 19, 412 (1985).

Coehn, Y., H. Taghavi and P.A. Ryan, J. Environ. Qual., 17, 2, 198 (1988).

Ehlers, W., W.J. Farmer, W.F. Spencer and J. Letey, Soil Sci. Soc. Amer., 34, 28 (1970).

Enfield, C., D. Walters, J. Wilson and M. Piwoni, Hazardous Waste & Hazardous Materials, 3, 57, (1986).

Farmer, W.J. and C.R. Jensen, Soil Sci. Soc. Amer., 34, 28 (1970).

Farmer, W.J., K. Igue, W.F. Spencer, and J.P. Martin, Soil Sci. Soc. Am. Proc., 36, 443-447 (1972).

Freeman, R.A. and J.M. Schroy, Env. Prog. 5, 1, 28-33 (1986).

Hamaker, J.W., In Organic Chemicals in the Soil Environment Vol. I, eds. C.A.I. Goring and J.W. Hamaker, Marcel Dekker, New York (1972).

Jury, W.A., W.F. Spencer and W.J. Spencer, J. Environ. Qual., 12, 4, 558 (1983).

Jury, W.A., W.F. Spencer and W.J. Spencer, J. Environ. Qual., 13, 4, 567, (1984a).

Jury, W.A., W.F. Spencer and W.J. Spencer, J. Environ. Qual., 13, 4, 573, (1984b).

Jury, W.A., W.F. Spencer and W.J. Spencer, J. Environ. Qual., 13, 4, 580, (1984c).

Mayer, R., J. Letey, and W.J. Farmer, Soil Sci. Soc. Am. Proc., 38, 563-568 (1974).

Parmele, L.H., E.R. Lemon and A.W. Taylor, Water, Air, and Soil Pollution, 1, 433 (1972).

Ryan, P.A. and Y. Cohen, submitted to Env. Sci. & Tech.

Shearer, R.C., J. Letey, W.J. Farmer and A. Klute, Soil Sci. Soc. Amer., 37, 189 (1973).

van Brakel, J. and P.M. Heertjes, Int. J. Heat Mass Transfer, 17, 1083, (1974).

MATHEMATICAL MODELING OF AIR TOXIC EMISSION FROM LANDFILL SITES

Christian Seigneur*, Anthony Wegrecki+, Shyam Nair[x], and Douglas Longwell+

* Currently with ENSR Consulting & Engineering, Oakland, California
\+ Bechtel Environmental, Inc., San Francisco, California
x Bechtel Environmental, Inc., Oak Ridge, Tennessee

INTRODUCTION

Hazardous waste sites emit toxic contaminants. These contaminants may be volatile organic compounds that diffuse through the soil and escape into the atmosphere, or they may be heavy metals, asbestos, or organic compounds present in the soil that are carried into the air by wind erosion. Quantification of the emission rates of these contaminants is absolutely essential to ensure clean air quality and worker safety at the site, to estimate risks associated with the emissions, and to comply with air quality regulations.

The relationship between an air pollutant emission source and the resulting atmospheric concentration at a downwind receptor can be established by means of atmospheric dispersion modeling. Adequate mathematical models are available to estimate source-receptor relationships. However, large uncertainties exist in estimating the quantity of emissions released from a source, particularly when the source is a hazardous waste site. In this study, a computer model is developed that will allow us to (a) estimate rapidly and inexpensively the levels of toxic emissions from a landfill site, (b) predict the future site conditions corresponding to a variety of remedial action scenarios, and (c) assess health and safety conditions at the site, thereby enhancing worker protection. Such quantitative information is necessary during the remedial investigation/feasibility study (RI/FS) phase of hazardous waste site cleanups.

BACKGROUND

Toxic air pollutants are emitted from hazardous waste sites in two forms: (1) particulate pollutants that are in solid or dissolved phase (both organic and inorganic compounds), and (2) gaseous pollutants (primarily organic compounds).

Wind Entrainment of Contaminated Particulates

A summary of models, currently available for fugitive dust emissions is presented in Table 1. Most wind erosion models are based on empirical formulations. In AP-42 (EPA, 1985a), equations are provided to estimate

Table 1. Summary of Major Models of Fugitive Dust Emissions

REFERENCE	APPLICATION	MODEL FORMULATION
EPA (1985)	Particulate emissions by vehicular traffic on unpaved roads	o Empirical formulation based on field data
EPA (1978)	Particulate emissions from exposed fields	o Empirical formulation based on field data o Entrainment windspeed threshold of 12 mph
Gillette (1981)	Respirable particulate emissions from exposed fields	o Empirical formulation based on field measurements o Particulates smaller than 10 microns in diameter o "Unlimited reservoir"
Cowherd (1983)	Respirable particulate emissions from exposed fields	o Empirical formulation based on field measurements o Particulates smaller than 10 microns in diameter o "Limited reservoir"

particulate emissions from vehicular activity on unpaved roads, i.e., mechanical disturbance of the soil. It is possible that during hazardous waste cleanup, unpaved roads may become contaminated. EPA (1985,b) also provides default values for various parameters for use when site specific data are not available.

Gillette (1981) and Cowherd (1983) separate soil surface types into two erodibility classes: (1) Unlimited reservoir and (2) limited reservoir. Unlimited reservoir surfaces are nonhomogeneous and contain nonerodibles such as stones, clumps of vegetation, etc.

The model of Gillette (1981) for respirable particulate emissions is empirical in nature and is based on field measurements of highly erodible surfaces (unlimited reservoir). The model is represented by the following equation:

$$E_{10} = 0.036\ (1-V)\ \left(\frac{U}{U_t}\right)^3\ F(X) \qquad (1)$$

Where E_{10} is annual PM_{10} (particles smaller than 10 microns in diameter) emission rate per unit area of contaminated surface ($g/m^2 hr$); V is fraction of contaminated surface with vegetative cover (equals 0 for bare soil); U is mean annual wind speed (m/s); X is 0.886 U_t/U = dimensionless ratio; U_t is threshold value of windspeed at 7m (m/s), [F(X) is 1.91 for $0< x <0.3$; 2.04 exp (-0.16 X) for $0.3 < x <1.8$;

$-1.4\ x+3.02$ for $0.9 < x < 1.8$; $-105\ x+2.39$ for $1.8 < x < 2.0$; $0.18\ (8x^3 + 12x) \exp(-X^2)$ for $X > 2.0$]

The threshold windspeed, Ut is based on site survey and it is a measure of the erodibility of the soil (EPA, 1985,b). There is no moisture related parameter in Equation (1) since highly erodible surfaces do not retain moisture. Estimation of U and U_t are discussed in greater detail in EPA (1985,b). EPA (1985,b) also provides worst-case emission rate estimate by using a simplified form of Equation (1).

Cowherd (1983) developed a model to estimate respirable particulate emissions from "limited reservoir" type surfaces. This model is based on field measurements from surface mines. This model is represented by the following equation:

$$E_{10} = 0.83 \frac{f\ P(U^+)\ (1-V)}{(PE/50)^2} \qquad (2)$$

Where E_{10} is annual PM_{10} emission rate per unit area of contaminated surface (mg/m^2-hr); f is frequency of disturbance per month; U_t is observed (or probable) fastest mile of wind for the period between disturbances (m/s); $P(U_t)$ is erosion potential, i.e., quantity of erodible particles present on the surface prior to the onset of wind erosion (g/m^2); V is fraction of contaminated surface area covered by continuous vegetation cover (equals 0 for bare soil); PE is Thorntwaite's Precipitation Evaporation Index used as a measure of soil moisture content. EPA (1985,b) discusses the estimation of various parameters in Equation (2) in greater detail. The implementation of Equations (1), (2), and the equation for vehicular traffic contribution to particulate emissions in a FORTRAN-based computer code is discussed later. The source terms are used to perform atmospheric dispersion simulation using ISCLT (EPA, 1987) model.

Volatile Emissions From Landfills

A summary of major models which have been developed to-date to simulate air emissions of toxic volatile organic compounds is presented in Table 2. The basic model for estimating volatile emissions from covered landfills (with no internal gas generation) was developed by Farmer et al. (1980). This model is based on Fick's Law of steady-state diffusion. The volatile gases move from regions of higher concentrations to regions of lower concentrations (i.e., from within the landfill to the surface) through small pore spaces within the soil. The effect of landfill liquids and solids on the diffusion processes, however, is ignored. These liquids and solids actually reduce the amount of pore space available for diffusion. As a result, highly compacted, very wet solids act to retard the diffusion process.

Shen (1982) and Farrino et al. (1983) modified the model of Farmer et al. (1980) to account for landfills containing a variety of hazardous chemicals. Their models assume that the chemicals are in saturated vapor equilibrium, soil is dry and all the pore spaces are available for diffusion and, therefore, represent a worst-case scenario of the earlier model.

In some instances, toxic waste material and biodegradable waste material are buried together. Decomposing biodegradable waste materials produce various gases such as methane and carbon monoxide. These gases increase the landfill internal gas pressure which results in the convective transport of toxic gases towards the surface/air interface.

Thibodeaux et al. (1982), Findikakis et al. (1979), and El-Fadel et al. (1986), developed models to predict the emission rates of gases generated from biodegradable processes. Thibodeaux used a single-cell, gradientless landfill model, whereas Findikakis et al. (1979), and El Fadel et al. (1986) used vertical resolution of the landfill to compute the pressure changes during the biogas generation process. El Fadel et al. (1986) used improved gas generation terms based on actual chemical and microbiological processes that lead to methanogeneration and included them in the model developed by Findikakis et al. (1979). Thibadeaux et al. (1982) also used the pressure gradient between the landfill and atmosphere to calculate the emission rates of other contaminant species that may not contribute to the biogas generation process.

All of the above models, however, address only the gas-phase transport of contaminants and ignore the sorptive processes in the landfill. In addition to the biogas generation, other non-biodegradable and not-yet-degraded volatile contaminants can also be released into the void spaces of the soil. During their transport to surface, they can be readsorbed/reabsorbed by surrounding solid/liquid phases. In such cases, emission rates of these contaminants can be a time-dependent process. It is possible to develop a simple model to account for the release of such chemicals from landfill sites, with and without biodegradation process, using material balance and thermodynamic considerations.

MODEL DESCRIPTION

For the estimation of emission rates of particulates by wind erosion, the models described in the Background section were implemented into a FORTRAN-code called DSTSRC. DSTSRC can calculate emission rates for up to 700 area sources and 100 unpaved roadways. It can be run both interactively and in batch modes. The output from DSTSRC are compatible with the input requirements for the ISC model.

For the estimation of soil gas emission rates, a one-dimensional contaminant transport model is developed to describe the transport of gases in the presence of liquid and gaseous phases in the porous spaces. For the gas-phase the material balance equation for a contaminant species can be described as:

$$\phi(1-s)\frac{\partial c}{\partial t} = \frac{\partial c}{\partial z}\left[D_{eff}\,\phi(1-s)c^{*}\,\frac{\partial}{\partial z}\left(\frac{c}{c^{*}}\right)\right] - \frac{\partial(u\phi(1-s)c)}{\partial z} - G + S_g \qquad (3)$$

where ϕ is the porosity of the medium, s is the moisture content in the soil expressed as volume of waste per unit volume of voids, c is the gas-phase contaminant concentration in the landfill (kmol/m^3 air or biogas mixture), D_{eff} is the effective diffusion coefficient of the contaminant species in air or biogas mixture (m^2/day), c* is the molar density of air or biogas mixture (kmol/m^3 of air or biogas mixture), u is the convective velocity of air or biogas mixture in the landfill (m/day), G is the mass transfer rate of contaminant species to the liquid-phase (kmol/m^3 of landfill volume-day), and S_g is the source term of contaminant to the gas phase (kmol/m^3 landfill-day). In Equation (3), u is calculated using Darcy's Law as,

Table 2. Summary of Major Models of Volatile Toxic Emissions From Landfill Sites

REFERENCE	APPLICATION	MODEL FORMULATION
Farmer et al. (1980)	Volatile emissions from covered landfills	o Fick's law of steady-state diffusion o One dimensional o No landfill gas production
Shen (1982) and Farino et al. (1983)	Worst case volatile emissions from covered landfills with a variety of chemical wastes	o Fick's law of steady-state diffusion o One dimensional o No landfill gas production o All pore space is available for diffusion o Chemical is at saturation equilibrium with respect to vapor
Thibodeaux (1981)	Volatile emissions from covered landfills with internal gas generation	o One cell, one dimensional o Diffusion o Mean internal gas velocity (advection) o Prandtl's mixing length theory
Thibodeaux et al. (1982)	Volatile emissions from covered landfills with internal gas generation	o One cell, one dimensional o Diffusion o Gas generation term (advection) o Atmospheric pressure fluctuations
El-Fadel et al. (1986)	Internal gas production	o One dimensional o Diffusion o Advection o Gas generation

$$u = -\frac{k}{\phi(1-s)\mu}\left(\frac{\delta p}{\delta z} + \rho g\right) \qquad (4)$$

where k is the permeability of the medium (m^2), μ is the visocity of air or biogas mixture ($N\text{-}day/m^2$), p is the barometric pressure in the landfill (N/m^2), ρ is the density of air or biogas mixture (kg/m^3), and g is the acceleration due to gravity (m/sec^2). The corresponding material balance equation for the liquid phase can be written as,

$$\phi s \frac{\delta c_1}{\delta t} = \frac{\delta}{\delta z} (D_1 \phi s \frac{\delta c_1}{\delta z}) + G + S_1 \qquad (5)$$

where c_1 is the contaminant concentration in the liquid (kmol/m^3 liquid), D_1 is the diffusion coefficient of contaminant species in liquid, and S_1 is the source term of contaminant to the liquid phase (kmol/m^3 of landfill volume-day).

The following assumptions are made in developing these equations: (1) water in the landfill is stagnant, (2) water content in the landfill is in quasi-steady state. This allows treatment of s as independent of time and independent of precipitation, infiltration, and evapotranspiration processes. Effects of concentration increase and decrease due to these physical processes can still be handled through the source terms S_1,(3) sorption processes between vapor and solid phases, and between liquid and solid phases are neglected and (4) the landfill is primarily in the unsaturated zone.

The mass transfer term G in Equations (3) and (5) can be expressed as a product of mass-transfer coefficient and the concentration difference between the bulk gas-phase contaminant concentration and the gas-phase contaminant concentration at the interface of water and gas. However, field values for the mass transfer coefficients not being readily available, and in accordance with the treatment of the solid/liquid interface in current groundwater transport models, dynamic equilibrium of contaminant concentrations is assumed at the gas-liquid interface. Assuming further that Henry's Law is valid at the contaminant concentrations encountered in the landfill, Equations (3) and (5) can be coupled together to yield,

$$[\phi(1-s) + \frac{\phi s}{H}] \frac{\delta c}{\delta t} = \frac{\delta}{\delta z} [D_{eff} \phi (1-s) c^* \frac{\delta}{\delta z} (\frac{c}{c^*}) + D_1 \frac{\phi s}{H} \frac{\delta c}{\delta z}]$$

$$- \frac{\delta (u \phi (1-s) c)}{\delta z} + S_g + S_1 \qquad (6)$$

where H is the Henry's Law constant expressed as (kmol/m^3 of air or biogas mixture) / (kmol/m^3 of water).

Equation (6) can be solved for the contaminant concentration using the zero gas-phase flux boundary condition at the landfill bottom, Drichelet boundary condition at the landfill top representing the ambient air contaminant concentration, and a known initial contaminant distribution in the landfill. Equation (6) can be solved either dynamically, as done by Findikakis et al. (1979), or kinematically, by imposing a prescribed pressure distribution. It is more appropriate to solve equation (6) kinematically for contaminants that do not contribute significantly to the biogas generation. The pressure profile can be generated using dynamic models or from direct measurement of vertical pressure distribution in the soil.

A finite element code was developed in FORTRAN language to solve Equation (6). For pure diffusion and diffusion dominated problems Galerkin method with linear elements is implemented, and for convection dominated problems (Pe>2), an upwinding scheme using quadratic asymmetric weighting functions and linear shape functions is implemented.

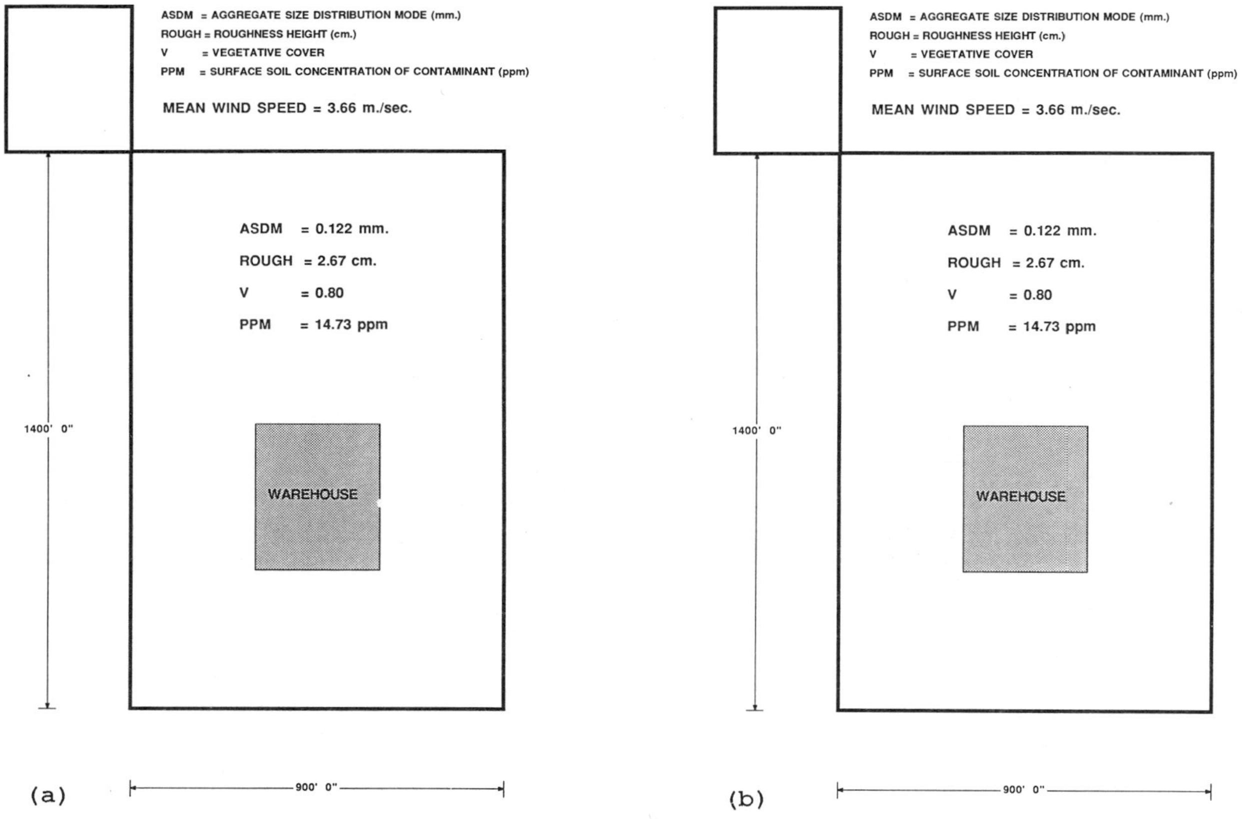

Figure 1. Site characteristics and input data for DSTSRC particulate emission model for (a) Approach 1 and (b) Approach 2.

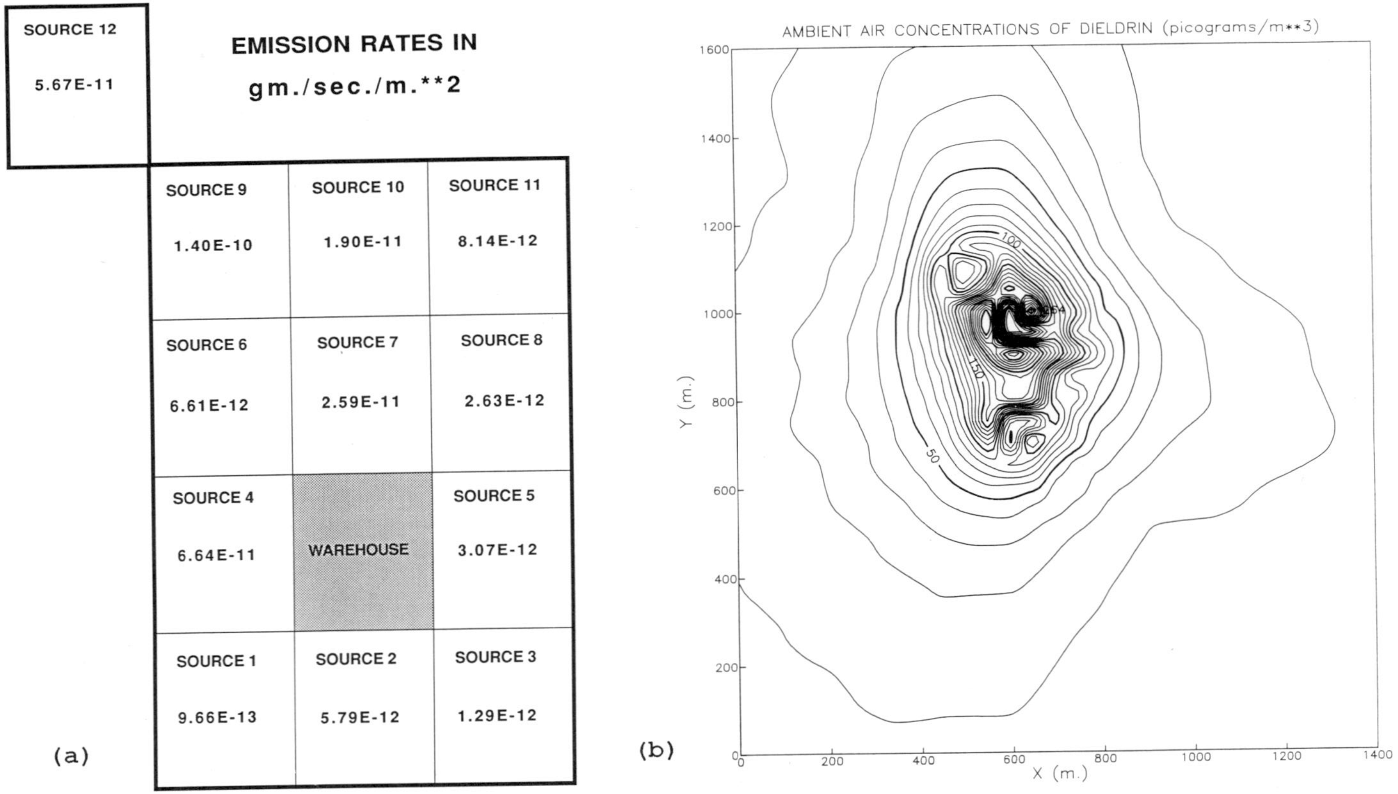

Figure 2. (a) Results from DSTSRC using Approach 1 and (b) results from ISCLT air dispersion model using the output of DSTSRC for Approach 1.

RESULTS AND DISCUSSIONS

Wind Entrainment of Contaminated Particulates

In this study, long-term exposure estimates of contaminated particulate emissions from a hypothetical hazardous waste facility is studied, using two different approaches. The surface soil at the site is assumed to be contaminated with a stable carcinogenic compound in particulate form. For risk assessment purposes, it is desirable to estimate the contaminant emission rate and the resulting long-term ambient air concentrations both on- and off-site. The hazardous waste site is 1400 ft in length and 900 ft in width. A 350 ft by 300 ft warehouse is located near the south-central section of the site, and no vehicular traffic exists at the site.

Input parameters to estimate the source term (EPA, 1985b) are presented in Figure 1a for this site. In the Modeling Approach 1, the site is divided into 12 area sources of 350 ft x 350 ft dimension. The site characteristics of each source are averaged (Figure 1a) and used as input to the wind erosion model DSTSRC. The aggregate size distribution mode for each source is such that the limited reservoir type surface can be applied to all 12 sources.

In Modeling Approach 2, the site is assumed to be one source with site-wide average characteristic data as input to DSTSRC. Figure 1b shows this approach and input data.

For the Approach 1, the DSTSRC results are summarized in Figure 2a. As expected, the largest emission rates occur to the north and west of the warehouse where maximum contaminant concentrations in soil were encountered. For the Approach 2, the emission rate was calculated as $3.72 \times 10^{-11} gm/m^2/s$.

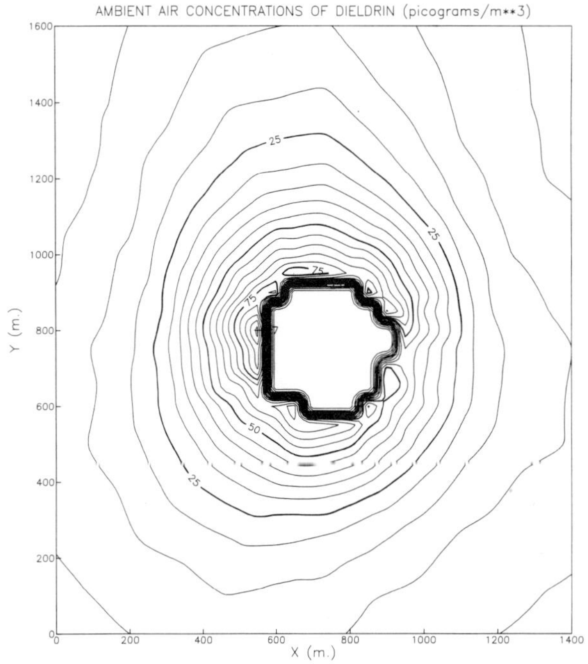

Figure 3. Results from ISCLT for particulate transport in the atmosphere using results of DSTSRC for Approach 2.

The ISCLT model was used to estimate the site-contribution to the annual average (long-term) ambient air concentration of contaminant for both approaches. To perform the simulations, meterological data recorded at a northern California site was used as input to ISCLT. The mean wind speed taken from this data is approximately the same as the 3.66 m/s value used in estimating the emission rates for the ISCLT simulation of multi-source approach. Each rectangle in Figure 1a was modeled as perfect square of equivalent area, as required by ISCLT. Using the DSTSRC emission rates in Figure 2a as input, the ISCLT generated output was plotted to produce Figure 2b, which shows the highest concentration of about 260 picogram/m^3 onsite.

The ISCLT simulation of the result of one-source approach using a perfect square-source of equal area as shown in Figure 1b generated the output shown in Figure 3. Highest concentration of 97 picogram/m^3 was obtained offsite.

The maximum concentration using the multi-source approach is 2-5 times that obtained using one single-source approach. However, as with most numerical schemes, the differences in the predicted concentration values from the two approaches decrease with distance from source.

Volatile Emissions

In the absence of site-specific information, transport and emission of 1,2-dichloroethylene is studied from a landfill 27m in depth, having a uniform porosity of 50%. Both liquid (water) and gas-phase (air) are present in the landfill. Eight runs were made, six using hydrostatic conditions in the landfill (pure diffusion problem), and two using prescribed vertical pressure profiles. In all runs, a linearly varying initial liquid-phase concentration with height, varying from a maximum at landfill bottom to zero at the landfill top/cover-bottom interface was assumed; and the gas-and liquid-phase concentrations were assumed to be in equilibrium at the beginning of simulation. Similarly, the landfill moisture content varied linearly from 20% of void spaces at bottom to zero at the landfill top/cover-bottom interface in all runs except Run 6, where the maximum moisture content was 40%. Table 3 presents the details of the eight simulation runs. The Henry's Law constant for 1,2-dichloroethylene was taken as 2.3 ((g/L water)/(g/L of air)). Viscosity of air for Runs 7 and 8 were taken as 1.82×10^{-5} N-s/m^2.

In Run 6, the maximum initial contaminant concentration was 500 ppm; and in all other runs it was 1000 ppm. Ambient 1,2-dichloroethylene concentration was kept at zero.

Emission rates of 1,2-dichlorothylene for Runs 1, 2, and 3 are plotted in Figure 4a, for Runs 1, 4 and 5 in Figure 4b, and for Run 1 and 6 in Figure 4c. By decreasing the cover porosity (Figure 4a), the overall emission rates are lowered in the first 600 days of simulation. After that, because of slower transport and higher concentration within the landfill for runs with lower porosity, the flux increases. Theoretically, the area under the curve is a constant between t=o and t=oo, because eventually the landfill will be emptied completely. By decreasing the cover diffusivity (by increasing tortuosity of cover material), a similar trend is observed in Figure 4b. Because of added effect of lower cover porosities, the fluxes are lower than in Figure 4a. In both figures, presence of cover material introduces a peak in flux which is delayed with increasing resistence to flow of gases. Increasing cover thickness both delays the peak and lowers the flux in the initial period of up to 500 days. Result of Figure 4c is

Table 3. Details of Simulations for the 1,2-dichloroethylene Transport Through The Landfill

Run No.	Cover Thickness	Cover Porosity o	Landfill Porosity o	Cover D_{eff} (D_1)	Landfill D_{eff} (D_1)	Max. Moisture S	Landfill* Permeability k	Cover* Permeability k_c
	(m)			($\frac{m^2}{d}$)	($\frac{m^2}{d}$)		(darcy)	(darcy)
1	0	0.5	0.5	0.803 (1.05E-4)	0.803 (1.05E-4)	0.2	-	--
2	1	0.25	0.5	0.803 (1.05E-4)	0.803 1.05E-4)	0.2	-	--
3	2	0.25	0.5	0.803 (1.05E-4)	0.803 (1.05E-4)	0.2	-	--
4	1	0.25	0.5	0.4015 (1.05E-4)	0.803 (1.05E-4)	0.2	-	--
5	2	0.25	0.5	0.4015 (1.05E-4)	0.803 (1.05E-4)	0.2	-	--
6	0	0.5	0.5	0.803 (1.05E-4)	0.803 (1.05E-4)	0.4	-	-
7	0	0.5	0.5	0.803 (1.05E-4)	0.803 (1.05E-4)	0.2	1	1
8	2	0.25	0.5	0.803 (1.05E-4)	0.803 (1.05E-4)	0.2	1	0.1

* For runs without an entry in the Table, pure diffusion problem was solved.

interesting. It shows that by reducing the contaminant concentration in half by adding more moisture, the flux is halved. This underscores the importance of precipitation events. It should also be noted that reconcentration of precipitation water by leaching processes is ignored in this analysis; but can be added through source term S_1 in Equation (6).

In Figure 5b, the emission rates for Runs 1, 7 and 8 are presented. The imposed pressure profile, roughly approximating the results of Findikakis et al. (1979) for similar landfill conditions, are shown for Runs 7 and 8 in Figure 5a. The complete depletion of soil gases within the first 400 days underscores the importance of convective processes accompanying biodegradation. It is interesting that Run 8 shows higher peak flux than Run 7, even though the cover has a lower permeability in Run 8 and Run 7 has no cover. The convective term is a function of both cover permeability and pressure gradient. The low permeability in Run 8 is more than offset by increased pressure gradient so that resulting

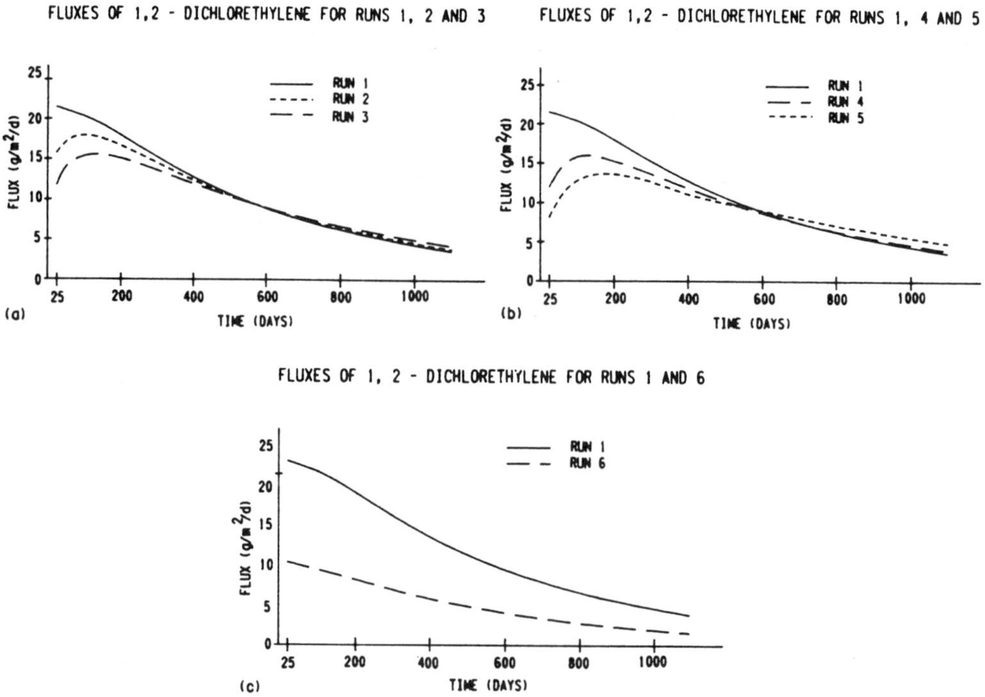

Figure 4. Emission rates of 1,2-dichloroethylene for Runs 1 though 6; (a) for Runs 1, 2, and 3, (b) for Runs 1, 4, and 5, (c) for Runs 1 and 6.

velocity in Run 8 is greater than in Run 7. Therefore, the choice of liner thickness has to be based on both cover permeability and gas-generation potential of the landfill, if it is to act as an effective barrier to the release of contamination to the atmosphere.

Figures 5c and 5d show the variation of concentration profiles with time for the diffusion-controlled transport of Run 1 and convection controlled Run 7, respectively. Run 7 shows that significant movement of contamination to the surface has already taken place by 25 days. By 500 days the landfill is nearly devoid of contaminants. Concentration profiles of Runs 2 through 6 show trends similar to Run 1, and Run 8 concentration profile follows Run 7 trend closely.

The emission rate of 7-22 g/m 2/day of 1,2-dichloroethylene for the pure diffusion case during the first year compares well with the results of Thibodeaux et al. (1982), where they report values of 11-30 g/m ^{2}/day of benzene for pure diffusion controlled transport. However, realistic input data collected from field experiments will have to be used to confirm this apparent agreement.

Again, for the hypothetical source region of Figure 2, an ISCLT simulation run was made using Approach 2 for Run 1 with an average first year source emission rate of 18.3 gm/in 2/day of 1,2-dichloroethylene from Run 1 for the same meteorlogical conditions that were used in the particulate dispersion simulations. The generated plot is shown in Figure 6, which shows highest concentrations near source regions.

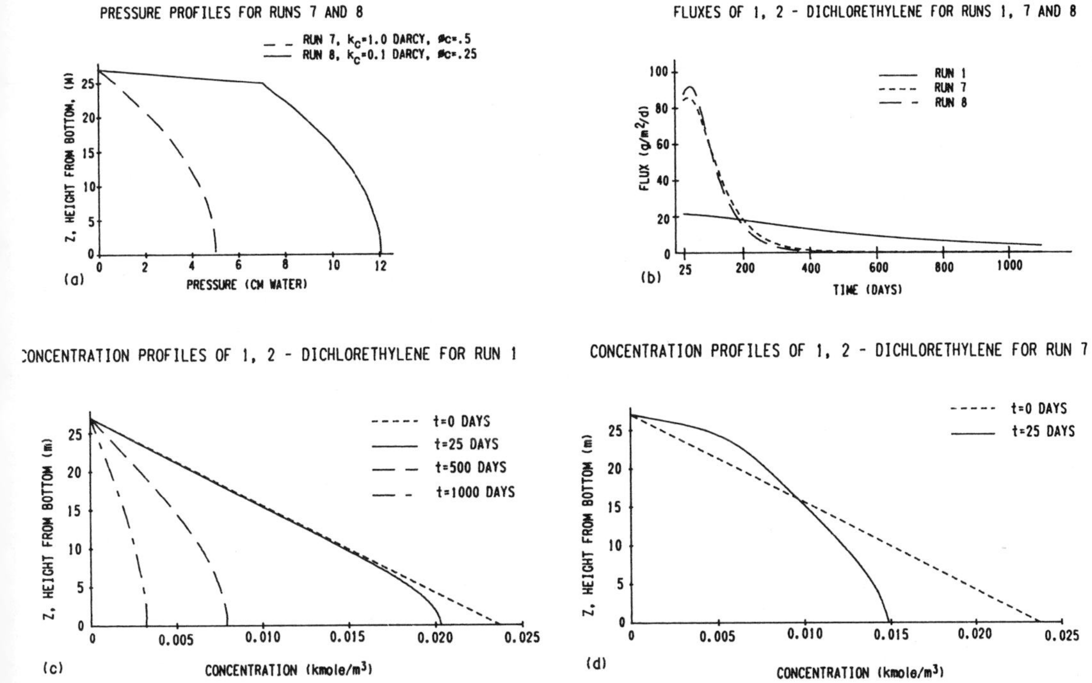

Figure 5. (a) Imposed pressure distributions for Runs 7 and 8; (b) emission rates for Runs 1, 7, and 8; (c) concentration profiles for Run 1; and (d) concentration profile for Run 7.

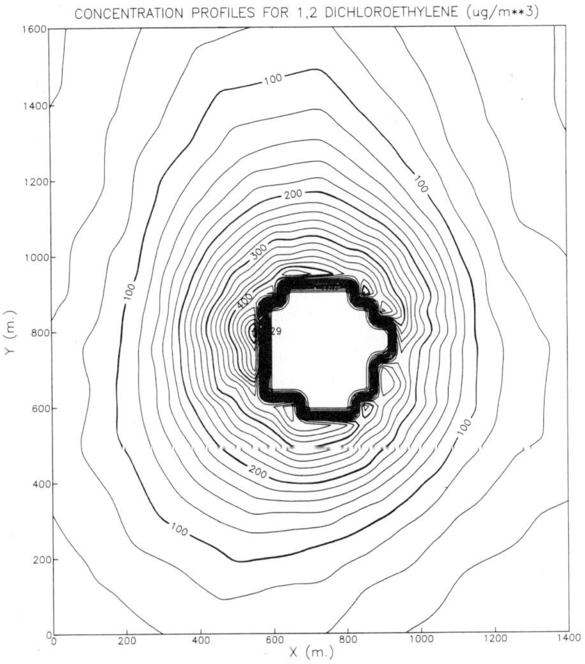

Figure 6. Results from ISCLT for 1,2-dichoroethylene transport through the atmosphere using Approach 2.

CONCLUSIONS

Two FORTRAN-based codes were developed, one to estimate particulate emission rates from landfill sites, based on existing models, and the other to estimate soil gas emission rates in the presence of both gas and liquid phases. The results of these were used to predict airborne concentrations of contaminants using ISC model.

Even though the runs considered for the soil gas emission estimation were theoretical in nature, the model can be applied easily to a real hazardous waste site. For pure diffusion problems, the model shows good agreement with observations of other researchers (Thibodeaux et al. (1982)). The model can be further improved by including sorption processes between gas-solid and liquid-solid interfaces. A sound framework has been developed, on which further improvements can be made.

ACKNOWLEDGMENTS

The authors would like to acknowledge the funding provided by Bechtel Group through a Technical Grant that made this research possible.

REFERENCES

Cowherd, C. A New Approach for Estimating wind-generated emissions from Storage Piles. Proceedings of the APCA Speciality Conference in Fugitive Dust Issues in the Coal Use Cycle, Pittsburgh, PA. 1983.

El-Fadel, M., A. N. Findikakis, and J. O. Leckie. Modeling Gas Production in Managed Sanitary Landfills - 2, Department of Civil Engineering, Stanford University, California, 1986.

EPA. Industrial Source Complex (ISC) Dispersion Model User's Guide Second Edition (Revised) Vol. 1, EPA-450/4-88-002a. Office of Air Quality Planning and Standards, Research Triangle Park, North Carolina, 1987.

EPA. Compilation of Air Pollutant Emission Factors, Volume I. AP-42. Office of Air Quality Planning and Standards, Research Triangle Park, North Carolina, 1985a.

EPA. Rapid Assessment of Exposure to Particulate Emissions from Surface Contamination Sites, EPA/600/8-85/002, National Technical Information Service, U.S. Department of Commerce, Springfield, VA, 1985b.

EPA. Fugitive Emissions from Integrated Iron and Steel Plants. Office of Research and Development, Research Triangle Park, North Carolina, 1978.

Farino, W., P. Spawn, M. Jasinski, and B. Murphy. Evaluation and Selection of Models for Estimating Air Emissions from Hazardous Waste Treatment, Storage, and Disposal Facilities. GCA Corporation, Bedford, Masachusetts, 1983.

Farmer, W. J., M. S. Yang, J. Letey, and W. F. Spencer, Hexachlorobenzene: Its Vapor Pressure and Vapor Phase Diffusion in Soil. Soil Sci. Soc. Am. J., Vol. 44, pp 676-680, 1980.

Findikakis, A. N. and J. O. Leckie, Numerical Simulation of Gas Flow in Sanitary Landfills, J. of Environ. Eng. Div., pp 932, 1979.

Gillette, D. A. Production of Dust that May Be Carried Great Distances. In Desert-Dust: Origin, Characteristics, and Effects on Man, edited by Troy Pewe, Geol. Soc. Am. Special paper 186, pp 11-26. 1981.

Shen, T. Air Quality Assessment for Land Disposal of Industrial Wastes, Environmental Management, Vol. 6, pp 297-305. 1982.

Tribodeaux, L. J., C. Springer, and L. M. Riley. Models of Mechanisms for the Vapor Phase Emission of Hazardous Chemicals from Landfills, J. Haz. Mat., Vol. 7, pp 63-74. 1982.

Thibodeaux, L. J., Estimating the Air Emissions of Chemicals from Hazardous Waste Landfills, J. Haz. Mat., Vol. 4, pp 235-244. 1981.

THEORETICAL CHEMODYNAMIC MODELS FOR PREDICTING VOLATILE EMISSIONS TO AIR FROM DREDGED MATERIAL DISPOSAL

Louis J. Thibodeaux
Department of Chemical Engineering
Louisiana State University
Baton Rouge, LA 70803

ABSTRACT

Some bottom sediment in both fresh and marine waters are contaminated with hazardous organic chemicals that are classified as volatile and semi-volatile. An example is the New Bedford Harbor and Acushnet River Estuary sediment which contains quantities of the polychlorinated biphenyls Aroclor 1242, 1248 and 1254. Dredged material contaminated with these and other volatile organic chemicals (VOCs) can be released to the atmosphere during and after disposal by volatilization. There is a need for methods to predict these volatilization losses in order to develop design, operating and management guidelines for controlling VOC emissions.

Volatilization rates for hydrophobic organic compounds from a confined disposal facility (CDF) containing contaminated dredged material are presently unknown. The primary purpose of this manuscript was to assess the availability of theoretical models for the evaluation of volatile emissions to air during the process of dredge material disposal in a CDF. The first objective was to identify the primary vapor phase transport mechanism for various CDF designs and stages of filling. This provides the theoretical basis for assessing relative volatilization rates. The second objective was to review available laboratory and field procedures for obtaining information needed to measure volatile losses.

Four VOC generating locales were identified. Emission locales are defined as specific locations within a CDF which exhibit common behavioral or operational characteristics that result in the release/generation of VOCs to air. The four are: the sediment relocation locale, the exposed sediment locale, the ponded sediment locale and the vegetation covered sediment locale. The word sediment is used above in place of the phrase dredged material.

Following a section which considers the thermodynamic basis of chemical vapor equilibrium and contaminated sediment, rate equations are presented and reviewed. These equations represent the quantitative results of models of emission mechanisms for each of the four locales. Computations using the equations will yield the chemical flux in mass per unit time. The rate equations are based on transport phenomena fundamentals and have the further advantage of inputs that require concentrations and surface areas of the contaminated sources. The models are sophisticated in the sense that they contain all the complexities of the physiochemical phenomena but some license is taken in assigning the thermal state, concentration gradients, source terms, geometric dimensions, etc. with simple mathematical approximations. This yields equations that are time and space averaged and the predictions of emission rates are therefore limited to time averages from area sources and not point values for specific time. Models for some locales are very crude and additional research is needed to develop more realistic predictive equations.

Emission rates, in the mass of specific or total VOCs per unit time, are primarily dependent on the chemical concentration at the source, the surface area of the source and the degree to which the dredged material is in direct contact with the air. The relative magnitude of these three parameters provides a basis upon which a tentative ranking of emission rates from the various locales can be given. On this basis the exposed sediment locale ranks first. The ponded sediment locale with a high suspended solids concentration in surface waters ranks second. Low in the rankings are bed sediment below a relatively quite water column such as exist in some ponded sediment locales and the vegetation covered sediment locale.

This report contains preliminary calculations of the emission rates of Aroclor 1242 and 1254 from a hypothetical CDF operation in the Upper Acushnet River Estuary of the New Bedford Harbor. The calculations appear in Appendix B and represent sites in two locales of the CDF.

INTRODUCTION

A confined disposal facility (CDF) is a diked area for gravity separation and storage of dredged material solids. When contaminated dredged material is placed in a CDF, the potential exists for volatile organic chemicals (VOCs) associated with the solid fraction of the sediment to be released to the air during and after disposal. The process is termed volatilization and under certain conditions may involve release of significant quantities of hydrophobic compounds such as polychlorinated byphenyls (Thomas, Pleasant and Maslansky 1979). The emission rates of PCBs and other more volatile organic compounds due to volatilization processes associated with operating a CDF are presently unknown. Methods for predicting these losses are needed in order to fully evaluate the confined disposal alternatives for dredged material from Superfund sites such as the New Bedford Harbor. Appendix B contains VOC emissions calculations for the proposed New Bedford Harbor pilot CDF.

The primary objective of this manuscript was to present a general theoretical framework for the development of methods, both mathematical and experimental, for predicting volatile chemical emissions to air from a CDF and its associated operations. The emphasis here is on the development of methods that address the VOC source generation terms in the water and dredged material compartments of the CDF. The information on source strength must be coupled with the appropriate air dispersion model to arrive at downwind concentration levels, contaminated air plume sizes, etc. These aspects of air modeling are not within the scope of this report.

This report presents the results and findings as a consequence of completing the above tasks. No attempt was made to review or present all the information in the open literature on the subject of volatiles from water, dredged material or soils. The subject area is very mature and the information is extensive however, that body of knowledge did provide the technical foundation for this report. Relevant portions of that body of knowledge as it was perceived to apply to CDFs was either selected whole or modified for presentation in this report. Other portions needed to be developed from first principles.

The theoretical chemodynamic models presented in this report reflect the need for mathematical algorithms that are practical but never-the-less contain sound scientific underpinnings. In many cases, the result is a set of vignette models. These models are sophisticated in the sense that they contain all the complexities of the physio-chemical phenomena but some license is taken in assigning gradients, source terms, geometric dimensions, etc. with simple approximations. This trade-off results in mathematical forms that are concise and explicit but allow the incorporation of all the known and relevant chemical behavior characteristics that are inherently difficult to model otherwise.

One or more partial differential equations can be presented to describe the minutia of chemical behavior in the various locales of a CDF. The result is mathematical overkill in the face of the degree of randomness and uncertainty that exist from site-to-site and time-to-time in nature. Development of this type of theoretical model framework was not within the scope of this study. In this light, the degree of compromise provided for by vignette models is justified.

EQUILIBRIUM PROCESSES

Background

All waterways contain particles, and sedimentation is a natural process resulting in deposition on the bottom. Particles enter urban waterways with water runoff and from controlled and uncontrolled water discharges. Chemical pollutants enter waterways by the same routes. The particles are predominantly soil solids. Fine-grained soils, such as silts and clays have a high affinity for many pollutants. VOCs is a general class of pollutants with finite vapor pressures and water solubilities that are known to be associated with waterborne soil particles. These VOCs include hydrophobic contaminants, such as PCBs, which have an especially high affinity for sediments containing organic matter (both natural and anthropogenic). As a result, deposited sediments have seen a significant sink for pollutants discharged to waterways.

Cleanups that involve dredging result in the removal and relocation of in-place polluted sediments. The process of sediment removal and relocation creates conditions that enhance the release of VOCs to air. VOCs enter the air primarily as individual molecular species in the gaseous state from water or sediment surfaces. The entry of VOCs to air from droplets created by bursting bubbles and wind blown sediment dust or other means of particle generation will not be considered in this work.

Chemical equilibrium between the various phases is a branch of the science of thermodynamics. The general criteria for chemical equilibrium as it applies to pollutants in the natural environment are presented elsewhere (Thibodeaux 1979). In the case of VOCs associated with sediment, three phases of matter are involved. These are the solid particles that constitute the sediment and include the sub-phases of organic matter and mineral matter. The organic matter can be both natural and anthropogenic in origin. The mineral matter is inorganic and includes the sand, silt and clay fractions. The two other primary phases are the fluids-air and water. In general, these are less complex in physical make-up than is the solid phase.

The emission of VOCs to air must commence with the proper theoretical chemical equilibrium laws between the three primary phases. A complete description in the case of the locales within a CDF will by necessity involve three binary phase chemical equilibrium conditions. These are air-water, water-sediment and sediment-air. The following section gives the details of each and presents the appropriate theoretical framework for arriving at equations for relating chemical concentrations of VOCs in the air, water and sediment.

Chemical Equilibrium Between Air and Water

Henry's Law The distribution of a volatile chemical between air and water can be represented by a simple ratio of concentrations if true solutions exist in both phases. Historically the ratio has become known as Henry's constant and the equilibrium condition for volatile chemical A is represented as

$$\rho_{A1} = H_\rho \rho_{A2} \tag{1}$$

where ρ_{A1} is concentration in air, gA/cm^3 air,

ρ_{A2} is concentration in water, gA/cm^3 water, and

H_ρ is Henry's constant, cm^3 water/cm^3 air.

The literature is rich with data and computational methods for obtaining this constant. A general source is the work of Lyman, Reehl and Rosenblatt (1982).

Oil-type Film on Water It has been observed both in the field and in the laboratory that an oil-type film can develop on the surface of water as contaminated sediments are agitated or otherwise violently disturbed. Apparently, the natural process of depositing sediment on bottom is a more gentle process and the free-phase oil attached to individual soil-particles remains. It is also possible that once deposited on bottom the contaminated sediment undergo some diagenetic processes whereby the adsorbed oil migrates under interfacial tension forces and coalesces. This results in the formation of oil microdroplets within the sediment. Agitation of this sediment with water during dredging can create conditions for the formation of an oil film.

This oil-type film floats on the water and its juxtaposition between the phases gives rise to a third fluid phase that separates the air from direct contact with the water. The films appear to be a few monomolecular layers in thickness and exist in patches so that complete coverage of the water surface does not occur. Nevertheless, its presence may complicate the normal Henry's Law treatment.

The equilibrium retationship for organic chemical A between air and an oil phase is (p. 196 in Thibodeaux 1979):

$$P_A = x_{A4}\delta_{A4}P_A^* \quad (2)$$

where P_A is the partial pressure in the air, atm

P_A^* is the pure component vapor pressure, atm

x_{A4} is the mole fraction A in the oil phase, and

δ_{A4} is the activity coefficient of A in the oil phase.

The equilibrium relationship between the oil and the underlying water is (p. 196 in Thibodeaux 1979):

$$\rho_{A2} = \rho_{A2}^* x_{A4}\delta_{A4} \quad (3)$$

where ρ_{A2} is the concentration in water, gA/cm^3 water, and

ρ_{A2}^* is the solubility of A in water, gA/cm^3.

If the oil film is thin so that it responds quickly and is in equilibrium with both air and water simultaneously then the product $x_{A4}\delta_{A4}$ may be eliminated and Equation 2 and 3 combined to give Henry's Law. Using the ideal gas law $\rho_{A1} = p_A M_A/RT$ completes the transformation so that the result is identical to Equation 1 where the constant is

$$H_p = M_A P_A^*/RT\rho_{A2}^* \quad (4)$$

where M_A is the molecular weight of A, g/mol,

R is the gas constant, 82.1 atm. cm^3/mol K, and

T is the absolute temperature, K.

If the oil film is thick it cannot be assumed to be in equilibrium with both phases simultaneously. In this case, it provides an additional resistance to the VOCs movement from water to air and the partition coefficient of the VOC between oil and water is needed (Springer, Valsaraj and Thibodeaux, 1985). Since very thin oil film appear to be the norm at CDF operations the above exception will not be considered further in this report.

Chemical Equilibrium between Water and Sediment

Distribution Coefficient The equilibrium distribution of a volatile chemical between water and sediment is typically represented by a constant which is the ratio of the concentration in (or on) the sediment to the concentration in water:

$$K_d = \omega_A/\rho_{A2} \quad (5)$$

where ω_A is the concentration on sediment gA/g sediment and K_d is the distribution coefficient, cm^3 water/g sed. (or l/Kg). The work of Huang et al. (1985) is concerned with VOCs on soils and contains limited desorption data. This is a much studied parameter, particularly for adsorption in the laboratory, and enjoys a voluminous literature heritage beyond the point of realistic applications to the natural environment in many cases. Hill, Myers and Brannon (1988) present the essential aspects of the theory of these distribution coefficients as they apply to CDFs. Pavlou (1978) demonstrated that several published correlations are adequate in predicting the range of the field observed partition coefficients. These field values were derived from analysis of porewater and sediment samples from contaminated sites.

For the purposes of this work it will be assumed that K_d can be adequately estimated by the most appropriate (i.e., chemical match-up) literature correlation or can be measured in the laboratory under desorption conditions. It will be assumed that in the case of dilute solutions up to the solubility limit the organic matter content of the sediment is the primary independent variable. This quantity along with a relationship between K_{oc} and the solubility of the VOC in water will yield K_d. The relation between the coefficient is:

$$K_d = K_{oc}\omega_c \tag{6}$$

where K_{oc} is the chemical partition coefficient normalized to the mass organic content in the sediment, L/KgOM, and
ω_c is the mass fraction organic content of the sediment, KgOM/Kg. OM is the organic matter.

The specific correlation for K_{oc} is the choice of the user. The compilation in Lyman, Reehl and Rosenblatt (1982) is a readily available source of current correlations.

<u>Solubility Limit</u> Extremely high concentrations of PCBs have been measured in the New Bedford Harbor sediment. Weaver (1982) reported concentrations from a few parts per million to over 100,000 mg/kg. If 0.13 mg/L is accepted as the solubility of Aroclor 1242 in freshwater at 11°C (Dexter and Pavlou 1978) then reasonable values of K_{oc} range from 10,000 to 100,000 L/Kg (see Lyman, Reehl and Rosenblatt 1982). For a 5% organic matter, 70% of the total PCBs being 1242, and a sediment PCB concentration of 36,000 ppm Equation (5) and (6) yields an equilibrium concentration in water of 2.5 and 0.25 mg/L. Each, based on a different K_{oc} value, exceed the solubility of 0.13 mg/L. This is thermodynamically impossible for a pure aqueous phase.

The use of distribution coefficients to compute equilibrium concentrations in water, based on concentrations in sediment, has its limits. Beyond that critical chemical concentration in sediment,

$$\omega_A^c = K_d\rho_{A2}^* \tag{7}$$

where ρ_{A2}^* is the solubility in water, the distribution coefficient model is inappropriate. Beyond this limit, the concentration in water is constant and equals the solubility. Figure 1 illustrates the solubility limit concept. However, batch leach data have been found to

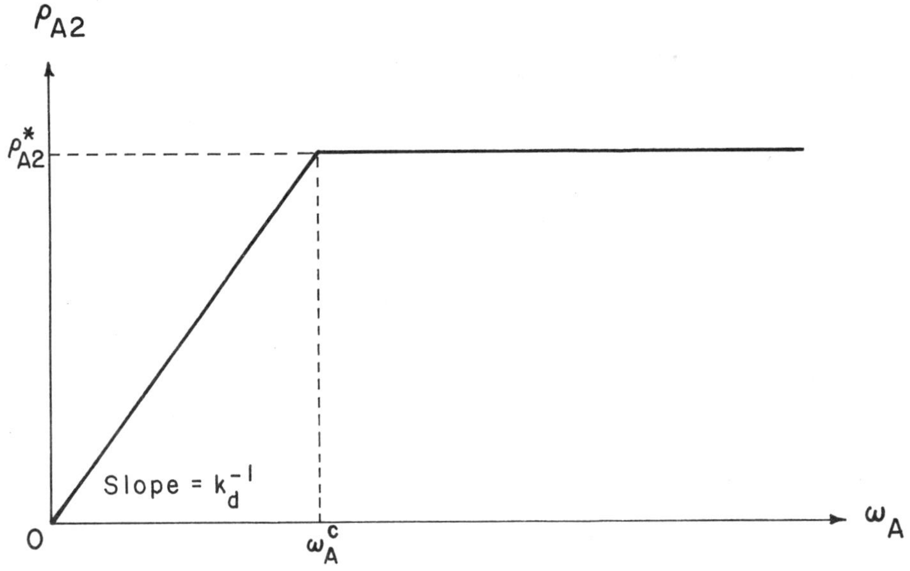

Figure 1. Limits of applicability of distribution coefficient Law and the solubility limit.

exceed such theoretical limits for "operationally defined" concentrations in water. Aroclor 1242 and 1254 concentrations in whole and centrifuged water samples in the Standard and Modified Elutriate Tests exceeded the solubility limit (Otis 1987). This effect is likely to be due to microparticulates in water and therefore not a true solution.

In this work, the ultimate source of all VOCs will be the solid particles either suspended in the water column or as bottom sediment. In the region of chemical concentrations at and beyond ω^c_A evaporation, under equilibrium conditions, occurs at a constant rate. The concentration in air corresponds to the pure component vapor pressure. Substituting ρ^*_{A2} into Equation 1 and using Equation 3 for H_ρ reveals this fact. As the concentration in sediment falls below ω^c_A the equilibrium concentration in water falls below the solubility. The evaporation process is said to be in the falling rate period. The equilibrium concentration in air is below the pure component vapor pressure. The chemical concentrations in all three phases approach zero as the equilibrium evaporation process continues.

Chemical Equilibrium Between Sediment and Air

Adsorption on Dry Sediment The adsorption of VOCs from the air onto dry sediment apparently obey the characteristics of Brunauer, Emmett and Teller (BET) isotherm (Chiou and Shoup 1985; Poe, Valsaraj and Thibodeaux 1988). Dry refers to a water content on the sediment of less than 5% of that needed to form a complete monolayer on the surface of the sediment. In most cases, the Langmuir Modification is appropriate for the levels of chemical concentration observed on most sediment. The Langmuir isotherm is

$$p_A = \frac{P_A^*}{B_1}\left[\frac{\omega_A/\omega_A^*}{1-\omega_A/\omega_A^*}\right] \tag{8}$$

where ω_A^* is the quantity of the VOC required to form one monolayer on the sediment surface adsorption sites, gA/g sed., and B_1 is the BET isotherm constant, dimensionless. In the case of a dilute loading of A on the soil the isotherm is linear and of the form

$$P_A = K_a\omega_A \tag{9}$$

where $K_a \equiv p_A^*/B_1\ \omega_A^*$ is the dry sediment air-solids partition coefficient.

<u>Adsorption on Damp Sediment</u> Damp sediments are those that contain sufficient water to the extent that it occupies 5% to 95% of that required for a single monolayer of coverage. Within this range, the water molecules occupy significant quantities of the adsorption sites so that it effectively competes with the VOC. Valsaraj and Thibodeaux (1987) employed the competitive multicomponent version of the BET and obtained an expression for VOC adsorption on damp sediment. The working form of the equation is:

$$P_A = \frac{P_A^*}{B_1}\left[\frac{\omega_A/\ \omega_A^*}{1-\omega_A/\omega_A^*-\omega_B/\omega_B^*}\right] \tag{10}$$

where ω_B is the moisture content of the sediment, and ω_B^* is the moisture needed to form one monolayer on the sediment surface adsorption sites, gH_2O/g sediment. For dilute loading levels of A on the sediment Equation 10 becomes:

$$P_A = K_a'\omega_A \tag{11}$$

where $K_a' \equiv K_a/[1-\omega_B/\omega_B^*]$ is the damp sediment air-solids partition coefficients. Methodologies for estimating ω_A^*, ω_B^* and B_1 are given by Valsaraj and Thibodeaux (1987). It is apparent from Equation 11 that as the sediment moisture level increases so does the equilibrium partial pressure of the VOC for a fixed soil loading level.

<u>Adsorption Onto Wet Sediment</u> A wet sediment is one that contains at least a monolayer of water molecules on the active adsorption sites. Due to its small thickness the water film on the sediment particles is simultaneously in equilibrium with the solid surface and with the pore gas. In the region of soil loadings below the solubility limit, Henry's Law applies for the air-water equilibrium and the distribution coefficient model applies for the water-sediment equilibrium. Eliminating the chemical concentration in water by solving and equating it

in the Equations 1 and 5 yields:

$$\rho_{A1} = \frac{H_{\rho}}{K_d} \omega_A \tag{12}$$

The ideal gas law can be used to express the left hand side as a partial pressure to yield

$$P_A = K''_a \omega_A \tag{13}$$

where $K''_a \equiv RTH_{\rho}/M_A K_d$ is the wet sediment air-solids partition coefficient

Vapor Pressure Limit Just as in the case of the solubility limit, it is thermodynamically impossible to increase the VOC content within the sediment to such a level that the partial pressure in the air (or gas) filled pore spaces exceeds the pure component vapor pressure at the system temperature. This pressure is an upper bound and limits the applicabilities of Equations 8 through 13 to a range of ω_A values. Figure 2 illustrates the vapor pressure limit concept and gives general regions of applicability of the dry, damp and wet sediment-air VOC adsorption models.

VOC evaporation from exposed sediment to air is a very complex process. The water content of the sediment is the primary factor. Its competition with the VOC for adsorption sites plus the coevaporation with the VOC under low humidity air and moderate temperature conditions is the root of the problem. As an example, consider a wet sediment-air equilibrium condition represented by the point A in Figure 2. This point may be of recently deposited and exposed surface of dredged sediment in the delta zone near the slurry discharge pipe within a CDF. This is a wet sediment case. The VOC content is high and above the solubility limit so that the pure component vapor pressure is exerted in the air layers immediately above the wet sediment bed. Due to downward perculation of water and low air humidity, the water content in the upper sediment layer drops rapidly. This is now a damp sediment case. The loss of water exposes active adsorption sites on the solid surface which can now be occupied by the VOC molecules and this effectively lowers the partial pressure in the air. The value is lower than P^*_A and is represented by point B. The surface sediment remains at this water content, somehow, for a period and the VOC content is reduced by evaporation to air. The equilibrium partial pressure decreases linearly with ω_A to a value represented by point C. Upon reaching point C, a rain event wets the soil and the VOC pressure returns to its full saturation value (point D). This is once again a wet sediment case. The net effect has been that ω_A is reduced due to the vaporization to air. However, the vapor pressure above the sediment remains high at P^*_A because the sediment is wet and contains a VOC at a concentration above ω^P_A.

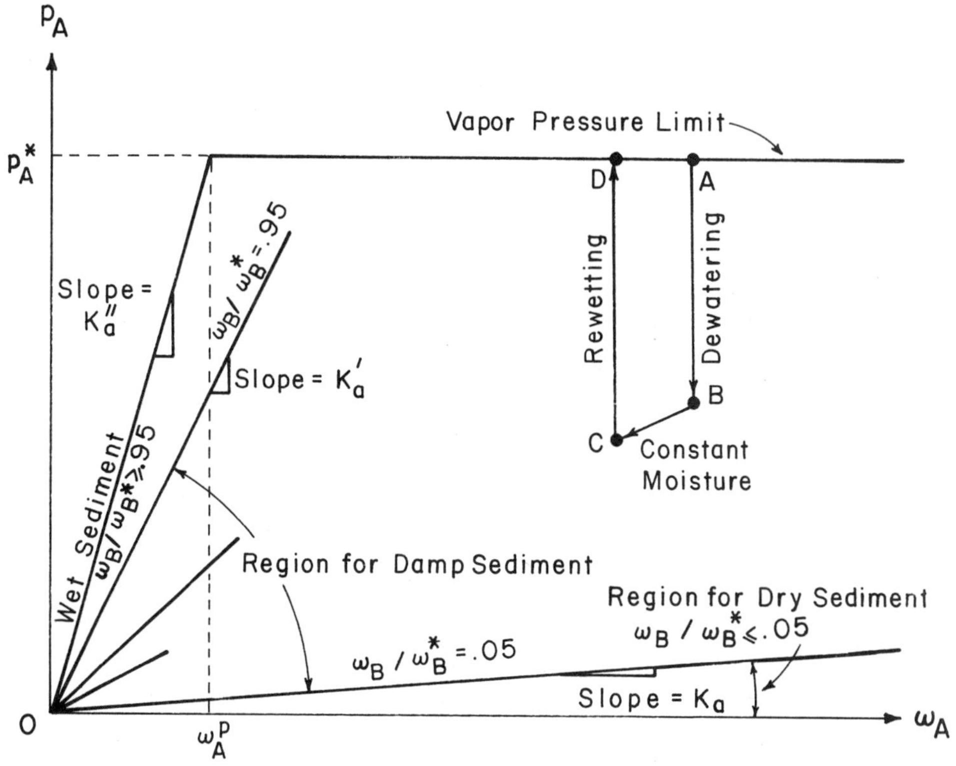

Figure 2. The vapor pressure limit and regions of applicability of sediment-air absorption models.

TRANSPORT PROCESSES

Volatile Organic Chemical Emission Locales

A CDF and its associated operations is a fairly complex unit from the standpoint of generating volatile organic chemicals emission to air. The procedure for developing a proper conceptual model, towards the end of providing realistic algorithms that will quantify emission rates, must address specific physicochemical mechanisms within the CDF as to how they control chemical transport to air. In this light, it is necessary to divide a CDF and its associated operations into four general VOC emission locales. As defined in the introduction, an emission locale is a specific location that contains common behavioral or operational characteristics which result in the release/generation of VOCs to air. Figure 3 illustrates the general locations of each major emission locale in a CDF.

Four locales have been identified. One involves those CDF operations which are concerned with sediment relocation. Specifically, they are dredging, transporting, discharging and other related sediment handling operations. The second emission locale is exposed and drying sediment beds void of vegetation. Typically, this may be the delta region formed from sediment laid down as the slurry emerges from a discharge pipe. This locale occupies the region from the waters edge to the dike or to the vegetation line that commences the marsh or upland region of the CDF. The third emission locale is that portion that contains water. This includes the area of sedimentation during disposal. The fourth and final emission locale is that portion of the CDF that is covered with vegetation. This may be grasses near the waters edge to trees in the older parts of the CDF.

The VOC emerging from the various locales described above do so through specific areas of the CDF surface. Plane surface areas need to be associated with each locale. As will be developed in the next part the individual locales will be characterized by a VOC emission flux rate which has dimensions of chemical mass rate per unit plane area. Fairly precise areas will be required in order to yield total chemical mass rates emitted.

Locale 1. Sediment Relocation Devices

Sediment Resuspension Due to Dredging During operation, all dredge plants, to differing degrees, disturb bottom sediment, creating a plume of suspended solids in the surrounding waters. The suspended solids plume can form relatively low concentrations in the upper water column, high concentrations near the bottom, or both, depending on the type of sediment and the amount of energy indroduced by the dredge.

Resuspension can be viewed as the difference between the amount of sediment loosened or disturbed from the bottom and the amount actually entrained and removed by the dredge. Therefore, the more efficient the dredging process is the less resuspension is likely to occur. There are two basic types of dredges - hydraulic and mechanical. Hydraulic dredges remove and transport sediment in slurry form. They are usually barge mounted and carry diesel or electric powered centrifugal pumps with discharge pipes. Cutterhead, suction, dustpan, hopper, and special-purpose dredges are types of hydraulic dredges. Mechanical dredges remove bottom sediment through the direct application of

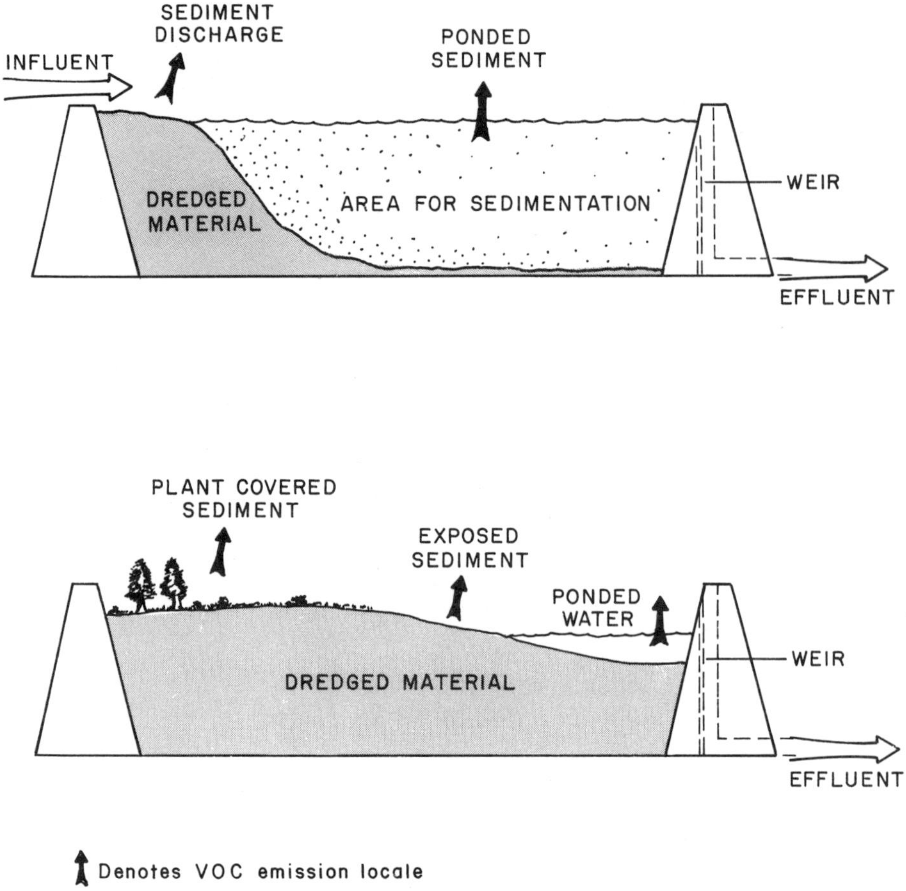

Figure 3. Major VOC Emission Locales: Sediment Relocation (discharge), Ponded, Exposed and Plant Covered.

mechanical force to dislodge and excavate the material at most in situ densities. Clamshell, dipper, dragline and ladder dredges are types of mechanical dredges.

VOC emissions may be enhanced due to the increase in suspended solids in the water column and the energy imparted that increases water turbulence. Contaminated sediment may release some constituents into the water column through resuspension of the sediment solids, dispersal of interstitial water, or desorption from the resuspended solids. Once resuspended, fine-grained sediments (clay and silt) tend to remain in the water column longer than sandy sediments due to slower settling velocities.

VOC are undoubtedly being emitted to air prior to the arrival and operation of the dredge. The contaminated sediment bed is a source of organic chemicals to the water column and they eventually move through the water surface to the air. The resuspension of bottom sediment by the dredge enhances the emission process. Contaminated sediment is brought into the water column near the air-water interface. Desorption of the contaminants from the suspended particles places them in solution from which evaporation to air occurs. Desorption from suspended particles short circuits the original pathway from the bottom and increase the concentration and emission rate.

Enhanced VOC emission will be assumed to be occuring from an area of the water surface, A (cm^2), that encompasses the suspended sediment plume generated in the waterway by the dredging operation. The concentration of suspended solid in the plume is ρ_{32} (Kg/l). If the rate of chemical vaporization is small relative to the mass of chemical associated with suspended solids then an equilibrium model can be used to estimate the VOC concentration in solution. The concentration in solution, ρ_{A2} (mg/l) is:

$$\rho_{A2} = \omega_A \rho_{32} / (K_d \rho_{32} + 1) \tag{14}$$

where ω_A is the concentration of the VOC on the original bed sediment, mg/kg. This equation applies provided $\rho_{A2} < \rho^*_{A2}$.

It is convenient to use chemical equilibrium models such as the above to arrive at concentrations in sediment porewater or the water column. Their use implies some assumptions that may not always be met in the particular locale. In this case, the critical assumption is that the water and solids are in contact a sufficient length of time so that equilibrium between the phases has been achieved. In the case of large particles or large K_d values, the release of the chemical is slow so that large times are needed for equilibrium to be achieved (Geschwend and Wu 1987). The process is particle diffusion controlled and therefore not at equilibrium. The assumption of equilibrium yields an elevated estimate of the VOC in solution and also an elevated emission rate.

The VOC flux rate through the air-water interface is

$$n_A = {}^1K'_{A2}(\rho_{A2} - \rho^{**}_{A2}) \tag{15}$$

where n_A is in gA/cm^2·s, ρ_{A2}^{**} is the hypothetical concentration in water for a VOC concentration in air of ρ_{A1}, gA/cm^3, and ${}^1K'_{A2}$ is the overall liquid phase mass-transfer coefficient, cm/s. This coefficient is computed from the two resistance theory (Thibodeaux 1979). The product of the flux rate and the sediment plume surface area yields the VOC emission rate, W_A in g/s.

Dredge Machinery Water Disturbances   The presence of dredge machinery (i.e., barge) will likely enhance ${}^1K'_{A2}$ above its background value as water turbulence is increased due to the operation. The degree of enhancement is unknown but likely to be no more than a few percent to a factor of two. The increased VOC emissions directly attributed to the operation of the dredge in the case of hydraulic devices are possibly small. On the other hand the operation of mechanical dredges will likely cause significant point sources of VOC emissions.

Mechanical dredges cause significant water turbulence at the point where the bucket, shovel, etc., breaks through the water surface in the process of being hoisted or lowered. This creates intense turbulence in the surface water and enhances VOC emissions. The surface of the bucket or shovel is coated with sediment and this is a very direct pathway for VOC emission. The surface area is small which makes the total emission rate small but does create an intense "point" source each time it is raised above the water surface. The discharge of the bucket contents into a scow or hopper barge creates another intense point source as does the exposed surface of the sediment in the receiving vessel.

All total, there are three VOC point sources involved with a mechanical dredge: the water surface under the bucket, the surface of the bucket with sediment heap and the surface of the receiving vessel. Two are intermittent and depend upon the cycle time and exposure time to air of the bucket while the third is continuous. Emission rates from each can be quantified with rate equations such as those used above.

The energy imparted into the water surface (i.e., horsepower or Watts) each time the bucket is raised or lowered can be easily estimated from the mass, velocity of the mechanical device and event time. This coupled with information on the water surface area disturbed and an estimate of the transport coefficient due to mechanical surface agitation (Equation 4.1-13, p. 146 Thibodeaux 1979 is suggested) will yield the product ${}^1K'_{A2}A$, cm^3/s. The number of events is double the 20 to 30 per hour which are typical bucket cycle times. The concentration driving force is that given in Equations 14 and 15.

As the bucket is lifted through the surface, swings upward and over the scow, deposits its load, swings back and finally plunges through the water surface again it is for all practical purposes a contaminated, sphere bobbing about in the air. The surface area of this almost spherically shaped object is known. Air-side transport coefficients ${}^2k'_{A1}$, cm/s for such objects can be estimated from

correlations involving the object Reynolds number (Bird, Stewart and Lightfoot 1960, p. 647). The VOC emission rate equation is:

$$n_A = {}^2k'_{A1}\left[\rho_{A1}^{*} - \rho_{A1}\right] \qquad (16)$$

In this equation, ρ^{*}_{A1} is the effective VOC concentration in air originating from the sediment clinging to the bucket that is dependent on the concentration in the sediment. Equation 12 is the appropriate equilibrium expression for this so that $\rho^{*}_{A1} = \omega_A H_\rho / K_d$ in Equation 16.

The Receiving Vessel Source The receiving vessel, scow or barge is an open-top vessel and a continuous VOC source either full or empty as long as it is mud splattered. Reible (1987) has developed a modification of the conventional turbulent boundary layer theory expression for the ${}^{2}k'_{A1}$ of flat surfaces. The term is the modification that accounts for the evaporating surface being a distance Z (ft) below the top of the vessel with diameter D(ft). The expression for the air-side transport coefficient is

$$Nu_{AB} = 0.036(1-Z/2D)\ Re^{0.8} Sc^{0.33} \tag{17}$$

where $Nu_{AB} \equiv {}^{2}k'_{A1}\ D/\ _{A1}$, $Re \equiv DV_\infty/\nu_1$, $Sc \equiv \nu_1/\ _{A1}$

$_{A1}$ is the molecular diffusivity of the VOC in air cm^2/s, V_∞ is the background wind speed cm/s and ν_1 is the kinematic viscosity of air, cm^2/s.

Summary The above development demonstrates the mechanisms by which VOC emission in a waterway can be enhanced by the presence of a dredging operation. It was assumed at the onset that the enhancement created by hydraulic dredging was due only to the generation of a suspended sediment plume. This plume was also present in the case of mechanical dredging but in addition, there were three more point sources. Rate equations and specific methodologies were suggested for each source of the dredging locale.

Sediment Discharge Emission Models Most hydraulic dredges such as the cutterhead, suction and dustpan are also equipped to pump all types of alluvial materials and compacted deposits, such as clay and hardpan, to CDFs. The sediment and water slurry is discharged from the end of a pipe either into the water or onto the sediment delta of the receiving CDF.

The discharge of a slurry from pipe is not unlike the natural process of water flowing over a dam. Reaeration occurs as the water flows over the dam, and empirical relationships for this have been developed and verified by field data. Among the empirical relationships available the equation, commonly referred to as the British formula, seems particularly well suited because it was calibrated with 54 small-stream headloss structures (Tetra Tech 1978). Since O_2 transport into water is liquid phase controlled and most VOC are desorbed by the same liquid phase mechanism, the fraction volatilized across the discharge can be estimated by

$$F = \frac{0.11ab(1+0.046T)H_d(\mathcal{D}_{A2}/\mathcal{D}_{B2})^{1/2}}{1+0.11ab(1+0.046T)H_d(\mathcal{D}_{A2}/\mathcal{D}_{B2})^{1/2}} \tag{18}$$

where b=0.60 for a round broad-crested curved face spillway. The half power of the diffusivity ratio coverts the correlation from an oxygen database to that for the VOC of interest. The a is a water quality factor and is equal to 1.00 in polluted water, T is water temperature (°C), H_d is the height through which the water falls (ft), $_{A2}$ and $_{B2}$ are the molecular diffusivities (cm^2/s) of the VOC and oxygen in water respectively. The quantity of the VOC approaching the discharge point that is volatilized is:

$$W_A = Q\rho_{A2}F \tag{19}$$

with W_A in g/s, Q is the volumetric rate of water (solids free) flow in the pipeline in cm^3/s and ρ_{A2} is the in-solution VOC concentration in g/cm^3. The use of Equation 14 to estimate the concentration is an excellent choice since the residence time and high turbulence levels in the pipe will assure near equilibrium conditions at the exit. The value of ρ_{32} is that of the pipeline sediment slurry.

The use of oxygen transfer coefficients as the data for a dredge slurry, as proposed above, will likely result in an over-estimate of the actual VOC emission rate. The recommended water quality factor a for clean water is 1.25. The recommended value of a = 1.00 for polluted water is a reasonably conservative choice for a sediment slurry, because effective viscosity of the slurry is much greater that that of water and the transport coefficient is typically a function of viscosity to the negative 2/3 power, Equation 18 will over predict actual VOC emission rates. The magnitude of the error cannot be estimated with information available.

If the discharge is submerged, Equation 18 is inappropriate for estimation of VOC emission because there is no contact between air and the dredged material slurry. For submerged discharge, VOC emission is hindered by the overlying water. VOC emission models for ponded sediment, discussed later in this report, are applicable to submerged discharge. The discharge of dredged sediment into the CDF may also be by bucket or shovel. In this case, the appropriate methodology recommended above for VOC emission estimates during bucket or shovel dredging should be used. Specifically see page 11.

It is possible that significant overland flow of water and sediment may occur from the discharge point to the waters edge of the CDF. Sediment deposition occurs along the entire stretch of the rivulets flow path providing a constant supply onto the bottom. This redeposited material will be considered to be a VOC source of constant concentration ω_A. Shallow rapid streams are known to be good VOC strippers (p. 156 Thibodeaux 1979). In the following paragraphs a vignette model is developed that accounts for these and other major sediment and chemical transport processes within the rivulet. A detailed derivation appears in Appendix A. The development is general enough that it can also be applied to the ponded locale of a CDF.

A sketch of a section of a rivulet stream is shown in Figure 4. As indicated in the sketch, the VOC evaporation occurs across the air/water interface. Other transport processes effecting the VOC within the differential element are advection by water flow, advection by suspended particle flow, deposition (net) of particles onto the bed, dissolution from the bed surface and dynamic equilibrium between suspended particles in the water column. This vignette model is steady-state with uniform concentration in the vertical and lateral direction. A simple net deposition particle transport and exchange model is used to describe the suspended solids concentration:

$$\rho_{32} = \rho^{o}_{32}\exp(-v_d xw/Q) \tag{20}$$

where ρ°_{32}, g/cm³, is the concentration of particles at the head of the rivulet streams, v_d is the net velocity of particles deposition, cm/s, x is the distance from rivulet head (cm), w is rivulet width (cm), and Q is the flow rate of water (cm³/s). This is one case of several particle transport and exchange models proposed in the EPA Technical Guidance Manual, Book II, Streams and Rivers, Ch. 3, Toxic Substances (Delos, et al., 1984). A mass balance on the VOC yields the following equation for the concentration, ρ_{A2} (g/cm³) with reach, x (cm)

$$\rho_{A2} = \frac{{}^{3}k'_{A2}\omega_A/K_d + {}^{1}K'_{A2}\,\rho^{**}_{2A}}{{}^{3}k'_{A2} + {}^{1}K'_{A2}} + \left[\frac{\rho^{o}_{A2} + {}^{3}k'_{A2}\omega_A/K_d + {}^{3}K'_{A2}\,\rho^{**}_{A2}}{{}^{3}k'_{A2} + {}^{1}K'_{A2}}\right] \tag{21}$$

$$\exp\left[-({}^{3}k_{A2} + K'_{A2})\ xw/Q + \left[\frac{{}^{3}k'_{A2} + {}^{1}K'_{A2}}{v_d}\right] \ln\left[\frac{1+K_d\,\rho^{o}_{32}}{1+K_d\rho 32\ \exp(-v_d xw/Q)}\right]\right]$$

where ${}^{3}k'_{A2}$ is the water-side transport coefficient above the bed (cm/s) and ρ^{o}_{A2} is the initial VOC concentration at x=0 in g/cm³. Appendix A contains the derivation of Equation 21.

The relative importance of the various transport parameters in Equation 21 is unknown. For example, if large particle fall out

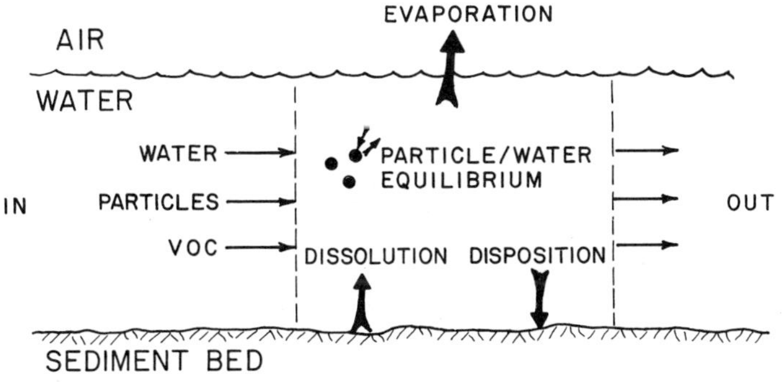

Figure 4. Major VOC Transport Pathways in a Rivulet.

quickly and $v_d = 0$ for the remainder of the length plus ${}^3k'_{A2} >> {}^1K'_{A2}$ then $\rho_{A2} = \omega_A/K_d$. These conditions force the evaporation to be much slower than dissolution resulting in a VOC concentration in water at equilibrium with the concentration in the bed. Nevertheless, it will be necessary to integrate ρ_{A2} over the length of the rivulet to get the average in order to use Equation 15 for the flux rate. Consult Delos, et al. (1984) for means of estimating v_d and Thibodeaux, Chang and Lewis (1980) for calculating ${}^3k'_{A2}$ based on stream and chemical properties.

Summary This concludes the section on models for VOC emissions resulting from the discharge of sediment into the CDF. As in the previous section on dredging operations the mechanisms that drive the VOC emission process are fairly energy intensive. Although the sources are intense in a comparative sense, the areas are small and only by calculation can the quantitative significance be determined. This is the case for the waterfall from the discharge pipe, the bucket/shovel discharge and the rivulet of water that runs down the face of sediment pile. Models for submerged discharge are considered in the section on ponded sediment; see page 19. The remaining emission locales are influenced by natural processes to a greater extent. The energy sources are subtle, however, the emission areas are large and the net effect could be the generation of a significant quantity of VOC's to the surrounding air boundary layer.

Locale 2. Exposed Sediment

This VOC emission locale is characterized by sediment that is exposed directly to air and void of any vegetation cover. It is the delta region formed by sediment discharged from a pipeline or from the bucket/shovel placement of sediment within the CDF and generally occupies the region from the waters edge to the vegetation line. Figure 3 illustrates this locale both in the early stages of filling and at the time the CDF is full. From a chemical transport point of view this is the most complex of the four VOC emission locales. The complexity is caused primarily by the presence of water. Attempts at complete mathematical descriptions of the coupled diffusion of thermal energy, with a phase change due to the evaporation of water, VOC, water vapor and liquid water in the three phases of a porous media with realistic closure and equilibrium formulations have been proposed (Linstrom and Piver 1985). The data input for such models is enormous and the output, without site-to-site calibration, is more doubtful than a series of vignette models where the inputs are few. The following development is such a set of vignette models that describe the major aspects of the VOC emissions from the exposed sediment locale.

Water Transport In The Upper Soil Layers The VOC emission rate from the upper soil layers is dependent on both the moisture content and the water transport rate. The dredged material in this locale will also be termed soil. The theoretical aspects of the role of water on the equilibrium partitioning of VOC's between soil particles and the vapor phase was presented in Part II. Water transport in the upper soil layers is driven by advection processes, both downward and upward as liquid, by capillary forces, evaporation and diffusion as vapor. In the case of soils that were once sediment contaminated with VOC's, it can be assumed that the VOC's are transported in the upper layers at much slower rates than water. This is true primarily for three reasons. The VOC's are strongly sorbed onto/into the soil particles, they enjoy less direct transport pathways and their vapor pressures are in general orders of magnitude lower than that of water. In the case of less direct transport pathways, it must be realized that liquid water can move as an essentially pure phase whereas the VOC must move with it and then only in solution. It will also be assumed that the upper soil layers in the CDF will exist in a predominantely unsaturated state with respect to water. Water that arrives at the exposed sediment locale, either by rain or runoff, is assumed to evaporate upward and/or percolate downward rapidly so that during the vast majority of years time the upper soil layers are unsaturated.

Since water is so mobile, its evaporation from the surface or from within soil enhances the transport of the VOC in the vapor state. The coupled process for transport through the soil-air interface is described by the following two equations (p. 79, Thibodeaux 1979):

$$N_A = {}^3k_{A1}(y_{Ai}-y_A)+y_{Ai}(N_A+N_B) \tag{22}$$

and

$$N_B = {}^3k_{B1}(y_{Bi}-y_B)+y_{Bi}(N_B+N_A) \tag{23}$$

where N_A an N_B are the molar flux rates of the VOC ($\equiv$A) and water ($\equiv$B) vapor respectively (mol/s.cm^2), y_{Ai} and y_{Bi} are mole fraction concentrations at a surface, y_A and y_B are concentrations far from the surface and ${}^3k_{A1}$ and ${}^3k_{B1}$ are the respective mass transfer coefficients on the air side of the soil surface (mol/s.cm^2). It is argued above that $N_B \gg N_A$. If the air is dry, contains none of the VOC and the mass-transport coefficients are inversely proportional to the square root of molecular weight then the equations can be combined to yield:

$$N_A = \left[\frac{y_{Bi}\left((M_A/M_B)^{1/2}-1\right)+1}{-y_{Bi}+1}\right]{}^3k_{A1}y_{Ai} \tag{24}$$

where M_A and M_B are the molecular weights of the VOC and H_2O respectively. The terms in the bracket is the enhancement factor that the water vapor transport imparts to the VOC transport. For example, the 40° the vapor pressure of water is 0.0732 atm so that for a VOC of M_A = 300 the enhancement factor is 1.32. This is not an insignificant factor.

<u>Vapor Phase VOC Transport In Upper Soil Layers</u> Contaminated sediment which are wet and exposed directly to the atmosphere is the case that results in the highest instantaneous VOC emission rates, however, the process is short lived. Volatile organics sorbed onto/into particles at the soil surface have a very short pathway to

the air. The mechanism of evaporation includes:

a. desorption from the particle surface (or pore) into the water film,

b. diffusion through the water film,

c. desorption from the water film into the air boundary layer and

d. finally vapor phase diffusion.

This is a series of four resistances. The water film is very thin and usually provides minimum resistance. If it is assumed as before that chemical equilibrium exist between the particle and the water film and since this interface and the water-air interface present no resistance then Equation 16 can be used. The assumption of particle-water chemical equilibrium is very weak in this case because at some point in this rapid surface evaporation process chemical diffusion within the particle phase will become the controlling factor. Nevertheless, Equation 16 will yield an upper bound VOC emission rate. Algorithms based on pesticide flux rate data above surface soils are available for estimating the air-side mass transfer coefficient (Thibodeaux and Scott 1985). This coefficient needs to be corrected by the water vapor transport enhnacement factor in Equation 24.

At the above VOC loss rate from the soil surface the top layers will quickly become depleted of the chemical. Continuing losses will come from within the soil pore spaces. At this point the process of evaporation is transient and changes from being air-side resistance controlled to soil-side vapor phase diffusion controlled. This occurs when the effective plane of VOC evaporation recedes to a depth h (cm):

$$h = 19\mathcal{D}_{A1}\varepsilon_1^{10/3}/3k'_{A1}\varepsilon^2 \qquad (25)$$

where ε_1, is the air filled porosity (cm^3/cm^3) and ε is the total soil porosity. Typically h is a few millimeters.

A model has been developed for the case of volatile chemical losses from soils that receive petroleum waste in so-called landfarming treatment process (Thibodeaux and Hwang 1982). This model has been verified in a laboratory microcosm (Dupont 1986). This investigator reported that the model "...consistently overestimated flux rates for the pure constituents by only a factor of approximately 2 to 10, and consequently appear quite effective for the estimation of volatile organic emissions from complex waste/soil systems". Since soil formed from dredged material is very similar, this model will be used as the basis for the case of VOCs originating from within the soil.

The emission pathway for VOC originating from particles positioned below the soil/air interface has an additional resistance to the four detailed in above. The vapor phase molecules must diffuse through the air (or gas) filled soil pore spaces prior to emerging into the air boundary layer. This is the fifth step in the overall evaporation process and is apparently the rate limiting step (Dupont 1986). The soil zone containing VOCs in the landfarm situation was a finite depth of 10 to 15 cm. whereas that for dredge material in a CDF is infinite for all practical purposes. Fick's second law with an effective diffusivity that reflects the so called retardation factor is appropriate. Due to the flat nature of the surface and the depth of the dredged material the semi-infinite solution is appropriate. In this case the instantaneous flux is:

$$n_A = \left[\frac{D_{A3}\,(\varepsilon_1 + K_d\rho_B/H_\rho)}{\pi t}\right]^{1/2} \left[\frac{\omega_A H_\rho}{K_d} - \rho_{Ali}\right] \tag{26}$$

where t is the time (seconds) the soil system has been inplace, ρ_{Ali} is the VOC concentration at the soil surface (gA/km^3), D_{A3} is the effective diffusivity (i.e., $\mathcal{D}_{A1}\varepsilon_1^{10/3}/\varepsilon_2$), and ρ_B is the bulk soil density (g/cm^3). Equation B4 in Appendix B combines the air-side and soil-side resistance into one relationship.

It is convenient to define the phrase "age of a sediment-soil". First, the word "sediment soil" will be used to denote the dredged material in the CDF that is exposed directly to the air so that with time it becomes a true soil. The age of a sediment-soil is the time it has been in-place in the CDF. In general, the soil near the waters edge of a CDF is younger than that at the vegetation line. This occurs mainly because of the delta forming process in which dredged sediment may be placed within a CDF. Therefore, the variable t in Equation 26 is the age of the sediment-soil and it appears in the denominator.

As a sediment-soil ages the VOC emission rate decreases (see Equation 26). If a simple moving boundary evaporation plane and pore diffusion model is assumed (p. 337, Thibodeaux 1979) the VOC lost from the soil surface creates a "dried-out" zone. The term "dried-out" is meant to confer that the VOC content of the sediment-soil has been depleted and does not refer to moisture. The depth-time relationship for creating this dried-out zone is approximately:

$$t \sim h^2(\varepsilon_1 + K_d\rho_B/H_\rho)/2D_{A3} \tag{27}$$

where h is the depth of the zone in cm. Considering the aging process as described in this and the previous two paragraphs, it is obvious that a realistic quantification of the VOC emission rate from the exposed sediment locale at any particular time will require a knowledge of the age of the sediment-soil in addition to the surface area. The total emission rate is the sum of the products of flux rate and surface area for the number of age categories chosen.

Caveats of the VOC Transport Model   The basic assumptions in the above VOC transport model are that the process is rate limited by the molecular diffusion of chemical species through the open vapor pores within the soil and that the particles, with their moisture covering, is the chemical mass source always maintaining an equilibrium concentration in the pore gas. A simple calculation with Equation 27 for Aroclor 1242 using the average solubility and vapor pressure shows that approximately 50 years are required to create a 30 cm. deep dried-out zone if vapor phase diffusion is assumed and 50,000 years if liquid phase chemical diffusion is assumed. Some laboratory data suggest the vapor phase process is more realistic. It has been observed in the laboratory that for an 8 cm. layer of Indiana Harbor Sediment exposed to air an 83% decrease in total Aroclor 1248 congener concentration occurred in 6 months (Palermo and Miller 1987).

The mechanisms which dominate the VOC emission process over the long haul is likely to be as demonstrated above. Water movement due to wet/dry cycles will likely have a secondary effect on the VOC loss rate from the upper soil layers. In general, the effect will be to enhance the rate. Wet and dry cycles will cause the VOC to move upward in clay and low permeability soils. Water evaporation at the surface and its capillary rise transports the soluble chemical upward. Here it readsorbs onto the cleaner surface soil and awaits dry-out so it can revaporize. In sandy, low permeability soils downward leaching is more likely to occur.

Soil surface cracking will likely enhance the VOC loss rate. Cracking increases the overall sediment-soil porosity and also shortens the diffusion pathway to the surface. As a first approximation the solid geometric shapes that make up the cracked soil surface can be modeled as vertical spines with cross section area A_c, cm^2, and perimeter P_c, cm. Heat transfer flux expressions for such shapes have been developed (Welty, Wicks and Wilson 1984). By use of analogy theories these expressions can be transformed to mass-transfer flux expressions and used for estimating VOC emission rates to air. This and other secondary effects such as capillary water movement effect on the VOC emission process in the upper soil layers await vignette model development.

Summary   The exposed sediment locale is likely to be a significant source of VOCs from the CDF to the surrounding air. In general, the pathway is short and the surface area is extensive so one would expect a relatively large quantitative rate. Tofflemire, Shen and Buckley (1981) give both laboratory and field data as evidence that PCBs volatilize from exposed sediment.

Locale 3. Ponded Sediment

The VOC emission locale is characterized by a water body which contains contaminated sediment in suspension and on bottom. The emissions occur through the water surface. Figure 3 illustrates the ponded sediment locale during the time dredged material is being placed into the CDF and when disposal has been completed. Each of these modes must be considered separately.

The Filling Phase   On page 14 a generalized model for VOC emissions from the overland flow of a rivulet is given. At the onset the development was generalized so that it would apply to the filling phase of a CDF and reflect all the relevant transport and the particle sedimentation process that are occuring during this period.
The primary model equations are 15, 20 and 21. DiToro and O'Connor (1981) have developed a less general model for estimating the PCB flux to the atmosphere.

The significant differences in the VOC emission rates between lake and rivulet are due to the relative magnitudes of the transport coefficients, surface area of the water body and volumetric flow rate. In general both mass-transfer coefficients, ${}^3k'_{A2}$ and ${}^1K'_{A2}$, will be smaller for the lake than for the rivulet. In addition, the overall particle deposition velocity, v_d, will depend on the settling characteristics of the dredged material. The removal of solids in a CDF is by either flocculant or zone settling of the dredged material slurry and a function of many variables (Montgomery 1978; Palermo, Montgomery and Poindexter 1978). Laboratory procedures developed by Montgomery (1978) should be used to determine settling velocities. The product of the length x, and the width w, of the surface will be larger for the lake. The volumetric flow rate for both will be comparable. For calculating the transport coefficients the appropriate correlations that represent lake conditions should be used. In the case of unstratified waterbodies, the ${}^3k'_{A2}$ correlation developed by Thibodeaux and Becker (1982) is recommended. For the case of a stratified water column an overall coefficient (i.e., ${}^1K'_{A2}$) must be chosen that represents the three or so layers of relatively un-mixed water (see p. 402 Thibodeaux 1979).

The amount of water column turbidity generated by an open-water pipeline disposal operation or barge pumpout into a CDF can probably be minimized most effectively by using a submerged diffuser system. This system is designed to eliminate all interaction between slurry and upper water column by radially discharging the slurry parallel to and just above the bottom at a low velocity. Some degree of sediment entrainment in the water column will occur. The degree likely falls somewhere on a scale between the sediment resuspension due to hydraulic dredging and that due to mechanical dredging. It is also likely that higher suspended solids concentration will occur near the air-water interface in shallow water than in deep water. From a mechanistic point of view, the chemical transport processes up the water column and the VOC emission process through the air-water interface is similar to dredging operations. The procedures outlined on page 10 are appropriate for estimating flux rates of submerged discharges.

The Bed-sediment Source   When the discharge of dredged material into the CDF is terminated, for a period thereafter the bed-sediment surface is the VOC source. The chemical pathway steps are:

a. desorption from the bed surface particles,
b. molecular diffusion through the benthic boundary layer,
c. movement up the water column to the interface and
d. volatilize through the air boundary layer into the atmosphere.

From a steady-state mass balance the following expression results for the VOC concentration in water:

$$\rho_{A2} = \frac{\omega_A \, {}^3k'_{A2} / K_d + {}^1K'_{A2} \, \rho^*_{A2}}{{}^3k'_{A2} + {}^1K'_{A2}} \tag{28}$$

Equation 15 is then used to arrive at the flux rate. If the water in the CDF is stratified then ${}^1K'_{A2}$ must be appropriately modified as indicated in the previous section.

As the sediment bed surface ages and gradually looses its VOC content a natural chemical leaching process occurs from the particles within the bed. The process of VOC emission is now controlled primarily by the very slow process of desorption and diffusion, in the water filled pores, to the sediment surface. The leaching is transient at this point and behaves similar to the vapor phase transport model presented on page 16. This type of transient leaching model has been developed for solid particles in bed-sediment form discharged at sea during offshore drilling for oil (Thibodeaux, Reible and Fang 1986). Selected portions of this model along with Equation 15 will yield the VOC emission flux rate for this aging period of the ponded locale in a CDF.

Hwang (1987) has developed a similar transient approach for estimating air emissions for estuaries during high tide, low tide and for the exposed sediment. He also presents a steady-state emission rate equation for bed sediment. These are based on knowing the average diffusion path length for a contaminant in sediment. This parameter can be estimated from concentration vs depth measurements. He notes that the models need to be validated.

<u>Summary</u> The ponded sediment locale is likely to be a significant VOC source only near the point where a sediment/water slurry enters. At this point the VOC concentration in solution is high and is maintained by a relatively high suspended sediment concentration in the region. As the water moves along, the particles settle and vaporization lowers the concentration. On the discharge end of the ponded locale the VOC emissions are likely low. As the CDF fills the ponded locale surface area diminishes and so do the emissions from this source. Remanent water bodies containing clear water that cover aged bottom sediment are likely to be very low VOC emission locales.

There is both laboratory and field evidence that bottom sediment containing VOCs are released to the air. Laboratory experiments that simulate vaporization of VOCs from bed-sediment covered with water were performed to demonstrate the release of PCBs (Tofflemire, Shen and Buckley 1981) and two chlorinated benzenes (Karickhoft and Morris 1985). Studies with PCBs in the two artificial outdoor ponds demonstrate that the same process occurs in the field (Larsson 1985). Clophen A 50, a PCB, was blended into sediment and observed in both air (20 cm above the surface) and in water for a two year period. The author observed that his results show that contaminated sediment may act as a source of chlorinated hydrocarbons released to the environment.

Locale 4. Vegetation Covered Sediment

The fourth and final emission locale is that portion of the CDF that is covered with vegetation. It occupies the region that commences with the line of grasses at the edge of the exposed sediment locale and extends into that region of the CDF containing older sediment. It can be characterized in general as the region of the CDF with a vegetation cover.

The existence of a vegetation cover on this locale causes significant changes in the soil environment, as compared to the exposed sediment locale, which effects the VOC emission rate. The presence of vegetation usually renders the upper soil layers more porous. This and other bioturbation processes in the zone would tend to increase ε_1, and ε which in turn increases the effective diffusivity of the VOC in soil (see page 16). With time the natural organic matter content of the soil will slowly increase. The source of this organic matter is decayed vegetation, microorganism remains, etc. Its presence will tend to retard the VOC transport by providing additional adsorption sites near the soil surface. The protrusion of plants into the air boundary layer increases the resistance of this chemical transport pathway. The net effect is a decrease in ${}^3k'_{A1}$. This reduces the water vapor and VOC transport rates directly but it also reduces these rates indirectly by maintaining a cooler (i.e., lower temperature) soil surface thermal environment. A reduction in the moisture evaporation rates has a secondary effect; it maintains the soil water content high thereby reducing ε_1. As indicated on page 17, this decreases the effective diffusivity of the VOC through the soil and should reduce the emission rate.

It appears from the above that the presence of vegetation comfounds the process of modeling the VOC emission process from this locale. The net effect of the above factors is a reduction in the VOC emission rates when compared to a similar exposed sediment locale that has the same age and depth to the sediment-soil contaminated zone. Nevertheless, the models developed for the exposed sediment locale generally apply. Equation 16 applies for the air-side, however, a factor to account for extent of vegetation cover in reducing the ${}^3k_{A1}$ mass-transfer coefficient must be incorporated. An equation similar to 26 applies on the soil side. It is likely more appropriated to restart the time (i.e., soil age) to zero and assume that a clean soil cover of depth h (cm) exists above the remaining contaminated sediment-soil. A model for the analogous problem of subsurface injection of liquid waste has been developed (Thibodeaux and Hwang 1982) and verified in the laboratory (Dupont 1986).

Since this locale likely has the lowest VOC emission rates, existing vignette models may be appropriate until such time that field data indicates otherwise. Theoretically, the other locales should have higher emission rates and therefore deserve closer attention.

LABORATORY AND FIELD EXPERIMENTS

Theoretical models must be tested against and adjusted to both laboratory and field data prior to their acceptance and use as predictive tools. In some, if not most, cases it is necessary to perform limited laboratory and/or field experiments. For example, this will need be done in order to test the significance of a particular VOC source term, to verify some hypothesized VOC generation process or to obtain critical adjustable parameters in the final mathematical algorithm of the model. In this regard, this section of the report is devoted to a brief review of existing laboratory and field test procedures for the technical information they can provide in support of the modeling objectives.

There are three levels of experimental protocols that are necessary to arrive at a complete understanding of VOC emission processes. Equilibrium experiments to quantify the partitioning of VOC's between sediment and air are necessary. Some protocols have been developed. The techniques developed by Spencer and co-workers (Spencer, Farmer, and Jury 1982) for quantifying the relative vapor pressures of pesticides that have been incorporated into soil are directly applicable. The above literature citation is a review article that includes the investigators work in this area for the last two decades. The recent works of Chiou and Shoup (1985) and of Poe, Valsaraj and Thibodeaux (1988) are also relevant to the subject.

Critical laboratory experiments simulating the transport of VOC's from water surfaces and exposed sediment surfaces will be necessary. Tofflemire, Shen and Buckley (1981) developed an apparatus and appropriate experimental procedures to measure VOC emission rates from water covered bottom sediment. The adaption and modification of this apparatus and procedure should lead to a more realistic laboratory simulation of the ponded sediment VOC emission locale. In the case of simulating the transport of VOC's from exposed sediment the techniques developed by Farmer and co-workers (Farmer, Yang, Letey and Spencer 1980) are most appropriate. The device was more recently applied by Karimi (1983) to measure VOC's from contaminated soils.

Field measurement techniques that allow the direct quantification of VOC emission rates from the exposed sediment and ponded locales are needed. Similar techniques exist for measuring volatiles from surface impoundments (Thibodeaux, Parker and Heck 1984), landfills (Eklund, Balfour and Schmidt 1985), and land treatment facilities (Eklund, Nelson and Wetherold 1987) have been developed and tested. The application of these apparatus and procedures to the CDF emission locales should be direct. Emission from the sediment relocation devices are, in effect, point sources. Techniques for quantifying emissions from such sources are available. That developed by Kolnsberg (1976) seems most appropriate for the sediment relocation operation associated with a CDF.

CONCLUSIONS AND RECOMMENDATIONS

There is much general information in the technical literature on the subject of volatile chemicals in water and on solids relative to contact with the gas phase. The basic theory of chemical volatilization from such sources is inplace. With a few minor exceptions the theory has been developed to the point that equations have been formulated so that the various rates can be quantified. Specifically on the subject of VOC's from CDF sources there is little of the above that is directly applicable.

This document contains an assemblage of vignette models and associated equations plus guidance for their use. The models are general so that they can be applied to any CDF, however features of the proposed dredging and disposal operation at New Bedford Harbor were used as the specific application. Four principal VOC emission locales were identified to exist in any CDF. The locales are: sediment relocation devices (dredging associated activities), exposed sediment, ponded sediment and vegetation covered sediment. Although it is theoretically possible for VOC's to be emitted from each locale it is very likely that some are more significant sources than others.

Emission rates, in mass of specific or total VOC's per unit time, are primarily dependent on the chemical concentration at the source, the surface area of the source and the degree to which the dredged material is in direct contact with the air. The relative magnitude of these three parameters provides a basis upon which a tentative ranking of emission rates from the various locales can be given. On this basis, the exposed sediment locale ranks first. The ponded sediment locale with a high suspended solids concentration in surface water ranks second. Low in the rankings are bed sediment below a relatively quiet water column such as exists in some ponded sediment locales. Emission from the vegetation covered sediment locale are also expected to be low.

The following are recommendations derived from the information generated in preparing this report. Preliminary model calculations can be made for the locales at this time, however some aspects are based on very crude equations and further development is needed. The recommendations in the model development section address these specific deficiencies. The recommendations in the sections on laboratory and field testing reflect general research activities that must be performed in order to build a higher degree of confidence in the predictive capability of the current generation of VOC emission models.

Preliminary Model Calculations

a) Perform detailed model calculations for all locales with quantitative information for New Bedford Harbor sediment. Appendix B contains two such detailed calculations.
b) Fine tune the tentative rankings of sources by performing a sensitivity analysis of critical model input parameters using the New Bedford Harbor information.

Model Development

a) Develop a vignette model and associated equations for the surface cracking/water evaporation coupled phenomena observed with much dredged material undergoing drying.
b) Develop an appropriate VOC emission model that accommodations the cracking/evaporation model.

Laboratory Testing

a) Develop laboratory test procedures to measure selected sediment/air chemical partition coefficients and related vapor pressure/adsorption parameters.
b) Simulate the VOC emission process from exposed sediment and ponded sediment locales for selected chemicals in a pilot-scale wind tunnel apparatus.

Field Tests

a) Perform field-tests at the New Bedford Harbor Superfund Site to measure VOC emission rates and concentrations in air.
b) Use this and other measured rates and concentrations in air for preliminary model verification, validation and calibration exercises.

ACKNOWLEDGEMENT

This manuscript is based on work performed for the Department of the Army Waterways Experiment Station, Corps of Engineers under contract No. DACW3987M2487. Mr. Tommy E. Myers, Water Supply and Waste Treatment Group (WSWTG), Environmental Engineering Division (EED), Environmental Laboratory (EL), WES was the technical monitor; general supervision was provided by Mr. Norman R. Francingues, Jr., Chief, WSWTG, Dr. Raymond L. Montgomer, Chief, EED, and Dr. John Harrison, Chief, EL.

Col. Dwayne G. Lee, CE, was Commander and Director and Dr. Robert W. Whalin was Technical Director of the WES during preparation of this report. The original WES report was entitled: Theroetical Models For Evaluation of Volatile Emissions to Air During Dredged Material Disposal with Applications to New Bedford Harbor.

REFERENCES

1. Bird, R. B., W. E. Stewart and E. N. Lightfoot 1960. Transport Phenomena, John Wiley & Sons, N. Y.

2. Chiou, C. T. and T. D. Shoup 1985. Environ. Sci. Technol., Vol 19, 1196.

3. Delos, C. G., J. V. DePinto, W. L. Richardson, P. W. Rodgers and K. Rygwelski 1984. Appendix B: Estimation and Use of Parameters in Modeling Toxics. Section 3.0 of EPA Tech. Guidance Manual, Book II, Streams and Rivers, Ch. 3 Toxic Substances. U.S. EPA, Washington.

4. Dexter, R. N. and S. P. Pavlou 1978. "Mass solubility and Aqueous Activity Coefficients of Stable Organic Chemicals in the Marine Environment: Polychlorinated Biphenyls", Marine Chemistry, Vol. 6, pp. 41-53.

5. DiToro, D. M. and D. J. O'Connor 1981. "Estimate of Maximum Probable PCB Flux to the Atmosphere from the Hudson River Sediment Disposal Basin", App. J in Unpublished report, Hydro Qual Inc. Mahwah, N. J.

6. Dupont, R. R. 1986. "Evaluation of Air Emission Release Rate Model Predictions of Hazardous Organics From Land Treatment Facilities", Environ. Prog., Vol 5, No. 3, p. 197.

7. Eklund, B. M., W. D. Balfour and C. E. Schmidt 1985. "Measurement of Fugitive Volatile Chemical Emissions Rates", Environ. Prog., Vol. 4, No. 3, pp. 199-202.

8. Eklund, B. M., Nelson, T. P., and Wetherold, R. G. 1987. "Field Assessment of Air Emissions and Their Control at a Refinery Land Treatment Facility," EPA/600/2-87/086a, Hazardous Waste Engineering Research Laboratory, U.S. Environmental Protection Agency, Cincinnati, OH.

9. Farmer, W. J., M. Yang, J. Letey and W. G. Spencer 1980. "Land Disposal of Hexachlorobenzene Waste: Controlling Vapor Movement in Soil", EPA, Office of Research and Development Final Report, No. EPA-600/12-80-11.

10. Hill, D. O., Myers, T. E. and Brannon, J. M. 1986. "Development and Application of Techniques for Predicting Leachate Quality in Confined Disposal Facilities; Part I: Background and Theory", Miscellaneous Paper D-88-1, U.S. Army Engineering Waterways Experiment Station, Vicksburg, MS.

11. Huang, J. C., B. A. Dempsey, S. V. Chang and H. Ganjidoost 1987. "Effects of Solid Concentration on the Partition Coefficient of Volatile Organic Compounds with Soils", Manuscript presented at AIChE Spg. Nat. Mtg., Houston, TX.

12. Hwang, S. T. 1987. "Multimedia Approach to Risk Assessment for Contaminated Sediment in a Marine Environment". Proceedings Superfund Conference, November, Washington, D.C.

13. Karickhoff, S. W. and K. R. Morris, 1985. "Impact of Tubificid Oligachaetes in Pollutant Transport in Bottom Sediment", Environ. Sci. Technol., Vol. 19, No. 1, pp. 51-56.

14. Karimi, A. A. 1983. "Studies of the Emission and Control of Volatile Organics in Hazardous Waste Landfills". Dissertation, Univ. So. Calif., Univ. Parks, L. A., CA.

15. Kolnsberg, H. J., 1986. "Technical Manual for Measurement of Fugitive Emissions: Upwind/Downwind Sampling Methods for Industrial Emissions", U.S. EPA, Ind. Env. Res. Lab., EPA-600/2-76-089a, 75p.

16. Larsson, P. 1985. "Contaminated Sediments of Lakes and Oceans Act as Source of Chlorinated Hydrocarbons for Release to Water and Atmosphere" Nature, Vol. 317, No. 6035, pp. 347-349.

17. Linstrom, F. T. and W. T. Piver 1985. "A Mathematical Model for the Transport and Fate of Organic Chemicals in Unsaturated/Saturated Soils", Environment Health Perspectives, Vol. 60, pp. 11-18.

18. Lyman, W. J., W. F. Reehl and D. H. Rosenblatt, 1982. Handbook of Chemical Property Estimation Methods: Chemical Behavior of Organic Compounds, McGraw-Hill, New York.

19. Montgomery, R. L. 1978. "Methodology for Design of Fine-Grained Dredged Material Containment Areas for Solids Retention," Technical Report D-78-56, US Army Engineer Waterways Experiment Station, Vicksburg, Miss.

20. Palermo, M. R. and J. Miller 1987. "Disposal Alternatives for PCB-Contaminated Sediments from Indiana Harbor, Indiana", Vol. 1. Environmental Laboratory Final Report, Dept. of the Army, Waterways Experiment Station, Vicksburg, Miss.

21. Palermo, M. R., Montgomery, R. L., and Poindexter, M. E. 1978. "Guidelines for Designing, Operating, and Managing Dredged Material Containment Areas," Technical Report D-78-10, US Army Engineer Waterways Experiment Station, Vicksburg, Miss.

22. Pavlou, S. P. and R. D. Kadeg 1987. "Preliminary Field Verification of the Equilibrium Partition Approach to Sediment Criteria Development". Paper No. 289, 8th An. Mtg. Soc. Env. Tox. Chem., Nov. 1987, Pensacola, FL.

23. Poe, S. H., K. T. Valsaraj and L. J. Thibodeaux 1988. "Equilibrium Vapor Phase Adsorption of Volatile Organic Chemicals on Dry Soils", J. Hazardous Materials, Vol. 9, p. 17-32.

24. Reible, D. D. 1987. Personal communication, Department of Chemical Engineering, Louisiana State University, Baton Rouge, LA.

25. Spencer, W. F., W. F. Farmer and W. A. Jury 1982. "Review: Behavior of Organic Chemicals at Soil, Air, Water Interfaces as Related to Predicting the Transport and Volatilization of Organic Pollutants", Environ. Toxicol. and Chem. Vol. 1, pp. 17-26.

26. Springer, C., K. T. Valsaraj, and L. J. Thibodeaux 1985. "The Use of Floating Oil Covers to Control Volatile Chemical Emissions from Surface Impoundments: Laboratory Investigations", Hazardous Waste and Hazardous Materials, Vol. 2, No. 4, pp. 487-501.

27. Tetra Tech 1978. "Rate Constants and Kinetic Formulations in Surface Water Quality Modeling". Prepared for the U.S. EPA.

28. Tofflemire, T. J., T. T. Shen, and E. H. Buckley 1981. "Volatilization of PCB from Sediment and Water: Experimental and Field Data", Technical Paper No. 63, PCB Workshop Toronto, Ontario, Canada.

29. Thibodeaux, L. J. 1979. Chemodynamics: Environmental Movement of Chemicals in Air, Water, and Soil, John Wiley and Sons, New York.

30. Thibodeaux, L. J. and S. T. Hwang 1982. "Landfarming of Petroleum Waste-Modeling the Air Emission Problem", Environ. Prog., Vol. 1, No. 1, p. 42.

31. Thibodeaux, L. J., L. K. Chang and D. J. Lewis 1980. "Dissolution Rates of Organic Contaminants Located at the Sediment Interface of Rivers, Streams and Tidal Zones". in Contaminants and Sediments Vol. 1, R. A. Baker, Editor. Ann Arbor Sci., p. 349.

32. Thibodeaux, L. J. and H. D. Scott 1985. "Air/Soil Exchange Coefficients" in Environmental Exposure from Chemicals, Vol. 1, W. Brock Neely and G. E. Blau Editors, CRC Press, Inc. Boca Raton, Florida, pp. 65-89.

33. Thibodeaux, L. J. and B. Becker 1982. "Chemical Transport Rates Near the Sediment in Wastewater Impoundments", Environ. Prog., Vol. 1, No. 4 pp. 296.

34. Thibodeaux, L. J., D. D. Reible and C. S. Fang 1986. "Transport of Chemical Contaminants in the Marine Environment Originating from Offshore Drilling Bottom Sediment" in Pollutants in a Multimedia Environment, Y. Cohen, Editor. Plenum Press, N.Y. pp. 49-64.

35. Thibodeaux, L. J., D. G. Parker, and H. H. Heck 1984. "Chemical Emissions from Surface Impoundments". Environ. Prog., Vol. 3, No. 2, pp. 73-78.

36. Thomas, R. F., R. C. Mt. Pleasant and S. P. Maslansky 1979. "Removal and Disposal of PCB-Contaminated River Bed Materials," Paper presented at the 1979 National Conference on Hazardous Material Risk Assessment, Disposal and Management, 25-27 April 1979, Miami Beach, Florida.

37. Valsaraj, K. T. and L. J. Thibodeaux 1988. "Role of Physical Adsorption in Determining the Vapor Pressure of Volatile Organic Chemicals Above Landfills and Landfarms", Paper No. 105, 8th An. Mtg. Soc. Env. Toxicol. and Chem., Nov., Pensacola, FL.

38. Weaver, G. 1982. "PCB Pollution in the New Bedford Massachusetts Area: A Status Report," Massachusetts Coastal Zone Management, Boston, Mass.

39. Welty, J. R., C. E. Wicks and R. E. Wilson 1984. Fundamentals of Momentum, Heat and Mass Transfer, John Wiley and Sons, New York, p. 266.

APPENDIX A

DERIVATION OF RIVULET AND PONDED VOC EMISSION MODEL

The major volatile chemical transport pathways for the rivulet and ponded sublocales within a CDF are illustrated in Figure 4. The following assumptions are involved in the model development for particle behavior:

- o plug flow of water and particles in the direction of flow,
- o unstratified and isothermal water column,
- o particle scour and resuspension from the sediment bed is characterized by net deposition velocity, and a
- o steady state flow process.

A particle mass balance on the volume element Δxhw in g/s is:

$$wh\, v_x\, \rho_{32}\Big|_x - wh\, \rho_{32}\Big|_{x+\Delta x} - v_d \rho_{32}\, \Delta x w = 0 \qquad \text{(A1)}$$

where the respective pathways are in by advection, out by advection and out by deposition. Divide by Δx and take limit as $\Delta x \to 0$ yields

$$\frac{d}{dx}\, Q\rho_{32} + wv_d \rho_{32} = 0 \qquad \text{(A2)}$$

where $Q = whv_x$. For constant flow, separation of variables and integration between (x=0) inlet solids of ρ^o_{32} to ρ_{32} at x yields

$$\rho_{32} = \rho^o_{32}\, \exp(-wv_d x/Q) \qquad \text{(20)}$$

The result is a simple exponential decrease of suspended solids concentration from the entrance with distance down the rivulet.

A portion of the volatile chemical is sorbed onto particles and behaves like the particles. However, it enjoys several other transport pathways. The following additional assumptions apply:

- o volatilization occurs across the air-water interface from solution only,
- o dissolution occurs from the bottom sediment to the water column,
- o advection of both soluble and sorbed fractions,
- o bottom sediment of constant concentration, and VOC
- o equilibrium between particle and solution phase in the water column.

A VOC mass balance on the volume element shown in figure 4 in g/s is:

$$(Q\rho_{A2}+ Q\rho_{32}\omega_A)|_x + {}^3k'_{A2}(\omega_A/K_d-\rho_{A2})w\Delta x \tag{A3}$$

$$-(Q\rho_{A2}+ Q\rho_{32}\omega_A)|_{x+\Delta x} -v_d\rho_{32}\,\omega_A w\Delta x-{}^1K'_{A2}(\rho_{A2} -\rho^*_{A2})w\Delta x=0$$

where the respective terms are in by particle plus solution advection, in by dissolution from the sediment bed, out by particle plus solution advection, out by net particle deposition and out by evaporation. The limiting process is repeated and the following equation results:

$$\frac{d}{dx}\,(Q\rho_{A2}+Q\rho_{32}\omega_A) + {}^3k'_{A2}(\omega_A /K_d -\rho_{A2})w \tag{A4}$$

$$-V_d\,\rho_{32}\omega_A w-{}^1K'_{A2}(\rho_{A2} -\rho^*_{A2})w=0$$

Equation A4 is used to eliminate ω_A and Equation 20 is used to eliminate ρ_{32} in Equation A4.

Performing the indicated differentiation yields to some simplifications. The following differential equations results:

$$-\frac{Q}{w}\,(1+K^o_d\rho_{32}\,\exp[-v_d\,xw/Q])\frac{d\rho_{A2}}{dx} = ({}^3k'_{A2}+{}^1K'_{A2})\rho_{A2}$$

$$-{}^3k'_{A2}\omega_A /K_d + {}^1K'_{A2}\rho^*_{A2} \tag{A5}$$

Separation of variables and integration of ρ_{A2} from its inlet x=0 value of ρ^o_{A2} to ρ_{A2} at x yields Equation 21.

APPENDIX B

PCB VAPORIZATION ESTIMATES FOR THE NEW BEDFORD HARBOR DREDGING AND DREDGED MATERIAL DISPOSAL PILOT STUDY

To demonstrate the application of the theoretical models for evaluating volatile emissions to air during dredged material disposal operations calculations were made based upon the conditions of the pilot scale CFD alternative proposed for New Bedford Harbor (NBH). Detailed operational, physical and chemical aspects of the site were obtained from Otis (1987). Three calculations were performed that represented possible operational stages of the pilot CDF.

The primary cell has a capacity to hold approximately 25,000 cubic yards of slurry. The surface area of the slurry is to be approximately 250 feet by 250 feet and this is the surface area of the PCB emissions. Aroclor 1242 and 1254 and total PCB are the volatile chemicals of concern. Table B1 contains the physicochemical properties of the Aroclor. The partition coefficients were estimated from the results reported on to the Standard Elutriate test with NBH sediment (Otis 1987) and therefore represent a desorption process. Solubility in water, pure component vapor pressure and molecular weight are from an EPA priority pollutant data list. Henry's constant (L_{H2O}/L air) was computed from the ratio of pure vapor density in air (mg/L air) to solubility in water (mg/L_{H2O}) as in Equation 4. The molecular diffusivities were estimated using the appropriate inverse molecular weight corrections (Thibodeaux 1979) as a first approximation. Phenol in water was used as the $\mathcal{D}_{A2}$ basis and benzene in air was used as the $\mathcal{D}_{A1}$ basis. Table B2 contains site specific information. The PCB and suspended solids concentrations are based on the test results reported by Otis (1987). Except for the density and porosities, which are estimates, the other data has the same source.

Emissions During Filling

Filling is assumed to occur by discharging the slurry into the CDF through a submerged diffuser. Due to the shallow depth of water the suspended solids concentration reported from the Standard Elutriate test was used. The appropriate model for this case is from the ponded sediment locale section of the report. See specifically paragraph two page 19. Equation 14 an 15 apply. Combining these yields:

$$n_A = {}^1K'_{A2} \left[\frac{\omega_A}{K_d + 1/\rho_{32}} + \rho^{**}_{A2} \right] \qquad \text{(B1)}$$

Since PCB emission from water is water-side resistance controlled and the fetch/depth ratio is 250/4 = 62.5 the following correlation from Lunney et. al. (1985) was appropriate:

$$^{1}k'_{A2} = 19.6\, V_x^{2.23}\, \mathcal{D}_{A2}^{2/3} \qquad \text{(B2)}$$

With V_x=25 mi/h and $_{A2}$ = 0.45E-5cm²/s for A-1242, $^{1}k'_{A2}$ =7.0cm/h. Assuming no PCB exist is the air above the CDF makes ρ^{**}_{A2} = 0. The effect concentration in water ρ_{A2} is:

$$\rho_{A2} \equiv \omega_A/(K_d + 1/\rho_{32}) \qquad \text{(B3)}$$

For A-1242 this is:

$$\rho_{A2}=0.48(432)\text{mg/kg}/(1.88E5+/490E-6)\text{L/kg}=.00109\text{mg/L}$$

This is less than the solubility of 1242 in water. Substituting the effective concentration and the transport coefficient values into Equation B1 yields:

$$n_A = 7\,\frac{\text{cm}}{\text{h}}\,[.00109-0\,\frac{\text{mg}}{\text{L}}]\,\frac{1}{1000\text{cm}^3} = 7.63E\text{-}6\text{mg/cm}^2.\text{h}$$

For A=5.8E7cm² this gives the emission rate of

$$W_A=n_A A=443\text{mg/h}.$$

The calculation for A-1254 is 311 mg/h for a total PCB rate of 754mg/h.

Exposed Sediment

Once the CDF is filled and the water removed, the solid dredged material will be exposed directly to the atmosphere. The section in the report body on the exposed sediment locale applies, specifically paragraphs on page 17. Early in the process the air-side resistance dominates. Later the soil-side processes dominate. Using the resistance-in-series concept yields:

$$n_A = \left[\frac{\omega_A H_\rho}{K_d} - \rho_{A1}\right] \Bigg/ \left[\left[\frac{\pi t}{D_{A3}(\varepsilon_1 + K_d\rho_B/H_\rho)}\right]^{1/2} + \frac{1}{^{3}k'_{A1}}\right] \qquad \text{(B4)}$$

Equation 26, the soil-side vaporization equation, is a part of Equation B4. Time appears explicitly and accounts for the transient nature of the evaporation process. Not included in this equation are enhancements due to water evaporation and soil surface cracking. Both are likely to increase the rate significantly.

The effective concentration of A-1242 in the soil pore air is ρ^*_{A1}:

$$\rho^*_{A1} \equiv \omega_A H_\rho/K_d=.48\text{x}432\,\frac{\text{mg}}{\text{kg}}\left|\frac{.0249\text{L}_{H_2O}}{\text{Lair}}\right|\frac{\text{kg}}{188{,}000\text{L}_{H_2O}} =2.75E\text{-}5\text{mg/L}$$

This is less than the pure vapor density of A-1242 which is: ($\rho^*_{A1} = p^o_A M_A/RT$) 0.058 mg/Lair (see paragraph on page 5 of report body). As in the previous calculation $\rho_{A1}=0$. The group of terms $\varepsilon_1 + K_d\rho_B/H_\rho$ is 9.06E6. From the relationship in paragraph two page 17:

$$D_{A3} = \mathcal{D}_{A1}\varepsilon_1^{10/3}/\varepsilon^2 = .036(.3)^{10/3}(.7)^2 = .00133\text{cm}^2/\text{s}$$

For the air-side mass-transfer coefficient the equation recommended by Thibodeaux and Scott (1985) is used

$$^3k'_{A1} = 0.036 R_e^{4/5} S_c^{1/3} \mathcal{D}_{A1}/L \tag{B5}$$

where $R_e = V_\infty L/\nu_1$ and $S_c = \nu_1/\mathcal{D}_{A1}$. For 25 mi/h wind ($V_\infty$) and 250 ft. fetch (L) R_e(dimensionless number) = 5.68E7, and S_c = 4.17 for ν_1= .15cm²s/ This yields from Equation A5:

$$^3k'_{A1} = .036(5.68E7)^{4/5}(4.17)^{1/3}(.036\text{cm}^2/\text{s})/(7620\text{cm})3600\text{s/h}$$

$$^3k'_{A1} = 1580\text{cm/h}$$

Substituting into Equation B4 yields:

$$n_A = (2.75E\text{-}8\text{mg/cm}^3 - 0)/(.00132\sqrt{t} + 1/1580),\ \text{mg/cm}^2.\ \text{hr and}$$

$$W_A = 2520/(2.08\sqrt{t}+1)\ \text{mg/h, with t in days}$$

where $W_A = n_A A$ as before. Table B3 contains the calculated A-1242 emission rates at selected days during the first two years for the un-capped, exposed sediment operation of the pilot CDF. The calculation is repeated using the A-254 properties in Table B1. The final flux equation is $W_A=2357/(1.08\sqrt{t}+1)$, mg/h with t in days. Calculated results appear in Table B3. The total PCB emission rate is the sum of the individual Aroclors rounded to three significant figures.

If the average emission rate for the first two years is 1000g/h then 0.9% of the PCB in the 5000 cubic yards of dredged material placed in the CDF evaporates. Capping with clean (i.e., nearly PCB free) sediment will reduce the emissions.

Capped PCB Contaminated Dredged Material

The emission rate theory for this particular case was not considered in the body of the report if a 6½ inch cap consisting of similar material is placed over the contaminated sediment the models used for steady-state emission from soil-covered landfills applies (Thibodeaux 1981). The same basic theory applies, however the pure component vapor density, used in landfill flux calculations, is replaced with $\omega_A H_\rho/K_d$. The appropriate flux equation is:

$$n_A = \frac{D_{A3}}{h}\left[\frac{\omega_A H_\rho}{K_d} - \rho_{A1}\right] \tag{B6}$$

where h is the cap thickness. As before $\rho_{A1}=0$. For A-1242 the calculations are:

$$n_A = \frac{.00133\text{cm}^2}{\text{s}} \left| \frac{1}{16.5\ \text{cm}} \left[2.75\text{E-5}-0\ \frac{\text{mg}}{\text{L}} \right] \frac{\text{L}}{1000\text{cm}^3} \right| \frac{3600\text{s}}{\text{h}} = 8.0\text{E-9mg/cm}^2.\text{h}$$

and W_A=.46mg/h. For A-1254 W_A =.44 mg/h The total PCB flux is 0.9 mg/h. As the time progresses, if the cap doesn't crack but retains its original seal, the rate will fall with the inverse of $\sqrt{t}$ just as with the exposed sediment case.

Summary

The preceding set of calculations are illustrative examples of how emission estimates can be performed. To the degree possible parameterization of condition for the proposed New Bedford Harbor pilot study dredging operation were used. Except for the exposed sediment case, the numerical values represent initial emission rates and are only approximate. This fact reflects the relative crude state-of-knowledge of the sediment bound volatile chemical desorption/evaporation process. However, the relative ratios of the predicted rates for each case are likely realistic. The exposed sediment rate is ~5000mg/h, ponded sediment is ~800mg/h and capped is 1 mg/h. These are relative ratios with the capped case as the base. Obviously capping will be an effective control methodology. Reducing the suspended solids level in the CDF during filling will reduce the emission rate during this period of operation.

TABLE B1

Aroclor Physicochemical Properties at 25°C

Aroclor	K_d	H_ρ	Solubility	Vap. Pres.	Mol.Wt.	$\mathcal{D}_{A2}$	$\mathcal{D}_{A1}$
	(L/Kg)	(L/L)	(ppm)	(mmHg)	(g/mol)	(cm²/s)	(cm²/s)
1242	188,000	.0249	.24	4.06E-4	267	.45E-5	.036
1254	304,000	.0337	.030	7.71E-5	238	.48E-5	.038

TABLE B2

NBH Site Specific Information and Data

Total PCB concentration in bed sediment: 432 mg/kg.
Aroclor ratios: 48% (wt) 1242 and 52% 1254.
CDF suspended solids concentration 490 ppm.
Temperature 25°C; Wind 25 mi/h.
Dredged material: air porosity (ε_1=) 0.3, total porosity (ε=) 0.7 and bulk density (ρ_B) 1.2kg/L
Water: depth 4 ft., surface area 250 ft. x 250 ft.

TABLE B3

PCB Emission Rates from Exposed Sediment vs Time

Time (day)	Aroclor 1242 (mg/h)	Aroclor 1254 (mg/h)	Total (mg/h)
0	2520	2360	4880
1/4	1240	1530	2770
1/2	1020	1340	2360
1	818	1130	1950
3	548	820	1370
10	333	534	867
100	116	200	316
730	44	78	122

REFERENCES

Lunney, P., C. Springer and L.J. Thibodeaux, 1985. Environ. Progress, Vol. 4, No. 3, 203.

Otis, M.J., 1987. "Pilot Study of Dredging and Dredged Material Disposal Alternatives - Superfund Site, New Bedford Harbor, Massachusetts". U.S. Army Corps of Engineers, New England Division, Waltham, Massachusetts.

Thibodeaux, L.J., 1979. Chemodynamics-Environmental Movement of Chemicals in Air, Water and Soil, Wiley, New York, p. 87.

Thibodeaux, L.J., 1981. "Estimating the Air Emissions of Chemicals from Hazardous Waste Landfills", J. of Hazardous Materials, Vol. 4, pp. 235-244.

Thibodeaux, L.J. and H.D. Scott, 1985. "Air/Soil Exchange Coefficients" in Environmental Exposure from Chemicals, Vol. 1, W. Brock Needly and G. E. Blau Editors, CRC Press, Inc. Boco Raton, Florida, pp. 65-89.

DIFFUSION EXPERIMENTS IN SOILS AND THEIR IMPLICATIONS ON MODELING TRANSPORT

David R. Shonnard and Richard L. Bell

Chemical Engineering Department
University of California, Davis
Davis, CA 95616

INTRODUCTION

The atmospheric emission rates of volatile organic compounds (VOC) from contaminated soils is of concern in many situations. Several examples of importance include gasoline contaminated soils, leaky underground storage tank soils, hazardous waste facility soils, and soils laden with volatile pesticides. It has been shown that VOC diffusive transport through unsaturated soils occurs almost exclusively in the gas phase (1-3) due to partitioning of the organic into the gas phase. This property was used to measure the effective diffusion coefficient, D_{eff}, of benzene vapor through moist soil as a function of soil bulk density and also of soil moisture content (4).

Partitioning of the VOC between soil vapor and soil mineral surfaces by adsorption and into soil organic matter by absorption controls the amount of VOC available for diffusion in the vapor phase. Linear adsorption isotherms have been measured in moist soils for most VOC's, however as moist soils dry sufficiently, adsorption becomes highly nonlinear and soil uptake capacity increases drastically (5-8). Laboratory diffusion experiments at UC Davis on gasoline contaminated soils have been conducted in order to measure atmospheric emission rates as a function of environmental conditions such as soil temperature, soil moisture, soil type, air relative humidity, and initial contaminant concentration in the soil. In this paper it will be shown that VOC emission rates out of the soil surface can be explained by understanding the adsorption behavior of both moist and air-dry soils using a published adsorption study (6).

ADSORPTION OF VOC IN SOIL

In most environmental situations, linear adsorption behavior is expected and has been observed for VOC's in moist soils (5). This occurs because solute-solute interaction are negligible in the soil water phase and adsorption to soil particles and organic matter occurs in the dilute limit. As soils dry, competitive adsorption with water causes adsorption isotherms to become progressively more nonlinear (6). At a soil moisture content of approximately 30 mg water per gram of dry soil, a step change in uptake of the pesticide Dieldrin was observed in a silty soil (8) and was accounted for by competitive adsorption between water and the pesticide. Dry soils have been observed to exhibit nonlinear adsorption, high uptake capacity, and mineral surface adsorption of VOC where moist soils are characterized by linear adsorption isotherms, a relatively low uptake capacity, and partitioning into the organic matter, as shown in Figure 1.

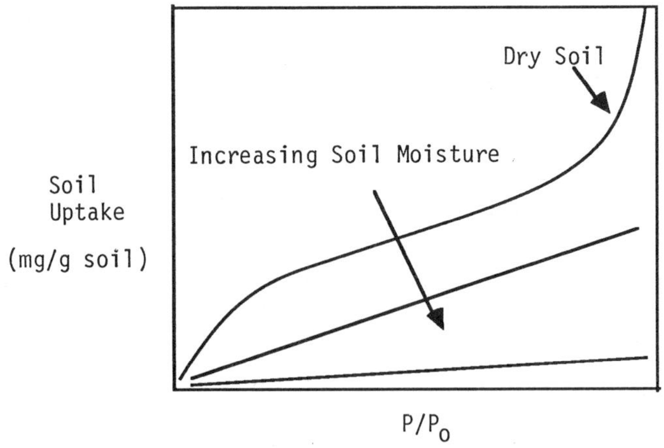

Figure 1. Adsorption behavior of a silty soil of composition 1.9% organic matter, 9% sand, 68% silt, and 21% clay taken from Chiou and Shoup (6).

EXPERIMENTAL

The apparatus used to obtain VOC emission measurements from soils is shown in Figure 2. It consisted of an air pretreatment section, a diffusion cell containing the contaminated soil, and a gas chromatograph interfaced to a microcomputer based data acquisition system for emission rate determination. The purpose of the air pretreatment section was to modify and control sweep air temperature, relative humidity, and volumetric flow rate. Air relative humidity was controlled by metering two air streams; one being saturated with water at the desired temperature of the experiment and the other which was completely dry. Air relative humidity and temperature and diffusion cell temperature were recorded every five minutes during each experiment using a Campbell Scientific CR10 measurement and control module.

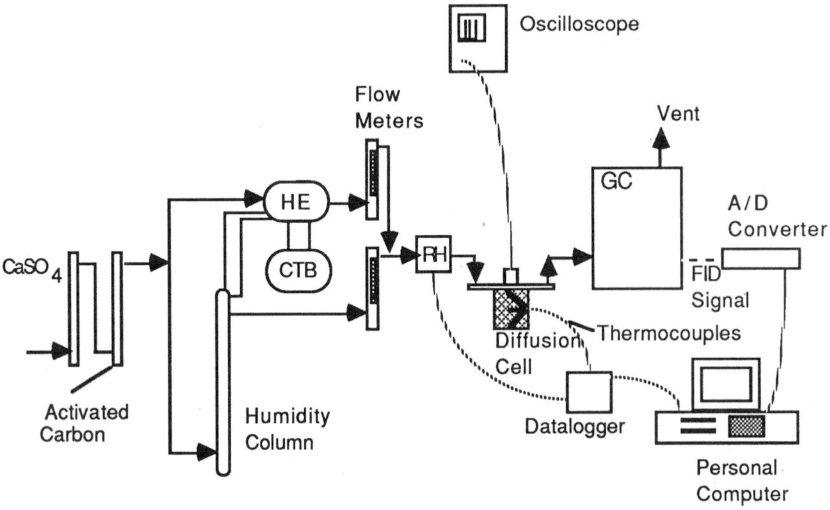

Figure 2. Schematic diagram of apparatus used to obtain VOC emission rate measurements from air-dry soils. HE = heat exchanger, CTB = constant temperature bath, RH = relative humidity probe, FID = flame ionization detector, A/D = analog to digital converter.

Prior to each experiment, enough organic contaminant (unleaded gasoline or n-nonane) was added to air-dry soil to result in a 1000, 500, or 250 ppm mixture. In order to achieve a uniform initial concentration, the mixture was allowed to tumble nondistructively in the diffusion cell for a minimum 24 hours. The diffusion cell soil column was designed from two cylinders of clear plexiglass and the mating surfaces of the inner and outer radius were sealed with rubber o-rings. One cylinder could slide within the other and the adjustable length of the soil column allowed for tumbling while the column was extended and later allowed for controlling the soil column length and therefore the soil bulk density before experiments began. Losses of VOC by soil handling were minimized by this proceedure. Two soil types, obtained from the UC Davis Soil Science Department, were used in this study; one was a fine silty loam (Yolo Silty Loam; 28% sand, 46% silt, 25% clay, 2.4% organic matter) and the other was a coarse, sandy soil (Auberry Sandy Loam; 76% sand, 16% silt, 8% clay, 2.6% organic matter).

The sweep air containing the VOC from the diffusion cell was sent to a sample loop in a Beckman 6800 Air Quality Chromatograph and was analyzed for total hydrocarbon with a flame ionization detector. During the first ten minutes of each experiment, measurements were obtained at one minute intervals and then at five minute intervals thereafter. No separations of the gasoline components was obtained and results were recorded as ppm of propane through a calibration of the FID signal. The output signal was converted from analog to digital form using a Keithley system 570 and the results were stored on an IBM XT personal computer. All trials were allowed to run for atleast 12 hours before termination.

RESULTS AND DISCUSSION

Typical Results. Typical results for experimental conditions and for emission rates as a function of time are shown in Figures 3 and 4 for gasoline contaminated air-dry soil at 41 C. Soil column temperature was controlled to within 2 C over the 12 hour experiment; with a similar precision for the sweep air relative humidity. The soil emission rates decreased monotonically by nearly two orders of magnitude over the course of 12 hours. Table I is a summary of experimental results using gasoline contaminated soils and using soils contaminated with one gasoline component; n-nonane (MW=128.26, b.p.=151 C).

Soil Moisture, Temperature, and Type. The set of results In Table I demonstrates the effects of soil moisture, soil temperature, and soil type on the rate of emission. Entries 1-3, when compared to entry 4 in Table I, show the effect of soil moisture and an observed

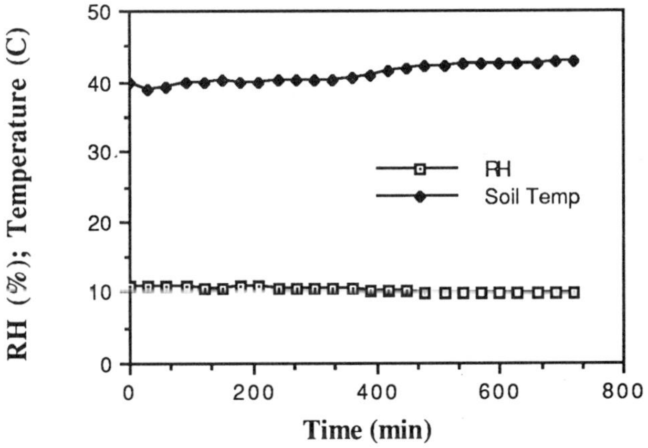

Figure 3. Typical results for soil temperature and sweep air relative humidity (RH) in emission rate experiments. Initial soil concentration = 1000 ppm gasoline. Yolo Silty Loam (air-dry).

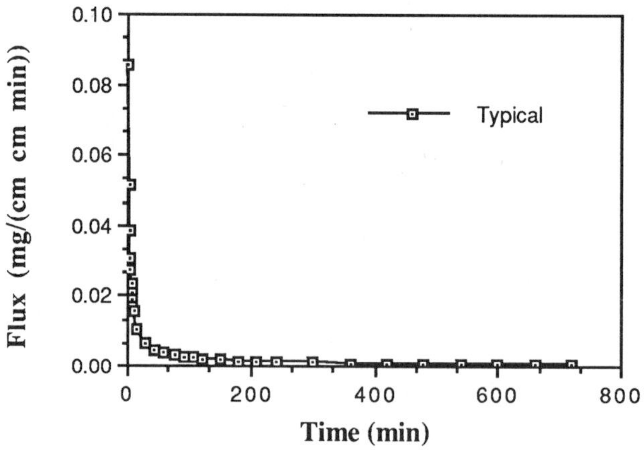

Figure 4. Typical emission rate measurements using a gasoline contaminated soil (1000 ppm, Yolo Silty Loam, air-dry) at 41 C.

increase in emission rate of a factor of nearly 6 over air-dry soil occurs when soil moisture was present at 30-35% of capacity. A comparison of entry 4 and entries 5-9 shows that an increase in soil temperature of 15 C causes an increase in emission rates of approximately a factor of 4. Soil type and its effect on emission rates is shown by the experiment using n-nonane in entry 14. At a soil temperature of 25 C, the emission rates from Auberry Sandy Loam are nearly a factor of 6 above the rates from Yolo Silty Loam under identical conditions of temperature and soil moisture. The observed emission rates can be qualitatively explained in terms of the adsorption behavior of VOC in soils as given in Figure 1. As the

Table I. Emission Rate Measurements from Contaminated Soils Showing the Effects of Soil Temperature, Soil Moisture Content, and Soil Type. Entries 1-13 are Yolo Silty Loam; Entry 14 is Auberry Sandy Loam.

	T_S (C)	Θ (%)*	RH (%)	M (mg/cm^2)	Initial Conc. (1000 ppm)
1.	24.7	35.2	51.1	2.175	Gasoline
2.	25.4	35.6	57.9	2.016	"
3.	26.3	31.7	53.8	2.305	"
4.	26.0	0.0	0.0	.379	"
5.	41.0	0.0	0.0	1.132	"
6.	42.0	0.0	0.0	1.028	"
7.	39.2	0.0	0.0	1.710	"
8.	41.0	0.0	0.0	1.430	"
9.	40.5	0.0	0.0	1.106	"
10.	23.5	0.0	0.0	.613	n-nonane
11.	24.5	0.0	0.0	.613	"
12.	24.0	0.0	0.0	.804	"
13.	25.5	0.0	0.0	.478	"
14.	25.0	0.0	0.0	3.810	"

* initially Θ (soil moisture content) = approx. 45 %

T_S = Soil Temperature (C); Θ = Soil Moisture Content (%); RH = Sweep Air Relative Humidity (%); M = VOC loss over 12 hours.

soil contaminant becomes more tightly bound to the soil, either through a reduction in temperature or a reduction in soil moisture, the soil vapor phase concentration decreases, and with it a decreased driving force for diffusion.

Initial Soil Concentration. A series of experiments were conducted on air-dry Auberry soil using 1000, 500, and 250 ppm initial concentration of n-nonane at 25 C. The results indicate that a nonlinear adsorption isotherm governs the diffusion behavior of n-nonane in the soil. The emission rates for replicate runs at each concentration level were averaged and then normalized to the 500 ppm level by dividing the 1000 ppm results by 2 and multiplying the 250 ppm results by 2. If a linear adsorption isotherm were in effect, then these normalized data sets should have coincided. Instead, as shown in Figure 5, the normalized data follow the trend; 1000 > 500 > 250 ppm. This trend can occur if the nonlinear adsorption behavior shown in Figure 1 for dry soils is true.

Relative Humidity. Increasing the relative humidity of the sweep air above air-dry soils has the effect of increasing emission rates. Figure 6 shows the emission rates of n-nonane from air-dry Auberry soil for a sweep air of 0 and 50% relative humidity. Since the air-dry soil was equilibrated with the humidity in the laboratory (approx. 50%), passing 0% air over the soil column will induce soil moisture diffusion as well as n-nonane diffusion. The loss of soil water at adsorption sites causes the n-nonane to be more tightly bound and thereby reduced emission rates. Over 12 hours, emission losses are reduced approximately 25% by comparison.

CONCLUSIONS

Emission rate measurements of VOC from the surface of contaminated soils were conducted in the laboratory on moist and air-dry soils. The effects of changes in soil moisture, soil temperature, initial soil concentration, sweep air relative humidity, and soil type were investigated. Changes in emission rates caused by the above mentioned changes in environmental conditions are accounted for qualitatively by the adsorption behavior of VOC in soil. Competitive adsorption between water and VOC accounts for the effects of soil moisture and sweep air relative humidity. Nonlinear adsorption behavior of VOC in air-dry soil accounts for the effect of initial VOC concentration. Differences in soil texture accounts for a factor of 6 increase in emission rates between two air-dry soils having nearly identical organic matter contents. An increase in soil temperature of 15 C caused a factor of nearly 4 increase in emission rates from an air-dry soil. The standard error associated with any

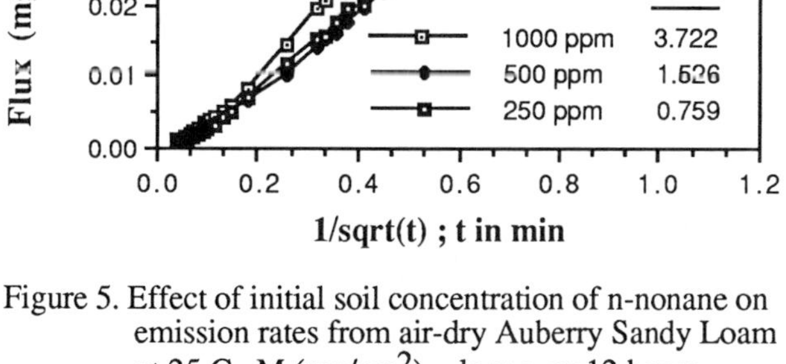

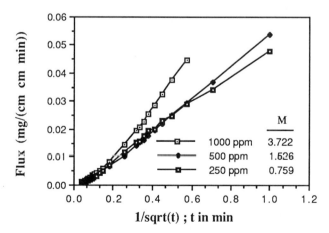

Figure 5. Effect of initial soil concentration of n-nonane on emission rates from air-dry Auberry Sandy Loam at 25 C. M (mg/cm^2) = loss over 12 hours.

measurtment of emission rates was calculated to be approximately 10%. This error estimate demonstrates not only the magnitude of uncertainty associated with any measurement but also the natural variability encountered when working with soils. It can be concluded from this work that in order to understand and predict emission rates from drying soils, competative adsorption between soil moisture and VOC must be understood more completely.

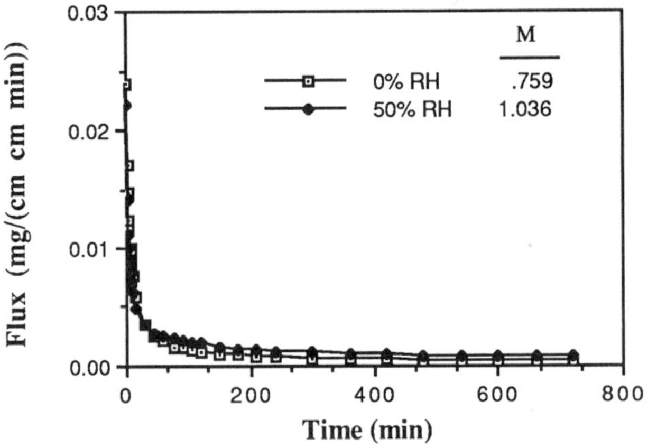

Figure 6. Effect of sweep air relative humidity (RH) on the emission rates of n-nonane from air-dry Auberry Sandy Loam. M (mg/cm^2) = loss over 12 hrs.

REFERENCES

(1) Goring, C.A.I. Theory and principles of soil fumigation. *Advanc. Pest Control Res.* **1962**, 5:47-84.
(2) Letey, J.; Farmer, W.J. Movement of pesticides in soil. In "Pesticides on Soil and Water"; Guenzi, W.D., Ed.; Soil Sci. Soc. Amer., Madison WI, **1974**; pp 67-97.
(3) Farmer, W.J.; Yang, M.S.; Letey, J.; Spencer, W.F. Hexachlorobenzene: Its vapor pressure and vapor phase diffusion in soil. *Soil Sci. Soc. Amer. J.* **1980**, 44:676-680.
(4) Karimi, A.A.; Farmer, W.J.; Cliath, M.M. Vapor phase diffusion of benzene in soil. accepted; *J. Envir. Qual.*
(5) Karichoff, S.W.; Organic pollutant sorption in aquatic systems. *J. Hyd. Eng.* **1984**; 110:707-735.
(6) Chiou, C.T.; Shoup, T.D. Soil sorption of organic vapors and effects of humidity on sorptive mechanism and capacity. *Envir. Sci Tech.* **1985**; 19:1196-1200.
(7) Spencer, W.S.; Cliath, M.M.; Factors affecting vapor loss of trifluralin from soil. *J. Agric. Food Chem.* **1974**; 22:987-991.
(8) Spencer, W.F.; Distribution of pesticides between soil, water and air. In "International Symposium on Pesticides in the Soil"; Michigan State University, **1970**; pp 120-128.

KEY WORDS

Diffusion experiments in soils, effects of soil temperature, soil moisture, initial soil concentration of contaminant, sweep air relative humidity, soil type,VOC, gasoline contaminated soils, n-nonane, adsorption.

EMISSION ESTIMATES FOR A HIGH VISCOSITY CRUDE OIL SURFACE IMPOUNDMENT: 1. FIELD MEASUREMENTS FOR HEAT TRANSFER MODEL VALIDATION

Barbara J. Morrison and Richard L. Bell

Department of Chemical Engineering
University of California at Davis
Davis, CA 95616

INTRODUCTION

Air pollution in California is an environmental problem which has led to much legislation and regulation in the past few decades. A pollutant of major concern is ozone, which is a product of the photochemical reactions of hydrocarbons and NO_x in the atmosphere. The primary sources of these reactants in metropolitan areas are mobile sources - automobiles, trucks, and buses. However, in the relatively rural area of the southern San Joaquin valley in Kern County the primary source of atmospheric hydrocarbons is the oil production industry. The scope of the pollution problem in this region is large and ozone standards set by both the EPA and the state are routinely exceeded.

Crude oil from this part of California is especially viscous. One method used to facilitate the recovery of the crude oil is steam injection. In this process steam is injected into the underground reservoir of oil to heat the oil and lower its viscosity; consequently, the wells produce up to eight times as much water as oil. The resulting mixture of water, oil, and steam is pumped to the surface into a sump or surface impoundment. Surface impoundments are defined as natural or man-made depressions, formed primarily of earthen materials, designed to hold an accumulation of liquid wastes or wastes containing free liquids (Eherenfeld, et al. 1986, p.17). In oil production fields, these impoundments or sumps are used to separate the mixture of oil, water, and sediment that arises from the steam recovery process for heavy crude oil. The mixture enters the sump from one or more inlet pipes and the phases separate, the crude oil forming a "pad" which floats on the water, as the mixture flows toward the outlet. At the outlet the oil and water phases exit the sump through separate outlet pipes.

During the period that the oil is exposed to the atmosphere some of the fraction of light hydrocarbons, or volatile organic compounds (VOCs), evaporate into the air above the oil. This is believed to be a significant source of atmospheric hydrocarbons in the San Joaquin Valley. Attempts have been made at estimating the emissions, both through empirical evaporation equations and direct measurement. The purpose of this study is to develop and evaluate a predictive model based on theoretical aspects of mass transfer; however, in this treatment we will only discuss the heat transfer aspects of the model in detail.

THEORY

Heat Transfer

Crude oil from Kern County falls under the definition of *heavy oil*. Heavy oil is any crude with an API gravity ranging from 10° to 20° at standard conditions and with a gas-free viscosity ranging from 100 to 10,000 centipoises (Guerard 1984). Because of its high viscosity, the organic phase is not modeled as a well-mixed phase, contrary to the previous work reported. Instead, we model the entire crude oil layer as a vertically unmixed or stagnant phase, and validate this assumption experimentally. (This may not hold true for very thin layers of crude.)

Both mass and heat transfer occur in the VOC evaporation process. It is important to know both the surface temperature of the oil and the temperature profile in the oil in order to determine the mass flux. A concentration gradient exists between the oil and air which provides a driving force for mass transfer. In addition, a temperature gradient exists between the oil and air due to the elevated temperature of the oil entering the sump and the much cooler ambient air temperature. The temperature difference between the oil and air of up to 60° C drives the heat transfer. Diffusion is the dominant mass transport mechanism in the oil. Since diffusion is strongly affected by temperature it is necessary to consider temperature effects on emissions. Surface temperature will also affect the concentration gradient through the VOC vapor pressures. The heat transport equation is described below.

Due to the physical dimensions of the system, small depth compared to large length and width, we assume that molecular diffusion and conduction occur only in the vertical direction. Figure 1 shows a schematic diagram of the three-phase system.

We use a method of estimating diffusivities of solutes in crude oil that relies upon the work of others in high viscosity solvents. We adopt the same dependence on temperature and molar volume as Wilke-Chang (1955) but use the Hiss-Cussler (1973) viscosity dependence as shown below

$$D_{AS} \propto \frac{T}{\eta^{2/3} V_A^{0.6}} \tag{1}$$

Conduction is the major mechanism of energy transport in the oil. We assume that the oil/water mixture enters the sump at some temperature T_0 and the water remains constant at this temperature for the duration of time that it resides in the pool. We further assume that temperature is constant across the oil/water interface. This assumption may not hold true if the oil layer is very thin but it suits our purposes in this problem. The oil/air interface is

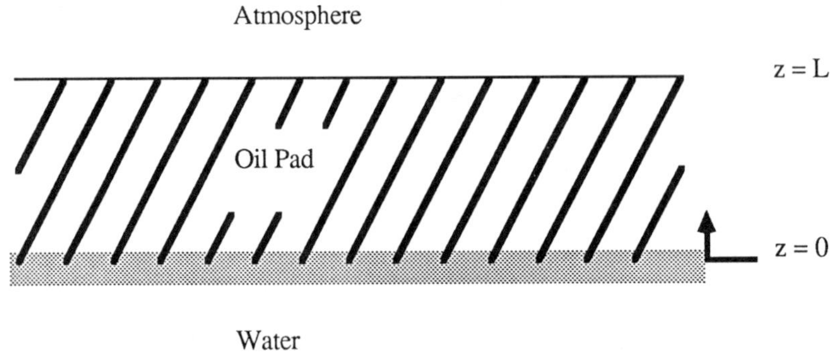

Figure 1. Schematic diagram of the three phase system.

subject to convection, longwave and shortwave radiation, and the compositional effect on the heat of vaporization. The heat transfer process is described by

GDE

$$\frac{\partial T}{\partial t} = \alpha \frac{\partial^2 T}{\partial z^2} \tag{2}$$

IC

$$T = T_o \quad \text{at} \quad t = 0 \tag{3}$$

BC 1

$$T = T_o \quad \text{at} \quad z = 0 \tag{4}$$

BC 2

$$-k\frac{\partial T}{\partial z} = h(T - T_\infty) + \sum_{i=1}^{n} N_i \, \Delta H_{vap} - Q_{rad} \quad \text{at} \quad z = L \tag{5}$$

where α is the thermal diffusivity, k is the thermal conductivity, h is the heat transfer coefficient, T_∞ is ambient air temperature, ΔH_{ivap} and N_i are the heat of vaporization and flux respectively of component i, and Q_{rad} is the net radiation flux toward the interface. Thermal diffusivity is defined as $\alpha = k/\rho c_p$. Vargaftik gives the thermal diffusivity for many different oils (Vargaftik 1975). The heavier oils have a value of about 9 x 10^{-8} m^2/s at warm temperatures (between 40° and 60° C.)

For petroleum fractions and oil mixtures, Cragoe's equation (Perry and Chilton 1974) for thermal conductivity can be used. This equation is

$$k = \frac{0.0677}{S}[1 - 0.0003(T_F - 32)](1.7307) \tag{6}$$

where S is specific gravity, T_F is temperature in °F, and the last constant is a conversion factor for k in W/m°K.

A heat transfer coefficient can be calculated for flat terrain with the equation

$$h = \frac{\kappa^2 \rho c_p u_{10}}{\ln\left(\frac{z_a}{z_o} + 1\right)} \tag{7}$$

where z_o is determined from a logarithmic velocity profile (Mitchell 1975).

The impoundment studied in this report could not be considered situated in open, flat terrain since it was built into a hill and was covered by a net. For this reason we assume that the windspeed at the surface is lower than would be predicted by a logarithmic wind velocity profile. This assumption was validated experimentally. For flow over a flat plate at nearly stagnant conditions, the relationship given by Bolz and Tuve predicts a value of h = 10 W/m^2K (Bolz and Tuve 1976, p. 538).

The heat of vaporization was calculated from the Watson correlation (Reid, Prausnitz, and Poling 1986, p. 228),

$$\Delta H_{vap} = \Delta H_{vap(bp)}\left(\frac{1 - T_r}{1 - T_{r(bp)}}\right)^{0.38} \tag{8}$$

where $\Delta H_{vap(bp)}$ is ΔH_{vap} of a compound at its boiling point (in J/mole), T_r is the reduced temperature of the compound, T/T_c, and $T_{r(bp)}$ is the reduced temperature of the compound at its boiling point, $T_{(bp)}/T_c$.

The three components in the radiation term include longwave radiation, both incoming and outgoing, and shortwave radiation. Incoming longwave radiation is estimated with the equation (Mitchell 1975)

$$Q_{LW} = \varepsilon\sigma T_{sky}^4 \tag{9}$$

where $T_{sky} = 0.0552(T_\infty)^{1.5}$
T_∞ = ambient temperature, K
ε = longwave emissivity (assumed to be 1)
σ = Stefan-Boltzmann constant, 5.669×10^{-8} W/m^2K^4

The outgoing black-body radiation is given by (Sellers 1965, p. 51)

$$Q_B = \varepsilon\sigma T_s^4 \tag{10}$$

where T_s is the surface temperature of the oil.

Shortwave solar radiation can be calculated with the equation (Pielke 1984, p. 211)

$$Q_s^* = S\left(\frac{d^*}{d}\right)^2 \cos(z^*) \tag{11}$$

where S is the solar constant, 1376 W/m^2, z^* is the zenith angle of the sun to the earth, d is the instantaneous distance from the sun to the earth, and d^* is the mean distance from the sun to the earth. The ratio of d^* to d is close to unity and is taken to be 1 for this problem; z^* is defined by

$$\cos(z^*) = \sin\phi \sin\delta + \cos\phi \cos\delta \cos h^* \tag{12}$$

where ϕ is the latitude (35°N or 0.611 radians in Bakersfield), h^* is the hour angle (0° at solar noon, -90° at sunrise, and 90° at sunset), and δ is the declination of the sun in radians, given by

$$\delta = 0.006918 - 0.399912 \cos d_o + 0.070257 \sin d_o - 0.006758 \cos 2d_o + 0.000907 \sin 2d_o - 0.002697 \cos 3d_o + 0.001480 \sin 3d_o \tag{13}$$

where $d_o = 2\pi m/365$
m = Julian day

The total solar radiation is then

$$Q_s = bQ_s^* \tag{14}$$

or

$$Q_s = b\,S \cos z^* \tag{15}$$

where b = 1-A, and A is the albedo of the oil. The albedo of fresh asphalt is 0.09 (Iqbal 1983, p. 290) and we will assume that this is also the albedo of crude oil.

The total radiation is

$$Q^{o}_{rad} = \sigma\left[\left(0.0552\, T_{\infty}^{1.5}\right)^{4} - T_{s}^{4}\right] + 0.9\, S \cos z^{*} \tag{16}$$

Cloud cover will affect the amount of radiation at the earth's surface. A modification to the calculated radiation is made with

$$Q_{rad} = Q^{o}_{rad}\,(1 - k^{*} n) \tag{17}$$

where n is the cloud cover in tenths and k* is a cloud cover correction given by Sellers (1965.)

Ambient temperature can be approximated with a sine function. An even better way of estimating ambient temperature is to use a Fourier series. McCutchan fit a four-term Fourier series to data which had been collected in Southern California in mountainous terrain (McCutchan 1979). This equation is

$$\begin{aligned} T = & -0.32815 + 0.96592\, T_{avg} - 0.43503\, T_{\Delta} \cos\frac{\pi t}{12} \\ & - 0.14453\, T_{\Delta} \sin\frac{\pi t}{12} + 0.09995\, T_{\Delta} \cos\frac{\pi t}{6} \\ & - 0.02450\, T_{avg} \sin\frac{\pi t}{6} \end{aligned} \tag{18}$$

where T_{avg} is the daily mean temperature = (maximum + minimum temperature)/2, T_{Δ} is the daily range of temperature = maximum - minimum temperature, and t is the local time. The ambient temperature at a specified time now can be substituted into the radiation calculation. The inputs to the radiation equation are average daily temperature, daily temperature range, day of year, time of day, cloud cover correction, and surface temperature. The surface temperature is not known but is calculated with the solution to the heat transfer equation, so an iterative process is required.

RESULTS

Experimental

Field work was performed at sump 36W, a primary sump at the Chevron-owned Cymric oil field in Kern County. The trips were made in March 1986, February 1987, and March 1987. The purpose of these trips was to: (1) gather meteorological data for comparison to the mathematical model input parameters; (2) measure temperature profiles for validation of the assumptions made in the model; (3) collect crude oil samples for information on viscosity and initial concentration.

The central piece of equipment used in the field tests was a Campbell Scientific model 21X datalogger which was programmed to store data from many different inputs. In addition, we used the following equipment for weather and temperature measurements: a wind direction vane; up to three anemometers for windspeed; a silicon pyranometer for solar radiation; a thermistor and relative humidity probe; and up to 30 Type T thermocouples for temperatures. The datalogger was programmed to take a reading from each instrument every minute and calculate and record the average of the readings every 15 minutes. In addition, the datalogger was programmed to convert readings from the instruments into engineering units, e.g. thermocouple voltages into degrees centigrade.

Theoretical

The governing differential equation was solved numerically using an implicit Crank-Nicolson finite difference technique. Finite differences are used to approximate differential increments of concentration in time and space coordinates. A FORTRAN program was written and run on a micro-Vax computer. An example of solar radiation data is compared to the predicted values of Equation 15 and shown in Figure 2. We see that the calculated values match fairly closely the measured values. Figure 3 shows ambient temperature data compared to the predicted values of Equation 18. The equation adequately predicts the ambient temperature measured at 2.4 m.

In March 1987 an experiment was set up to determine the temperature at various depths for different locations in the sump. Experimentally determined temperature profiles are compared to the theoretical model. Figure 4 shows the measured profile as temperature versus time at various depths. Figure 5 shows the predicted temperature profiles in the same format. More data revealed that the profile changed for a thinner oil pad (0.15 m) and the model reflected this change accurately. It is apparent from the temperature stratification that no vertical mixing takes place. The main assumption in the theoretical model is validated.

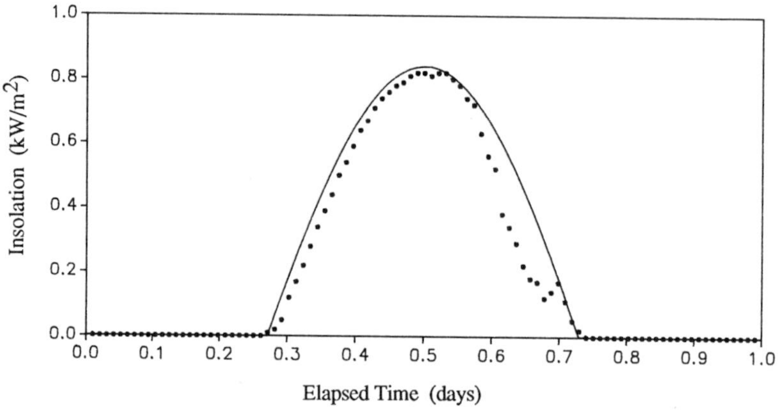

Figure 2. Data (•) and predicted (—) solar radiation for February 21, 1987. Midnight is at 0 days and noon is at 0.5 days.

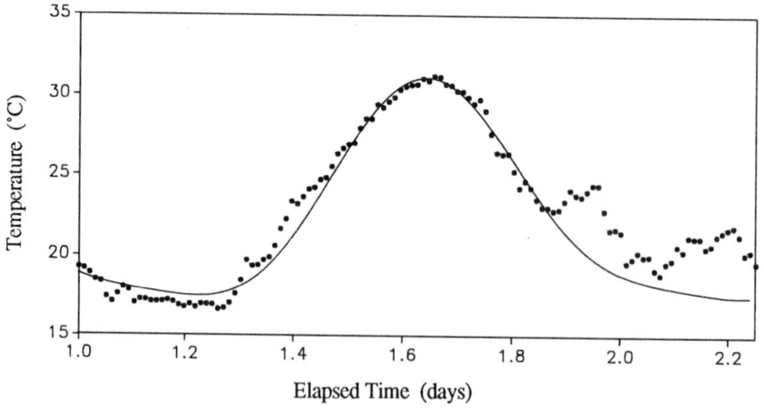

Figure 3. Data (•) and predicted (—) ambient air temperature for March 31, 1987. Midnight is at 1.0 days and noon is at 1.5 days.

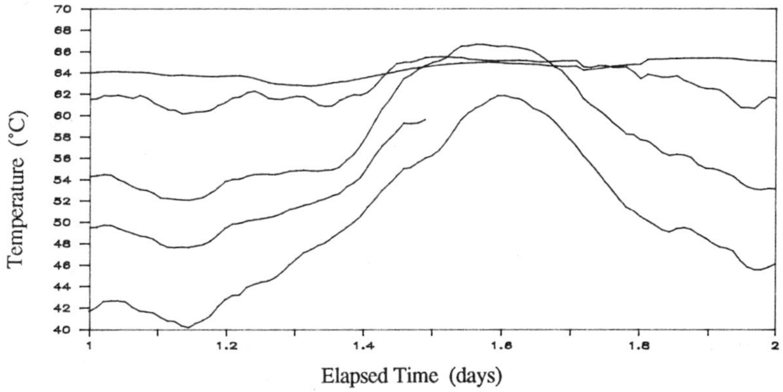

Figure 4. Measured temperature profile from float B at sump 36W, March 1987.

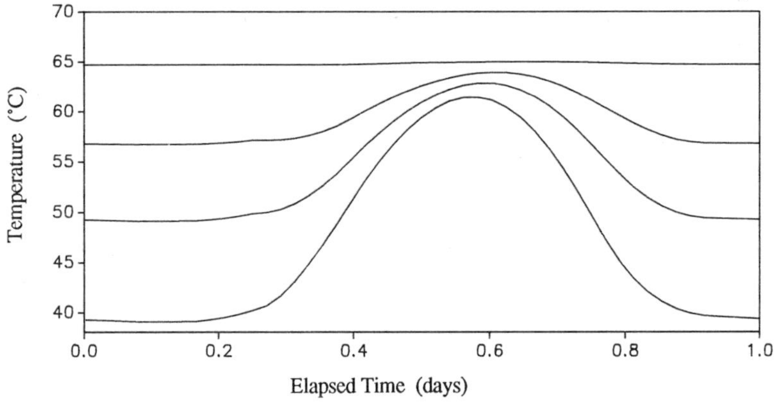

Figure 5. Predicted temperature profile for an oil pad depth of 0.2 m, from the theoretical meteorological conditions for March 30.

Data was recorded during an operations upset in March 1986. During this period oil was being recycled from the outlet back into the sump and the depth of the oil pad grew to over 2 feet. The temperature was found to stay constant over time for depths of 7 inches and 19 inches. The temperature data supports the assumption of constant temperature made for the boundary condition at the oil/water interface in the heat transfer model.

CONCLUSION

Evaporative emissions from a high viscosity crude oil surface impoundment can be estimated with transient one-dimensional heat and mass transfer models. Emissions are affected by heat transfer through the temperature dependence of the VOC diffusion coefficients, vapor pressures, and heats of vaporization. In this study the main assumption of both models was validated - that of no vertical mixing of the crude oil. In a future publication, mass transfer theory for evaporative emissions from the crude oil layer will be described and both theoretical and experimental results will be discussed.

LITERATURE CITED

Bolz, R.E., and Tuve, G.L. eds., "CRC Handbook of Tables for Applied Engineering Science," 2nd Edition. CRC Press, Cleveland, Ohio (1976).

California Air Resources Board, "Determination of Hydrocarbon Emissions from Oil Field Production Sumps." Report No. ARB/SS-87-05 (December 1986).

Ehrenfeld, J.R., Ong, J.H., Farino, W., Spawn, P., Jasinski, M., Murphy, B., Dixon, D., and Rissman, E., eds., "Controlling Volatile Emissions at Hazardous Waste Sites." Noyes Publications, Park Ridge, New Jersey (1986).

Guerard, W.F. Heavy Oil in California, 3rd ed., California Department of Conservation, Division of Oil and Gas, Publication No. TR28 (1984).

Hiss, T.G., and Cussler, E.L., Diffusion in High Viscosity Liquids, AIChE J., 19 (1973), 698-703.

Iqbal, M., "An Introduction to Solar Radiation." Academic Press, Canada (1983).

McCutchan, M.H., Determining the Diurnal Variation of Surface Temperature in Mountainous Terrain, J. Appl. Meteor., 18 (1979), 1224-1229.

Mitchell, J., Microclimatic Modeling of the Desert, in "Heat and Mass Transfer in the Biosphere, Part 1: Transfer processes in plant environment," D.A. de Vries and N.H. Afgan, eds., Scripta Book Co., Washington, D.C. (1975), 273-286.

Perry, R.H. and Chilton, C.H., eds., "Chemical Engineers' Handbook," 5th Edition. McGraw-Hill, Inc., New York (1974).

Pielke, R.A., "Mesoscale Meteorological Modeling." Academic Press, Orlando, Florida (1984).

Reid, R.C., Prausnitz, J.M. and Poling, B.E. "The Properties of Liquids and Gases," 4th Edition. McGraw-Hill, Inc., New York (1986).

Sellers, W.D., "Physical Climatology." The University of Chicago Press, Chicago (1965).

University of California at Davis, Department of Chemical Engineering, "Research and Development on Methods for the Engineering Evaluation and Control of Toxic Airborne Effluents," Report to the California Air Resources Board, Contract No. A4-159-32, Volume 2 (June 1987).

Vargaftik, N.B., "Tables on the Thermodynamic Properties of Liquids and Gases." Hemisphere Publishing, Washington DC (1975).

Whitaker, S., "Fundamental Principles of Heat Transfer," Robert E. Krieger, Malabar, Florida (1983).

Wilke, C.R., and P. Chang, Correlation of Diffusion Coefficients in Dilute Solutions, AIChE J., 1 (1955).

Keywords

EMISSION ESTIMATES FOR A HIGH VISCOSITY CRUDE OIL SURFACE IMPOUNDMENT: 1. FIELD MEASUREMENTS FOR HEAT TRANSFER MODEL VALIDATION

crude oil

diffusion

diurnal cycle

emissions

heat transfer

hydrocarbon

mass flux

numerical model

temperature profile

VOC

TRANSPORT OF POLLUTANTS FROM SOILS TO GROUNDWATER

SOLUTE TRANSPORT IN HETEROGENEOUS FIELD SOILS

Martinus Th. van Genuchten and Peter J. Shouse

U.S. Salinity Laboratory
USDA-ARS
Riverside, California

INTRODUCTION

The purpose of this paper is to briefly review current approaches to quantifying (modeling) solute transport in the unsaturated (vadose) zone of field soils. Much progress has been attained in the analytical and numerical description of vadose zone transfer processes. A variety of mathematical models are now available to describe and predict water flow and solute transport between the land surface and the groundwater table. The most popular models remain the classical Richards' equations for unsaturated flow and the Fickian-based convection-dispersion equation for solute transport. While deterministic solutions of these equations remain useful tools in both fundamental and applied research, their practical utility for predicting actual field-scale water and solute distributions is increasingly being questioned. Problems caused by preferential flow through soil macropores, spatial and temporal variability in the soil hydraulic properties, various nonequilibrium processes affecting chemical transport, and a lack of progress in improving our field measurement technology, have contributed to some disillusionment with the classical models. A number of alternative deterministic and stochastic approaches have been proposed to better deal with field-scale heterogeneities. These models have greatly increased our quantitative understanding of field-scale flow and transport processes, and in some cases also resulted in better practical tools for management purposes. In this paper we shall briefly review those models, and also outline a number of areas in need of further research and development.

CLASSICAL APPROACHES TO MODELING WATER AND SOLUTE MOVEMENT

The importance of the unsaturated zone as an inextricable part of the hydrological cycle has long been understood. Detailed studies of the unsaturated zone have been further motivated by concerns about soil and groundwater pollution, as well as by attempts to optimally manage the root zone of agricultural soils for maximum crop production. These studies have greatly increased our conceptual understanding of the many complex and interactive physical, chemical and microbial processes operating in the unsaturated zone.

Quantitative descriptions of one-dimensional chemical transport in the unsaturated zone are generally based on the classical Fickian-based convection-dispersion equation

$$\frac{\partial(\theta Rc)}{\partial t} = \frac{\partial}{\partial z}(\theta D \frac{\partial c}{\partial z} - qc) \quad (1)$$

in which c is the soil solution concentration, θ is the volumetric water content, R is a retardation factor accounting for equilibrium sorption or exchange between the liquid and solid phases of the soil, D is the dispersion coefficient, q is the volumetric fluid flux density, z is soil depth and t is time. The fluid flux q is given by Darcy's law as

$$q = - K(h) \frac{\partial h}{\partial z} + K(h) \quad (2)$$

where K is the hydraulic conductivity and h the soil water pressure head. For transient flow, q can be calculated from solutions of the unsaturated-saturated flow equation (also known as Richards' equation)

$$C(h) \frac{\partial h}{\partial t} = \frac{\partial}{\partial z} [K(h) \frac{\partial h}{\partial z} - K(h)] \quad (3)$$

where C is the soil water capacity, being the slope of the soil water retention curve, $\theta(h)$. Sources or sinks of water and solutes in the system have been conveniently ignored in (1) and (2). For conditions of steady-state water flow in homogeneous soils, and assuming linear sorption, Eq. (1) can be reduced to the much simpler form

$$R \frac{\partial c}{\partial t} = D \frac{\partial^2 c}{\partial z^2} - v \frac{\partial c}{\partial z} \quad (4)$$

where $v = q/\theta$ is the pore water velocity.

Use of the retardation factor R in (1) assumes equilibrium-type interactions between the liquid and solid phases of the soil. If, in addition, simple linearity is assumed between the solution (c) and sorbed (s) concentrations, R becomes independent of concentration. Use of a linear equilibrium isotherm can greatly simplify the mathematics of a transport problem (Valocchi, 1984). Unfortunately, adsorption-desorption and exchange reactions are generally nonlinear and usually depend also on the presence of competing species in the soil solution, thus requiring the consideration of solution chemistry and/or cation exchange principles. Widely different solution techniques have been proposed to deal with the resulting multicomponent transport problem (e.g., Rubin and James, 1973; Miller and Benson, 1983; Kirkner et al., 1984; Cederberg et al., 1985; Charbeneau, 1988).

The assumption of instantaneous (equilibrium) sorption is also being questioned. A number of chemically-controlled kinetic and diffusion-controlled "physical" nonequilibrium models have been used to describe nonequilibrium transport (Wagenet, 1983). Among these, the most popular and simplest one is the first-order linear rate equation

$$\frac{\partial s}{\partial t} = \alpha(kc - s)$$

where k is a distribution coefficient characterizing equilibrium sorption, and α is a first-order rate constant. More refined nonequilibrium transport models invoke the two-site sorption or two-region (double-porosity) assumptions. Two-site models assume that sorption sites in a soil can be divided into two fractions each exhibiting different equilibrium and kinetic adsorption properties (Selim et al., 1976; Rao et al., 1979; Parker and Jardine, 1986). Two-region (or mobile-immobile) models, on the other hand, assume that the sorption rate is controlled by the rate at which solutes diffuse from relatively mobile (flowing) liquid regions to reaction sites in equilibrium with immobile (dead-end) water. Diffusion into and out of these immobile water pockets is generally modeled as an apparent first-order exchange process (van Genuchten, 1981; DeSmedt and Wierenga, 1984; Carnahan and Remer, 1984; and van Eijkeren and Loch, 1984). Nkedi-Kizza et al. (1984) recently showed that the two-site and two-region models can be put into the same dimensionless form using model-specific parameters, and then used this information to show that effluent curves from laboratory soil columns alone cannot be used to differentiate between chemical and physical phenomena that cause apparent nonequilibrium. This means that independent experiments (such as batch studies or displacement experiments with non-reactive tracers) are needed for verification of the two types of models.

Similar problems related to nonlinear and nonequilibrium sorption also pertain to the transport of organic solutes (MacKay et al. 1985; Wagenet and Rao, 1985; Pinder and Abriola, 1986). Depending upon the type of organic involved, models to predict their transport in the unsaturated zone may also need to account for volatilization, microbial, chemical and photochemical transformations, and possibly multiphase flow. Clearly, model sophistication in this case should be commensurate with the specific purpose of the modeling exercise. For example, Jury et al. (1983) developed a useful analytical model to screen or rank compounds as to their relative hazard for soil and groundwater contamination. Such a screening model can be far less complex than more comprehensive models needed to predict actual temporal and spatial distribution in the vadose zone (Abriola and Pinder, 1985).

While the above classical models based on Eqs. (1) and (3) have been, and undoubtedly will remain, indispensable tools in research and management of unsaturated zone transfer processes (Addiscott and Wagenet, 1985), growing evidence in the literature suggests that deterministic solutions of these models may not acurately describe transfer processes in natural field soils (Sposito et al., 1986a; Nielsen et al., 1986). Factors contributing to this failure to describe field-scale processes are (1) preferential movement of water and solutes through large, continuous macropores in the soil, and (2) spatial and temporal variability of field-scale flow and transport properties. These two problem areas, and alternative approaches to deal with them, are briefly discussed below.

TRANSPORT IN STRUCTURED SOILS

Recently, valid questions have arisen about the usefulness of Eqs. (1) and (3) for describing flow and transport processes in structured soils characterized by large continuous voids, such as natural interaggregate pores, interpedal voids, earthworm and gopher holes, decayed root channels, or drying cracks in clay soils. The movement of water and solutes in such soils (Fig. 1) can be substantially different from that in relatively homogeneous materials (Beven and Germann, 1982; White, 1985; Seyfried and Rao, 1987).

Attempts to describe water flow in unsaturated structured soils have generally centered on two-domain, two-region, or bicontinuum approaches. One domain consists of the soil matrix in which water flow is described with the conventional Darcian-based unsaturated water flow equation, while the other domain consists of either a single macropore, or of a statistical network of macropores, through which water flows primarily under the influence of gravity (Yeh and Luxmoore, 1982; Germann and Beven, 1985; Davidson, 1985). A more elaborate numerical model for drainage of water from structured soils was developed by Wang and Narasimhan (1985). Their formulation not only considers flow along the walls of partially saturated rectangular macropores, but also flow between unsaturated soil matrix blocks.

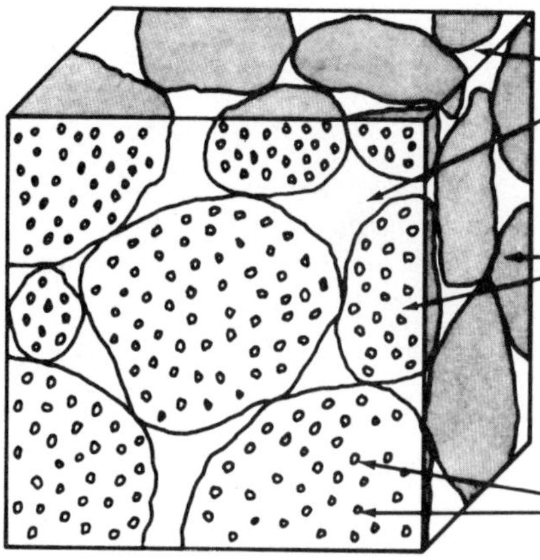

Figure 1. Schematic representation of a structured soil.

While macropore flow itself has important implications in subsurface hydrology in general, and on infiltration and unsaturated water flow in particular, its main implications are in the accelerated movement of surface-applied fertilizers or pollutants into and through the unsaturated zone. A large number of two-region type models have been developed over the years to describe this type of preferential solute transport. Like the older quasi-empirical first-order mobile-immobile equations, these models assume that the chemical is transported through a single well-defined pore or crack of known geometry, or through the voids between well-defined, uniformly-sized aggregates. Contrary to those previous models, however, Fickian-based diffusion equations are used to more rigorously describe the transfer of solutes from the larger pores into micropores of the soil matrix. A variety of analytical (Rasmuson, 1984; van Genuchten, 1985; Goltz and Roberts, 1986a) and numerical (Huyakorn et al., 1983) models, often focusing on transport in fractured groundwater systems, currently exist. While these types of models have been successfully tested in the laboratory, their field verification has started only recently (Gillham et al., 1984; Maloszewski and Zuber, 1985; Goltz and Roberts, 1986b).

Although two-region models are conceptually pleasing and have resulted in improved prediction capabilities, the question remains whether or not geometry-inspired two-region solute transport models are too complicated for routine use in research or management. They require a large number of parameters which are not easily measured independently, especially in the field. In contrast, the classical Fickian-based transport equation is much simpler. Moreover the classical model may well be applicable under certain conditions dictated by the spatial scale of the transport problem. Several attempts have been made to define conditions for which the much simpler classical model (as well as the first-order mobile-immobile model) may be valid (Parker and Valocchi, 1986), in which case the effects of soil matrix (intra-aggregate) diffusion can be lumped into an effective dispersion coefficient, D (Valocchi, 1985; van Genuchten and Dalton, 1986).

STOCHASTIC APPROACHES

Two-region models simulating preferential movement of water and solutes through soils represent attempts to deal with pore structure heterogeneity at spatial scales somewhat intermediate between the usual laboratory measurements and the large field scale. As such they are useful for predicting predominantly vertical transport in structured but areally homogeneous field soils. Unfortunately, few soils are areally homogeneous. This is illustrated in Fig. 2 which shows a map of measured bromide concentrations 399 days after application of an areally uniform, instantaneous Br-pulse to the surface of an undisturbed field soil in Switzerland. The unequal redistribution of the tracer, especially horizontally along the transect, raises serious questions about how to most efficiently sample these types of distributions, and how to best simulate the field-scale transport process. Clearly, some type of stochastic approach is needed.

A large number of stochastic approaches are currently being pursued (Dagan, 1984; Sposito et al., 1986a). Following Jury (1984), we conveniently group the stochastic approaches into scaling theories, Monte Carlo methods, and stochastic-continuum models. A common assumption of all field-scale stochastic transport models is that parameters are treated as random variables with discrete values assigned according to a given probability distribution. In practice, the stochastic approach cannot be used without several simplifying assumptions, including (1) the stationarity hypothesis which assumes that a random transport parameter has the same probability density function (pdf) at every point in the field (the mean and variance are constant), and (2) the ergodicity hypothesis which states that ensemble averages can be replaced by spatial averages, and that spatial replicates can be used to construct the appropriate pdf's for the transport parameters.

Current scaling theories applied to field-scale flow and transport problems have evolved from the early work of Miller and Miller (1956) on microscopic geometric similitude. The approach considers different regions of a heterogeneous field soil to be similar if their microscopic geometric structures are scale magnifications of each other. Transport parameters at any point within a given field soil are related to the parameters at an arbitrary reference point (*) through length scale ratios, or scaling factors α (= λ/λ^*, where λ represents the microscopic length scale).

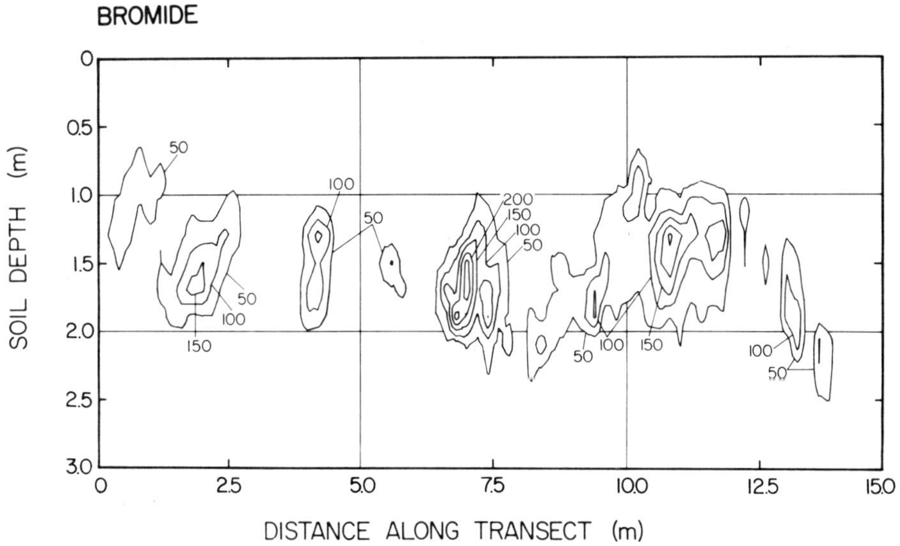

Figure 2. Observed gravimetric bromide concentrations (mg per kg dry soil) 399 days after application of a bromide pulse to the soil surface (after Schulin et al., 1987).

Hydraulic conductivity and soil water retention parameters of a particular region in the field are then calculated from those of the reference soil by means of a set of prescribed equations (Miller, 1980; Nielsen et al., 1983; Tillotson and Nielsen, 1984; Sposito and Jury, 1985). Recent work by Jury et al. (1987) suggests that two scaling factors may be needed for soils which are not strictly similar. Applications of scaling theory to field problems normally assume that α is a random variable characterized by a certain probability density function (Peck, 1983). The method has been a central part of the stochastic flow and transport models of Bresler and Dagan (1983) and Dagan and Bresler (1983).

Monte Carlo simulations assume that the flow and transport parameters are random variables with values assigned from a joint pdf. The water flow or solute transport models are repeatedly run with coefficient values from the assumed pdf until a large number of possible outcomes have been generated. These outcomes are then used to calculate sample means and variances of the underlying stochastic transfer process. Anderson and Shapiro (1983) used this type of simulation to study steady state unsaturated water flow in a heterogeneous soil. Amoozegard-Fard et al. (1982) used the same method to show that of all factors influencing field-scale solute transport, the pore-water velocity had the greatest influence, while local dispersion phenomena were only of secondary importance. A more detailed study by Persaud et al. (1985) reached similar conclusions.

Stochastic continuum models were initially used primarily in ground-water studies as exemplified by the works of Gelhar et al. (1979) and Gelhar and Axness (1983). In these models all random variables are represented by the sum of their mean values plus random fluctuations which, when substituted into the convection-dispersion equation (Eq. 4), lead to a new mean transport model with additional terms. The modified model is evaluated by deriving first-order approximations for the fluctuations and solved by means of fourier transforms. The approach, among other things, leads to a macro-scale dispersion coefficient which only applies asymptotically as distance and/or time increase. Spatial correlations of solute velocity variations, characterized by its autocorrelation function, have been shown to play important roles in the derivation of the asymptotic convection-dispersion equation (Sposito et al., 1986a). We note that a closely related formulation was used recently by Yeh et al. (1985a,b) to study unsaturated flow in isotropic and anisotropic soils.

A different continuum approach was followed by Simmons (1982) who neglected the dispersion coefficient D in Eq. 4, and developed a formal theoretical approach using the pore-water velocity and the travel time as random variables. Jury (1982) and Jury et al. (1986) similarly neglected D in the development of their transfer function model (TFM) of solute transport. The TFM leads to an estimation of the distribution of travel times from the soil surface down to some reference depth L. Solute transport is characterized by a travel time probability density function $f_L(t)$, which for many soil transport processes may be represented as lognormal. The flux concentration in the profile is represented with a convolution integral of $f_L(t)$ and the imposed flux concentration at the soil surface. Recent applications of the transfer function model are given by White et al. (1986) and Dyson and White (1987), whereas its relationship to the classical one-region (Eq. 4) and first-order (mobile-immoble) two-region transport models is discussed by Sposito et al. (1986b).

A closely related continuum formulation, somewhat resembling the works of Amoozegard-Fard et al. (1982) and Bresler and Dagan (1983), was also used by Parker and van Genuchten (1984). These authors consider the entire field to be composed of numerous independent, one-dimensional soil columns, each having unique flow and transport properties and subjected to specific local boundary conditions. The pore-water velocity, while taken to be constant with time and depth in each column, is assumed to vary lognormally among the columns. Finally, transport in each column is described with the classical transport equation (Eq. 4), thus including a dispersion coefficient which was linearly related to the pore water velocity. As with many of the other stochastic models discussed above, the formulation does not allow for any lateral movement between the neighboring columns, an assumption which is inconsistent with the observed distributions in Fig. 2.

Still other statistical approaches exist. For example, Knighton and Wagenet (1987) simulated solute transport using a continuous Markov process. Fractal distributions of soil heterogeneity (Hewett, 1986; Wheatcraft and Tyler, 1988), and their influence on water and solute movement, are also providing new opportunities for studying the unsaturated zone. More work in these areas of research can be expected in the near future.

CONCLUDING REMARKS

This review shows that a large number of widely different models are available for describing field-scale solute transport. While deterministic models have and continue to be invaluable tools in research and management, their usefulness for predicting actual field-scale distributions in time and space is increasingly being questioned. The stochastic approaches appear especially useful for estimating solute travel times in the vadose zone, as well as for predicting areally averaged solute transport loadings to underlying groundwater systems. Their evaluation in the field has thus far been limited to only a handful of data sets. Hence, a few carefully executed field experiments are needed to determine their validity for describing field-scale processes, especially for relatively deep unsaturated profiles.

The development and use of simulation models has significantly increased our quantitative understanding of the main physical, chemical and microbiological processes operative in the unsaturated zone. The proliferation of computer models in research and management likely will continue as computer costs keep decreasing and the need for more realistic predictions increases. Unfortunately, the simulation of field-scale processes requires considerable effort in quantifing spatially and temporally varying soil hydraulic and solute transport parameters. Thus, the completeness of experimental data, and the accuracy of the estimated model parameters, may eventually become the critical factors determining the usefulness of site-specific simulations. Many of our current methods for measuring relevant unsaturated flow and transport parameters are largely those that were introduced several decades ago. Thus, new methods and technologies of measurement are sorely needed to keep pace with our ability to simulate increasingly complex laboratory and field systems. A number of potentially powerful methods based on parameter estimation techniques have recently been introduced (Wagner and Gorelick, 1986, Kool et al., 1987; Kool and Parker, 1988). Other studies have contributed to a better measurement technology (Paetzold et al., 1985; Dalton et al., 1984; Topp et al., 1988). It is important that work in these and related areas of research continues.

REFERENCES

Abriola, L. M. and G. F. Pinder, 1985, A multi-phase approach to the modeling of porous media contamination by organic compounds, 1, Equation development, Water Resour. Res., 21:11-18.

Addiscott, T.M., and R. J. Wagenet, 1985, Concepts of solute leaching in soils: A review of modeling approaches, J. Soil Sci., 36:411-424.

Amoozegard-Fard, A., D. R. Nielsen, and A. W. Warrick, 1982, Soil solute concentration distributions for spatially varying pore water velocities and apparent diffusion coefficients, Soil Sci. Soc. Am. J., 46:3-9.

Anderson, J., and A. M. Shapiro, 1983, Stochastic analysis of one-dimensional and steady-state unsaturated flow: A comparison of Monte Carlo and perturbation methods, Water Resour. Res., 19:121-133.

Beven, K., and P. Germann, 1982, Macropores and water flow in soils, Water Resour. Res., 18:1311-1325.

Bresler, E., and G. Dagan, 1983, Unsaturated flow in spatially variable fields, 3, Solute transport models and their application to two fields, Water Resour. Res., 19:429-435.

Carnahan, C. L., and J. S. Remer, 1984, Nonequilibrium and equilibrium sorption with linear sorption isotherm during mass transport through an infinite porous medium: Some analytical solutions, J. Hydrol., 73:227-258.

Cederberg, G. A., R. L. Street, and J. O. Leckie, 1985, A groundwater mass transport and equilibrium chemistry model for multi-component systems, Water Resour. Res., 21:1095-1104.

Charbeneau, R. J., 1988, Multicomponent exchange and subsurface solute transport: Characteristics, coherence and the Riemann problem, Water Resour. Res., 24:57-64.

Dagan, G., 1984, Solute transport in heterogeneous porous formations, J. Fluid Mech., 145:151-177.

Dagan, G., and E. Bresler, 1983, Unsaturated flow in spatially variable fields, 1, Derivation of models of infiltration and redistribution, Water Resour. Res., 19:413-420.

Dalton, F. N., W. N. Herkelrath, D.S. Rawlins, and J.D. Rhoades, 1984, Time-domain reflectometry: Simultaneous measurement of soil water content and electrical conductivity with a single probe, Science, 224:989-990.

Davidson, M. R., 1985, Numerical calculation of saturated-unsaturated infiltration in a cracked soil, Water Resour. Res., 21:709-714.

DeSmedt, F., and P. J. Wierenga, 1984, Solute transfer through columns of glass beads, Water Resour. Res., 20:225-232.

Dyson, J. S., and R. E. White, 1987, A comparison of the convection-dispersion equation and transfer function model for predicting chloride leaching through an undisturbed, structured clay soil, J. Soil Sci., 38:157-172.

Gelhar, L. W., and C. L. Axness, 1983, Three-dimensional stochastic analysis of macrodispersion in aquifers, Water Resour. Res., 19:161-180.

Gelhar, L. W., A. L. Gutjahr, and R. N. Naff, 1979, Stochastic analysis of macrodispersion in aquifers, Water Resour. Res., 15:1387-1397.

Germann, P. F., and K. Beven, 1985, Kinematic wave approximation to infiltration into soils with sorbing macropores, Water Resour. Res., 21:990-996.

Gillham, R. W.,E. A. Sudicky, J. A. Cherry, and E. O. Frind, 1984, An advection-diffusion concept for transport in heterogeneous unconsolidated geologic deposits, Water Resour. Res., 20:369-378.

Goltz, M. N., and P. V. Roberts, 1986a, Three-dimensional solutions for solute transport in an infinite medium with mobile and immobile zones, Water Resour. Res., 22:1139-1148.

Goltz, M. N., and P. V. Roberts, 1986b, Interpreting organic solute transport data from a field experiment using physical nonequilibrium models, J. Contam. Hydrol., 1:77-93.

Hewett, T. A, 1986, Fractal distributions of reservoir heterogeneity and their influence on fluid transport, Paper SPE 5386, 61st Annual Technical Conf., Soc. Pet. Eng., New Orleans, Louisiana, Oct. 5-8, 1986.

Huyakorn, P. S., B. H. Lester, and J. W. Mercer, 1983, An efficient finite element technique for modeling transport in fractured porous media, 1, Single species transport, Water Resour. Res., 19:841-854.

Jury, W. A., 1982, Simulation of solute transport using a transfer function model, Water Resour. Res., 18:363-368.

Jury, W. A., 1984, Field scale water and solute transport through unsaturated soils, in: "Soil Salinity Under Irrigation", I. Shainberg and J. Shalhevet, eds., pp. 115-129, Springer-Verlag, New York.
Jury, W. A., W. F. Spencer, and W. J. Farmer, 1983, Behavior assessment model for trace organics in soil, I, Model description, J. Environ. Qual., 12:558-564.
Jury, W. A., G. Sposito, and R. E. White, 1986, A transfer function model of solute transport through soil, 1, Fundamental concepts, Water Resour. Res, 22:243-247.
Jury, W. A., D. Russo, and G. Sposito, 1987, Spatial variability of water and solute transport in unsaturated soil, II, Scaling models of water transport, Hilgardia, 55:33-55.
Knighton, R. E., and R. J. Wagenet, 1987, Simulation of solute transport using a continuous time Markov process, 1, Theory and steady state application, Water Resour. Res., 23:1911-1916.
Kirkner, D. J., Theis, T. L., and A. A. Jennings, 1984, Multicomponent solute transport with sorption and soluble complexation, Adv. Water Resour., 7:120-125.
Kool, J.B., and J. C. Parker, 1988, Analysis of the inverse problem for transient unsaturated flow, Water Resour. Res., 24:817-830.
Kool, J. B., J. C. Parker, and M. Th. van Genuchten, 1987, Parameter estimation for unsaturated flow and transport models, J. Hydrol., 91: 255-293.
MacKay, D. M., P. V. Roberts, and J. A. Cherry, 1985, Transport of organic contaminants in groundwater, Environ. Sci. Techn., 19:384-392.
Maloszewski, P., and A. Zuber. 1985, On the theory of tracer experiments in fissured rocks with a porous matrix, J. Hydrol., 79:333-358.
Miller, C. W., and L. V. Benson, 1983, Simulation of solute transport in a chemically reactive heterogeneous system: Model development and application, Water Resour. Res., 19:381-391.
Miller, E. E. 1980. Similitude and scaling of soil-water phenomena, in: "Applications of Soil Physics", D. Hillel, ed., pp. 300-318, Academic Press, New York.
Miller, E. E., and R. D. Miller, 1956, Physical theory of capillary flow phenomena, J. Appl. Phys., 27:324-332.
Nielsen, D. R., P. M. Tillotson, and S. R. Veira, 1983, Analyzing field-measured soil-water properties, Agric. Water Manage., 6:93-109.
Nielsen, D. R., M. Th. van Genuchten, and J. W. Biggar, 1986, Water flow and solute transport processes in the unsaturated zone, Water Resour. Res., 22:89S-108S.
Nkedi-Kizza, P., J. W. Biggar, H. M. Selim, M. Th. van Genuchten, P. J. Wierenga, J. M. Davidson, and D. R. Nielsen, 1984, On the equivalence of two conceptual models for describing ion exchange during transport through an aggregated Oxisol, Water Resour. Res., 20:1123-1130.
Paetzold, R. F., G. A. Matzkanin, and A. De Los Santos, 1985, Surface soil water content measurement using pulsed nuclear magnetic resonance techniques, Soil Sci. Soc. Am. J., 49:537-540.
Parker, J. C., and P. M. Jardine, 1986, Effects of heterogeneous adsorption behavior on ion transport, Water Resour. Res., 22:1334-1340.
Parker, J. C., and A. J. Valocchi, 1986, Constraints on the validity of equilibrium and first-order kinetic transport models in structured soils, Water Resour. Res., 22:399-407.
Parker, J. C., and M. Th. van Genuchten, 1984, Determining transport parameters from laboratory and field tracer experiments, Bull. 84-3, Virginia Agric. Exp. Sta., Blacksburg, Virginia.
Peck, A. J., 1983, Field variability of soil physical properties, Adv. Irrigation, 2:189-221.
Persaud, N., J. V. Giraldez, and A. C. Chang. 1985, Monte-Carlo simulation of non-interacting solute transport in a spatially heterogeneous soil, Soil Sci. Soc. Am. J., 49:562-568.

Pinder, G. F., and L. M. Abriola, 1986, On the simulation of nonaqueous phase organic compounds in the subsurface, Water Resour. Res., 22:109S-119S.

Rao, P. S. C., J. M. Davidson, and H. M. Selim, 1979, Evaluation of conceptual models for describing nonequilibrium adsorption-desorption of pesticides during steady flow in soils, Soil Sci. Soc. Am. J., 43:22-28.

Rasmuson, A., 1984, Migration of radionuclides in fissured rock: Analytical solutions for the case of constant source strength, Water Resour. Res., 20:1435-1442.

Rubin, J., and R. V. James, 1973, Dispersion-affected transport of reacting solutes in saturated porous media: Galerkin method applied to equilibrium-controlled exchange in unidirectional steady water flow, Water Resour. Res., 9:1332-1356.

Schulin, R., M. Th. van Genuchten, H. Fluhler, and P. Ferlin, 1987, An experimental study of solute transport in a stony field soil, Water Resour. Res., 23:1785-1794.

Selim, H. M., J. M. Davidson, and R.S. Mansell, 1976, Evaluation of a two-site adsorption-desorption model for describing solute transport in soils, Proceedings 1976 Summer Computer Simulation Conference, pp. 444-448, July 12-14, 1976, Washington, D.C.

Seyfried, M. S., and P. S. C. Rao, 1987, Solute transport in undisturbed columns of an aggregated tropical soil: Preferential flow effects, Soil Sci. Soc. Am. J., 51:1434-1444.

Simmons, C. S., 1982, A stochastic-convective ensemble method for representing dispersive transport in groundwater, Report CS-2258, Electric Power Research Institute, Palo Alto, California.

Sposito, G., and W. A. Jury, 1985, Inspectional analysis in the theory of water flow through soil, Soil Sci. Soc. Am. J., 49:791-797.

Sposito, G., W. A. Jury, and V. K. Gupta, 1986a, Fundamental problems in the stochastic convection-dispersion model of solute transport in aquifers and field soils, Water Resour. Res., 22:77-88.

Sposito, G., R. E. White, P. R. Darrah, and W. A. Jury, 1986b, A transfer function model of solute transport through soil, 3, The convection-dispersion equation, Water Resour. Res., 22:255-262.

Tillotson, P. M., and D. R. Nielsen, 1984, Scale factors in soil science, Soil Sci. Soc. Am. J., 48:953-959.

Topp, G. C., M. Yanuka, W.D. Zebchuk, and S. Zegelin, 1988, Determination of electrical conductivity using time domain reflectometry: Soil and water experiments in coaxial lines, Water Resour. Res., 24:945-952.

Valocchi, A. J., 1984, Describing the transport of ion-exchanging contaminants using an effective K_d approach, Water Resour. Res., 20:499-503.

Valocchi, A. J., 1985, Validity of the local equilibrium assumption for modeling sorbing solute transport through homogeneous soils, Water Resour. Res., 21:808-820.

van Eijkeren, J. C. M., and J. P. G. Loch, 1984, Transport of cationic solutes in sorbing porous media, Water Resour. Res., 20:714-718.

van Genuchten, M. Th., 1981. Non-equilibrium transport parameters from miscible displacement experiments, Res. Report No. 119, U.S. Salinity Lab., Riverside, California.

van Genuchten, M. Th., 1985, A general approach for modeling solute transport in structured soils, Memoires Int. Assoc. Hydrogeol., 17: 513-526.

van Genuchten, M. Th., and F. N. Dalton, 1986, Models for simulating salt movement in aggregated field soils, Geoderma, 38:165-183.

Wagenet, R. J. Principles of salt movement in soil, 1983, in: "Chemical Mobility and Reactivity in Soil Systems", D. W. Nelson et al., eds., pp. 123-140, Soil Science Society of America, Madison, Wisconsin.

Wagenet, R. J., and P. S. C. Rao. 1985, Basic concepts of modeling pesticide fate in the crop root zone, Weed Sci., 33(Suppl. 2):25-32.

Wagner, B. J., and S. M. Gorelick, 1986, A statistical method for estimating transport parameters: Theory and applications to one-dimensional advective-dispersive systems, Water Resour. Res., 22:1303-1315.

Wang, J. S. Y., and T. N. Narasimhan, 1985, Hydrologic mechanisms governing fluid flow in a partially saturated, fractured, porous medium, Water Resour. Res., 21:1861-1874.

Wheatcraft, S. W., and S. W. Tyler, 1988, An explanation of scale-dependent dispersivity in heterogeneous aquifers using concepts of fractal geometry, Water Resour. Res., 24:566-578.

White, R. E., 1985, The influence of macropores on the transport of dissolved matter through soil, Adv. Soil Sci., 3:95-120.

White, R. E., J. S. Dyson, R. A. Haigh, W. A. Jury, and G. Sposito, 1986, A transfer function model of solute transport through soil, 2, illustrative examples, Water Resour. Res., 22:248-254.

Yeh, G. T., and R. J. Luxmoore, 1982, Chemical transport in macropore-mesopore media under partially saturated conditions, in: "Symposium on Unsaturated Flow and Transport Modeling", E. M. Arnold, G. W., Gee and R. W. Nelson, eds., pp. 267-281, NUREG/CP-0030, U. S. Nuclear Regulatory Commission Washington, DC.

Yeh, T.-C. J., L. W. Gelhar and A. L. Gutjahr, 1985a, Stochastic analysis of unsaturated flow in heterogeneous soils, 1, Statistically isotropic media, Water Resour. Res., 21:447-456.

Yeh, T.-C. J., L. W. Gelhar and A. L. Gutjahr, 1985b, Stochastic analysis of unsaturated flow in heterogeneous soils, 2, Statistically anisotropic media with variable α, Water Resour. Res., 21:457-464.

CONTAMINANT TRANSPORT IN THE SUBSURFACE: SORPTION EQUILIBRIUM AND THE ROLE OF NONAQUEOUS PHASE LIQUIDS

Dermont C. Bouchard

Robert S. Kerr Environmental Research Laboratory
U.S. Environmental Protection Agency
Ada, OK 74820, U.S.A.

ABSTRACT

In this discussion the sources of nonequilibrium during contaminant transport are reviewed and the effects of contaminant solubility and geologic material organic carbon content on sorption equilibrium are reported. For experimental work with three neutral organic compounds (NOC's) and geologic material from the saturated zone (TOC = 0.33 g kg^{-1}), unsaturated zone (TOC = 2.6 g kg^{-1}), and surface soil (TOC = 6.9 g kg^{-1}), sorption nonequilibrium during solute transport through columns was detectable for contaminant-geologic material combinations yielding retardation factors > 2.

The process of nonaqueous phase liquid (NAPL) transport and immobilization in porous media is discussed with emphasis on the immobilized fraction of NAPL, termed the residual saturation. Parameters affecting NAPL immobilization and mobilization are discussed and experimental information from a heptane-water-aquifer material system is presented. Phenols retardation through residually saturated heptane was predictable using independently measured contaminant heptane-water partitioning values. However, nonequilibrium partitioning during transport introduced error into the predictions.

INTRODUCTION

Decisions on hazardous waste disposal, site remediation, and groundwater protection often have far reaching environmental and economic consequences. It is imperative that these decisions be supported by an understanding of contaminant transport and transformation in the environment. This discussion will focus on contaminant transport in the subsurface; and, more specifically, on the effects of natural and anthropogenic organic carbon on sorption equilibrium during contaminant transport.

One of the parameters distinquishing contaminant transport in the subsurface from that in surface soil is the very low level of natural organic carbon associated with the mineral phase of subsurface geologic material. The low organic carbon content of subsurface materials results in very low sorption of many neutral organic compounds (NOC's) (1-3). In addition to affecting the magnitude of sorption, the low organic carbon content of subsurface material also affects sorption equilibrium (4) and exacerbates the effects of anthropogenic organic carbon (5). The following report discusses the role of sorbent organic carbon content in sorption equilibrium by examining contaminant transport through geologic material of varying organic carbon content. In addition, this report also discusses the effects of one type of anthropogenic organic carbon, hydrocarbonaceous nonaqueous phase liquids (NAPL), on contaminant transport.

NONEQUILIBRIUM DURING SOLUTE TRANSPORT

Sources of Nonequilibrium

Most early models for describing contaminant transport in porous media have assumed instantaneous and reversible sorption, linear and single valued sorption isotherms, and diffusion equilibrium during solute transport (6,7). In short, it was assumed that all processes affecting contaminant transport were at equilibrium. However, contaminant breakthrough curves (BTC's) from miscible displacement studies with soils and other geologic material have often differed from the symmetric sigmoids predicted by the equilibrium contaminant transport models (4,8,9). BTC asymmetry has usually been manifested as fronting, or early contaminant breakthrough; or tailing, where system output concentration (C) slowly approaches the input concentration (Co) for sorption, or C slowly approaches zero for desorption.

BTC asymmetry has been attributed to physical characteristics of the geologic material, such as pore geometry, that result in regions of immobile, intra-aggregate pore water that are only accessible by diffusion. Contaminant transport then is assumed to occur by convection and dispersion in mobile water regions, and soley by diffusion in the immobile regions. Nonequilibrium conditions and asymmetric BTC's result when there is slow contaminant diffusion into and out of the immobile water sinks. Since this process occurs entirely in the aqueous phase, it affects both sorbed and nonsorbed contaminants. For sorbed contaminants, sorption-desorption kinetics that are slow relative to contaminant-geologic material exposure time will also result in nonequilibrium conditions during solute transport. Combined equilibrium and kinetic models where sorption was assumed to be at equilibrium on some sorption sites and not at others have described some asymmetric BTC's fairly well (10-11). However, slow aqueous phase diffusion and slow sorption kinetics have very similar effects on contaminant transport; and are, therefore, easily confounded. In addition, combined equilibrium and kinetic models have been shown to be mathematically equivalent to the mobile-immobile water diffusion models (12,13).

Sorption Nonequilibrium

Sorption experiments utilizing both column and batch technics have indicated that attainment of sorption equilibrium can occur very slowly in some systems (4,14,15). The rate of sorption has often been observed to vary inversely with the magnitude of sorption. So solute-sorbent combinations having a high degree of sorption, such as pyrene and a surface soil with 1.0% organic carbon, will take much longer to reach sorption equilibrium than trichloroethene and a subsurface material with an organic carbon content of 0.05%. The objectives of the research described here were to separate and evaluate the effects of contaminant solubility and sorbent organic carbon content on sorption nonequilibrium during contaminant transport. To accomplish this, 25 mm i.d. by 50 mm long columns were packed with three different geologic materials (<250u fraction) of varying organic content, and miscible displacement studies were conducted under saturated conditions with three NOC's. See Tables 1 and 2 for relevant properties of the geologic material, columns, and NOC's.

Figure 1 contains the BTC's for ^3H_2O and the herbicide diuron on the Lula aquifer material. The very low sorption of diuron was consistent with the very low organic carbon content of the aquifer material. Some researchers have observed that NOC's sorption on low organic surfaces may be due to both organic and mineral components of the geologic material (1-2). However, for most NOC's, sorption to mineral surfaces will be minimal, and low sorbent organic carbon content results in low contaminant sorption.

Table 1. Chemical and physical properties of the geologic material and columns.

Geologic material	Organic carbon (g kg^{-1}) $\pm$95% C.L.	CEC (mmol kg^{-1}) $\pm$95% C.L.	ρ (Mg m^{-3})	θ (m^3 m^{-3})
Eustis surface soil	6.9 $\pm$ 0.6	22 $\pm$ 4	1.64	0.373
Unsaturated zone material	2.6 $\pm$ 0.1	110 $\pm$ 4	1.42	0.464
Lula aquifer material	0.33 $\pm$ 0.01	90 $\pm$ 2	1.52	0.432

Table 2. Some properties of the neutral organic solutes

Neutral Compounds	Molecular wt.	Aqueous sol. at 25C (M L^{-1})	log K_{ow}	pKa
Hexazinone	252.3	1.31×10^{-1}	1.03	1.2
Atrazine	215.7	3.25×10^{-4}	2.35	1.6
Diuron	233.1	1.80×10^{-4}	2.81	--

The rate of sorption is inversely proportional to the magnitude of sorption, so that low sorption also results in more rapid sorption-desorption kinetics (4,14-15). In Figure 1, the diuron BTC's are symmetric sigmoids where the diuron concentration at the column outlet (C) rapidly approached and then equalled the input diuron concentration (C_o). On the desorption side, diuron was rapidly displaced from the column as C approached and then equalled zero. The symmetric BTC's obtained for diuron were very similar to BTC's obtained for other NOC's and aquifer material

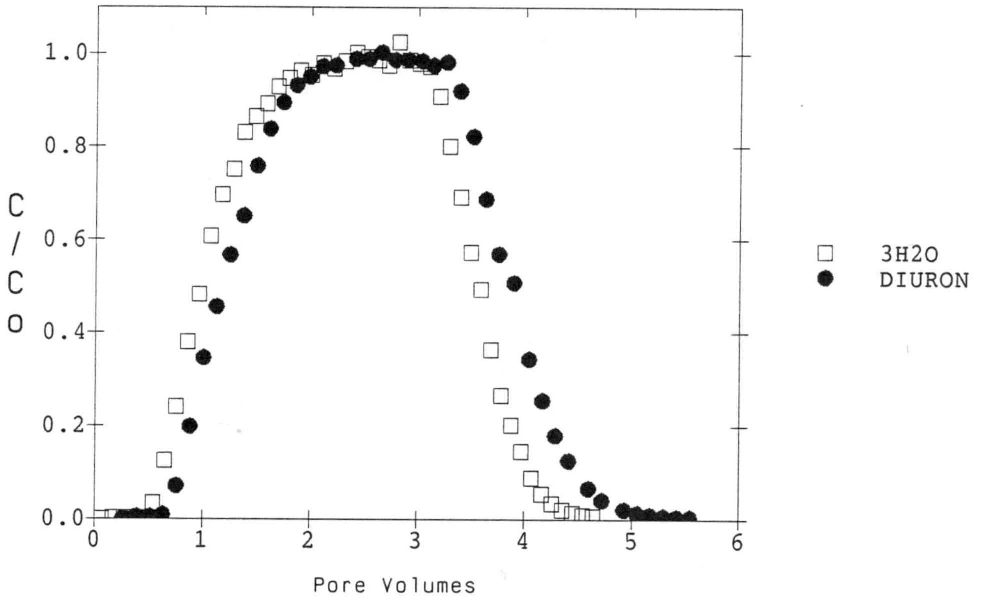

Fig. 1. ^3H_2O and diuron BTC's on Lula aquifer material.

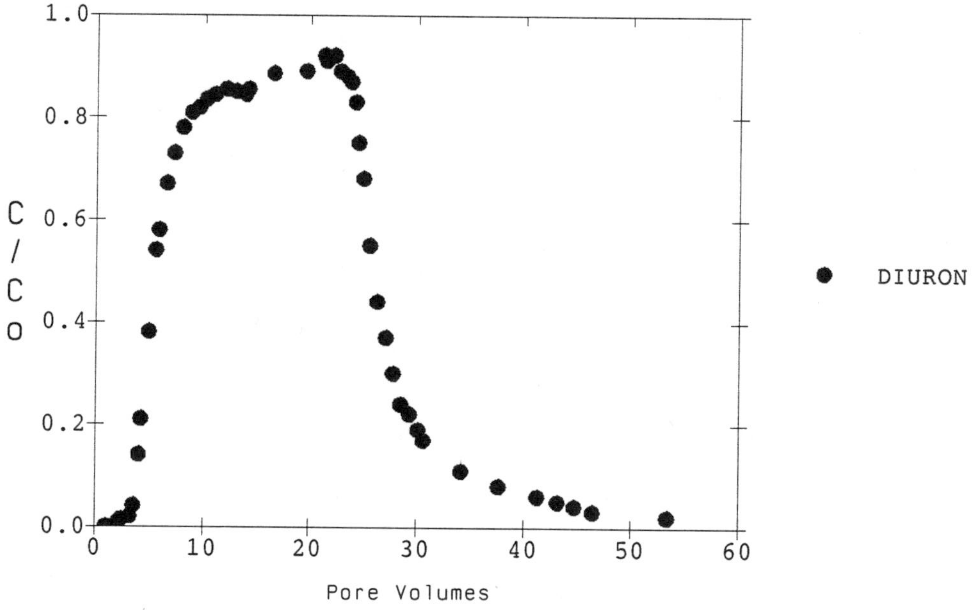

Fig. 2 Diuron BTC's on Eustis surface soil.

and indicated that rapid and reversible sorption was occurring (4). Figure 2 contains the BTC's for diuron through the higher organic carbon Eustis surface soil. As would be expected, diuron was more highly retarded on the higher organic carbon Eustis soil. In addition, the diuron BTC's were more asymmetric than the diuron-aquifer material BTC's. This asymmetry is reflected in the BTC flattening out and the subsequent slow approach to $C = C_o$, and from the tailing exhibited by the desorption side of the diuron BTC.

A measure of BTC asymmetry, and therefore the degree of sorption nonequilibrium, can be obtained by comparing the dimensionless retardation factors (R_T) computed by two different methods:

$$R_{T1} = V / V_o \text{ , when } C = 0.5C_o \qquad [1]$$

where V is the cumulative effluent volume (m^3), V_o is total column pore volume (m^3), and C and C_o are defined earlier. Or,

$$R_{T2} = \text{ area behind contaminant BTC / area behind } {}^3H_2O \text{ BTC} \qquad [2]$$

For symmetric BTC's, retardation factors computed by these two methods will yield similar values. The ratio R_{T2} / R_{T1} can be used as an index of nonequilibrium: as the ratio deviates from one, the greater the BTC asymmetry and nonequilibrium. Table 3 contains the R_T values computed by the two methods and their ratios for the three NOC's and three geologic materials.

There was little NOC's retardation and all NOC's BTC's were symmetric on the Lula aquifer material (Table 3). The low organic carbon content of the Lula material apparently allowed for rapid contaminant mass transfer between the organic carbon and solution phases relative to contaminant residence times in the columns. In fact, no nonequilibrium transport was detectable for contaminant-geologic materials combinations with $R_T < 2$. However, the data in Table 3 does indicate that nonequilibrium increased as geologic material organic carbon increased and as solute solubility decreased, thus demonstrating that both contaminant and geologic material were important in determining the rate of sorption.

As noted earlier, nonequilibrium conditions and BTC asymmetry may result from slow contaminant diffusion into and out of regions of "immobile" pore water. However, the nonaggregated geologic material and contaminant column residence times used in these experiments were chosen to exclude slow molecular diffusion in the aqueous phase from contributing to nonequilibrium

Table 3. Contaminant retardation factors (R_T) computed from miscible displacement data[1].

Geologic material-compound	R_{T1}	R_{T2}	R_{T2} / R_{T1}
Lula aquifer material-			
hexazinone	1.38	1.36	0.99
atrazine	1.10	1.14	1.04
diuron	1.26	1.26	1.00
Unsat. zone material-			
hexazinone	1.80	1.73	0.96
atrazine	1.65	1.66	1.01
diuron	3.40	3.95	1.16
Eustis surface soil-			
hexazinone	1.20	1.29	1.07
atrazine	2.00	2.45	1.22
diuron	5.35	8.76	1.64

[1]Adapted from ref. (4).

during solute transport, and symmetric BTC's for ^3H_2O transport in the columns (identical boundary conditions) verified that slow aqueous phase diffusion was not a significant limiting process in our columns. Also, batch sorption experiments yielded linear sorption isotherms, so BTC asymmetry was not due to curvilinear isotherms. Therefore, the source of the nonequilibrium could be attributed to the sorption-desorption process itself.

Environmental Consequences of Sorption Nonequilibrium

An important consequence of nonequilibrium sorption during solute transport is the slow desorptive release of contaminants from geologic material to the aqueous phase. This has important implications for the remediation of sites contaminated with hazardous wastes. One of the major groundwater remediation technologies in use today is "Pump and Treat", where contaminated groundwater is withdrawn from wells, and then pumped to the surface for treatment by adsorption on activated charcoal or other technics. The slow desorptive release of contaminants from the solid to solution phase may require that pumping times far in excess of those predicted assuming sorption equilibrium be used for contaminant removal.

Given the importance of organic carbon in NOC's sorption and transport in the environment, it is clear that any perturbation in the amount or type of organic carbon in a system will profoundly affect NOC's transport. The effects of anthropogenic organic carbon are magnified in the subsurface where the natural organic carbon content is usually very low. When organic carbon is mobile, it may increase, or facilitate, NOC's transport through dissolution in (16), or sorption on (17), the mobile organic carbon phase. When organic carbon is immobile, however, NOC's sorption on the stationary organic carbon will decrease NOC's mobility in the environment (5) and may also affect sorption-desorption kinetics. The ubiquity of carbon based fuels and lubricants has resulted in widespread environmental contamination by these fluids from leaking pipelines, underground storage tanks, and other sources. In the following section, the effects of this anthropogenic carbon on solute transport is discussed.

CONTAMINANT TRANSPORT IN THE PRESENCE OF NONAQUEOUS PHASE LIQUIDS: FOCUS ON THE RESIDUAL SATURATION

NAPL Transport and Immobilization in Porous Media

For the multiphase flow of immiscible fluids through porous media, the interfacial tensions existing between the fluids and between the fluids and the porous media must be considered. When a liquid is in contact with another immiscible liquid, gas, or solid, an interfacial tension (γ) exists between the two phases. The interfacial tension existing between two phases a and b (γ_{ab}) may be defined as the amount of work that must be done to separate a unit area of a from b. For air and water $\gamma = 7.288 \times 10^{-2}$ joules m^{-2} at 20 C.

When water and an immiscible fluid are in contact with geologic material, an interfacial tension (γ_{sf}) exists between the solid phase and each fluid. The chemical composition of the fluids and the geologic material will determine which fluid will have a greater tendency to spread over and coat, or "wet" the geologic material surface. Wettability of geologic material by fluids is determined by the contact angle formed between the solid surface and the fluid-gas (unsaturated conditions) or fluid-fluid (saturated conditions) interface, where the contact angle is measured through the denser fluid. A fluid is said to wet the solid, i.e. is the wetting fluid, when the contact angle is less than 90°. See Bear (18) for a diagram and additional explanation. There are conditions under which the NAPL may be considered to be the wetting fluid or conditions of intermediate or mixed wettability where more than one fluid wets the geologic material. For example, in the unsaturated zone the NAPL may be considered to be the wetting fluid relative to air, but non-wetting relative to water. Under most environmental conditions, when water and an NAPL (usually a hydrocarbon based fuel or halocarbon solvent) are present, it is reasonable to assume that water is the wetting fluid. It will be assumed for the remainder of this discussion that water is the wetting fluid.

Darcy's Law may be used to describe the laminar flow of a single fluid through porous media:

$$v = -k\rho g \;/\; u \;(dh \;/\; dl) \qquad [3]$$

where v is the specific discharge or Darcy velocity [$m\ s^{-1}$] of the fluid, k is the permeability (also called the intrinsic permeability) of the porous media [m^{2}], ρ is fluid density [$Mg\ m^{-3}$], g is acceleration due to gravity [$9.8\ m\ s^{-2}$], u is the fluid dynamic viscosity [$kg\ m^{-1} s^{-1}$] and dh / dl is the hydraulic gradient across the system. Porous media permeability is determined by properties such as particle size, geometry, and packing (or media porosity), that effect fluid flow. The ratio u / ρ, termed the kinematic viscosity, may be used to compare fluid velocities through a given porous medium. Schwille (19) presents a table of kinematic viscosities that illustrates that chlorinated solvents and gasoline have flow velocities 1.5-3.0 greater than water through dry porous media due to the lower kinematic viscosities of the solvents and gasoline. Hydrocarbon fuels characterized by higher molecular weight fractions than gasoline, such as diesel fuel, have kinematic viscosities higher than water and thus move at velocities lower than water.

For the simultaneous laminar flow of immiscible fluids through porous media, a permeability can be defined for each fluid (k_w, k_h) where the subscripts w and h denote the wetting phase, water, and the non-wetting hydrocarbon phase, respectively. A form of Darcy's Law can then be applied to describe the flow of the two phases.

$$v_w = -k_w \rho_w g \;/\; u_w \;(dh \;/\; dl) \qquad [4]$$

$$v_h = -k_h \rho_h g \;/\; u_h \;(dh \;/dl) \qquad [5]$$

Like k, k_w and k_h are independent of fluid properties and flow rate; however, k_w and k_h are functions of the porous media fluid saturation:

$$S_w = V_w / V_v \qquad [6]$$

$$S_h = V_h / V_v \qquad [7]$$

where S_w is the fractional volume of water [m^3 m^{-3}], V_w is the pore volume occupied by water [m^3], V_v is the total pore volume [m^3], and S_h and V_h are the corresponding values for the hydrocarbon fluid.

Relative permeabilities for the wetting fluid ($k_{rel,w}$) and non-wetting fluid ($k_{rel,h}$) may be defined:

$$k_{rel,w} = k_w / k \qquad [8]$$

$$k_{rel,h} = k_h / k \qquad [9]$$

It is clear that:

$$k_{rel,w} < 1 \qquad [10]$$

$$k_{rel,h} < 1 \qquad [11]$$

whenever two or more immiscible fluids are present. Typical relative permeability curves (20) indicate that:

$$k_{rel,w} + k_{rel,h} < 1 \qquad [12]$$

whenever two or more immiscible fluids are present.

On a macroscopic level, when two immiscible fluids, such as water and gasoline, flow simultaneously through porous media each fluid tends to establish a flow path through a different set of pores. The flow path for each fluid remains quite stable as long as S_w and S_h are constant. As S_h decreases as the distance from the fluid source increases, the non-wetting phase will become discontinuous provided dh/dl remains constant. This discontinuous, essentially immobile, fraction of the NAPL is termed the residual saturation (S_{hr}):

$$S_{hr} = V_h \text{ discontinuous} / V_v \qquad [13]$$

NAPL Residual Saturation

As a NAPL flows through the unsaturated zone under gravitational head as a continuous phase, some of the NAPL volume is consumed through capillary retention in the pore space. Assuming that the NAPL is non-wetting relative to water, and wetting relative to air, the NAPL residual saturation will exist in the geologic material as drops and droplet aggregates (blobs) bounded by water in the smaller pores, and as pendular rings bounded by air in the larger pores.

If the NAPL volume is small relative to the travel distance through the unsaturated zone to groundwater, all of the NAPL will be retained as residual saturation and no bulk fluid transport to groundwater will occur. Contaminant transport will then be governed primarily by contaminant volatility (gas phase transport in the unsaturated zone), solubility (leaching of contaminants from the immobilized residual saturation by water percolating through the unsaturated zone) and advective-dispersive transport. However, if the volume of the NAPL is large relative to the travel distance through the unsaturated zone to groundwater, the NAPL will reach the groundwater and either accumulate above the saturated zone along the capillary fringe (NAPL less dense than water - nonhalogenated hydrocarbon fluids) or penetrate into the aquifer where it will move laterally with groundwater flow and vertically until an impermeable layer is encountered (NAPL denser than water (DNAPL) - halogenated solvents).

From a site remediation standpoint, it is much easier to remove the continuous bulk fluid or "free product" NAPL than the fluid present as residual saturation. Interception of the NAPL with a recovery well and a dual pump system can result in recovery of large masses of the bulk fluid. However, large amounts of the NAPL will be left behind as residual saturation. Most research on the residual saturation of NAPL's in porous media has been reported in the petroleum engineering literature. Much of this work was motivated by the desire to increase the efficiency of tertiary crude oil recovery methods as residual crude oil typically occupies 25-50% of the pore space in geologic material following secondary oil recovery methods such as waterflooding (21). Laboratory residual saturation measurements have varied between 15 and 40% in saturated porous media (22-23) with smaller quantities being retained in unsaturated porous media. Conrad et al. (24) observed residual saturations of 29 and 9% for Soltrol - 130 in sandy media under saturated and unsaturated conditions respectively. Wilson et al. (25) have presented data showing that S_{hr} in air/water/hydrocarbon fluid systems increased with decreasing glass bead diameter. In water/hydrocarbon fluid systems, however, S_{hr} values were 2-5

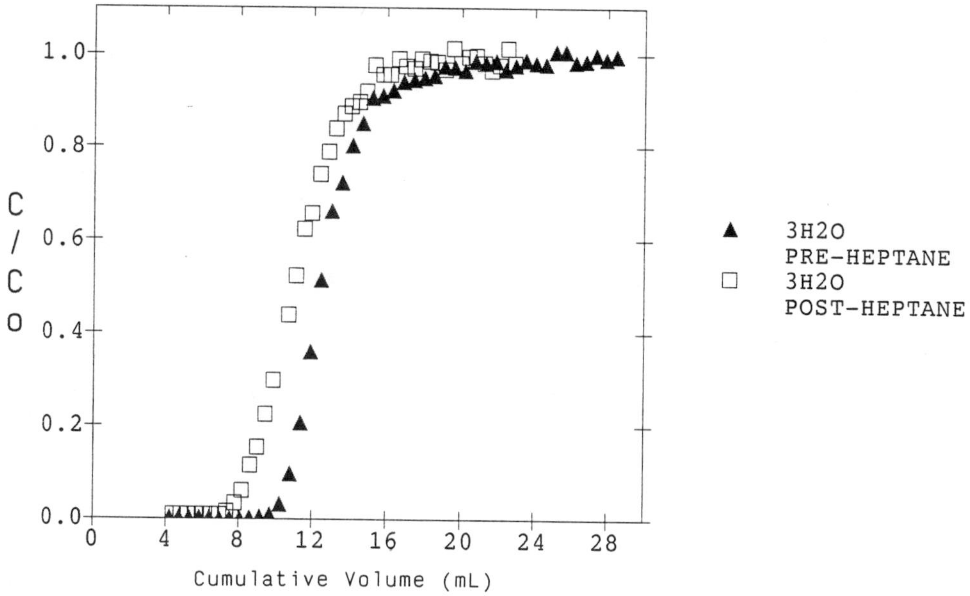

Fig. 3. ^3H_2O BTC's pre- and post-heptane on Lula aquifer material

times larger than in the unsaturated systems and were independent of glass bead diameter.

We used the prep-scale liquid chromatography columns packed with unconsolidated aquifer material described earlier to study NAPL entrapment. The columns were initially saturated with 0.01M $CaCl_2$. The 0.01M $CaCl_2$ was then displaced with heptane eluting from the bottom of the column upward. After passage of 13 pore volumes of heptane through the column, the heptane was displaced with 0.01M $CaCl_2$ at $v_w = 5.94 \times 10^{-6}$ m s^{-1} for approximately 20 pore volumes. The v_w was then reduced to 4.75×10^{-6} m s^{-1} for all BTC's. As discussed below in NAPL mobilization, increased viscous forces at higher v_w promote residual NAPL mobilization. The 5.94×10^{-6} m s^{-1} velocity was used to assure that the residual heptane in the column would remain immobile at $v_w = 4.75 \times 10^{-6}$ m s^{-1}. ^3H_2O BTC's were run prior to, and after, heptane introduction into the column (Fig. 3) The area between the BTC's is equal to the difference between V_w pre-heptane and V_w post-heptane, and represents the volume of residually saturated heptane, V_h . Finding V_h = 1.45 ml, the heptane residual saturation, S_{hr}, was calculated to be 15% using equation [13].

Before discussing NOC's transport in these heptane-water-aquifer material systems, it is first necessary to understand the mechanisms of NAPL entrapment and mobilization in porous media. As water displaces a NAPL in a porous medium, unstable water-NAPL interfaces are formed that result in the NAPL becoming a discontinuous phase of isolated blobs. It is the stability of the water-NAPL interface that determines both NAPL entrapment and mobilization. Interface stability is in turn determined by porous media pore geometry and pore topology (pore interconnectivity), and the capillary, viscous, buoyancy, and inertial forces operating in the system.

Two important mechanisms in NAPL entrapment in porous media are "snap-off" and "by-passing" (22,26-27). Snap-off results in NAPL entrapment in porous media when a NAPL droplet disconnects from a larger fluid mass as a result of the instability of the "neck meniscus" formed as the NAPL mass passes through a narrow pore throat into a larger pore space (pore body). The ratio of pore body to pore throat radius is an important factor in the snap-off mechanism. Considering a number of possible pore throat shapes, the critical pore body radius to pore throat radius (termed aspect ratio) was approximately 3 (26). That is, for aspect ratios >3 snap-off can occur. The residual saturation resulting from snap-off is characterized by a large number of individual droplets in single pores (singlets). A NAPL may be trapped via by-passing when the NAPL bulk mass is transported downstream, leaving the remaining NAPL trapped, as stable water-NAPL meniscii are formed on both the upstream and downstream side of the trapped NAPL.

For both the snap-off and by-passing mechanism, pore geometry and topology; rather than absolute pore size, are important porous media characteristics that determine the type and amount of NAPL retained as residual saturation. The amount of NAPL trapped via by-passing relative to trapping by snap-off increases as the porous media aspect ratio decreases. Also, larger droplet aggregates tend to result from by-passing compared to the singlets characteristic of snap-off. The end result of both snap-off and by-passing is that NAPL is trapped in the porous media and is immobile under existing water flow rates (viscous forces) and interfacial tensions.

Mobilization of the Residual Saturation

The highly curved menisci existing between the NAPL residual saturation and water in the pores are stable to system perturbations brought about by increased pore water velocities (increased viscous forces) attainable under most pump and treat remediation conditions. Also, the menisci effectively block pores to the flow of water reducing the saturated hydraulic conductivity.

As a result of the interfacial tension between two immiscible phases, a pressure discontinuity exists across the interface separating the two phases. The difference in pressure between the nonwetting fluid side of the interface and the pressure on the wetting fluid side is termed the capillary pressure (P_c).

$$P_c = P_h - P_w \qquad [14]$$

For a non-wetting fluid and a wetting fluid to co-exist in porous media, the pressure of the non-wetting fluid must exceed that of the wetting fluid. The Laplace equation relates pressure, interfacial tension, and interface curvature:

$$P_c = 2\gamma_{wh} \ / \ r \qquad [15]$$

where r is the radius of curvature for spherical interfaces.

From the Laplace equation it is evident that capillary pressure increases with interfacial tension and interface curvature (curvature = 1/r). Therefore, fluids exhibiting high interfacial tensions and curvature in porous media will be more strongly retained as residual saturation by capillary forces. In water-wet systems, the residual NAPL tends to be trapped in larger pores as the NAPL is unable to displace the highly curved water meniscii in smaller pores. It follows that NAPL existing in small pores will be more difficult to remove than NAPL in larger radius pores.

Residual NAPL trapped by capillary forces may be mobilized if opposed by sufficiently high viscous or gravity forces. The dimensionless ratios of viscous to capillary forces and gravity/buoyancy forces to capillary forces are termed the capillary number (N_c), and the Bond number (N_b), respectively.

$$N_c = k \ \rho_w \ g \ / \ \gamma_{wh} \ dh \ / \ dl \qquad [16]$$

where ρ is the density of water [Mg m^{-3}], γ_{wh} is the interfacial tension between water and the NAPL [N m^{-1}], and the other parameters are defined earlier. Or, alternatively:

$$N_c = v_w \ u_w \ / \ \gamma_{wh} \qquad [17]$$

$$N_b = k \ \Delta\rho_{wh} \ g \ / \ \gamma_{wh} \qquad [18]$$

where $\Delta\rho_{wh}$ is the density difference of the two fluids [Mg m^{-3}].

When the water velocity is in the same direction as the gravity/buoyancy forces (downward for DNAPL; upward for NAPL), the capillary and Bond numbers are additive. Under most environmental conditions, the magnitude of viscous and gravity forces are small relative to capillary forces, i.e., capillary and Bond numbers are small and NAPL entrapment as residual saturation is favored. Morrow and Songkran (28) found that NAPL trapping in porous media was dominated by capillary forces when $N_b < 5 \times 10^{-3}$ and $N_c < 10^{-6}$. With $N_b = 1.8 \times 10^{-9}$ and $N_c = 9.5 \times 10^{-8}$, buoyancy forces had no significant effect on heptane entrapment, and heptane entrapment was dominated by capillary forces in our heptane-water-aquifer material system.

In order to mobilize the residual NAPL, the water velocity (v_w) can be increased (increased viscous forces and higher capillary number) and/or the interfacial tension between water and the residual NAPL (γ_{wh}) can be reduced (decreased capillary forces and higher capillary and Bond number). The critical capillary number, (N_c)crit, is defined as the capillary number at which residual NAPL mobilization is initiated. For water-wet sandstones, (N_c)crit have ranged from 5×10^{-6} (21), to 2×10^{-5} (29) with N_c approaching 1.5×10^{-3} for complete NAPL mobilization. Though differences in porous media particle size and geometry will result in different (N_c)crit values, the (N_c)crit values reported for these sandstones are probably reasonably close to those for unconsolidated porous media of similar particle size and geometry.

Chatzis et al. (21) observed that NAPL blobs dispersed into smaller blobs at the initial stages of residual saturation mobilization. Thus, NAPL mobilization was not necessarily related to the initial blob-size distribution, but to the blob-size distribution at the prevailing capillary number. This is important as the viscous forces required for residual NAPL mobilization are expected to be inversely proportional to blob length in the direction of flow (30-31). NAPL mobilization involves NAPL displacement of water at the leading edge of blob movement (drainage), and water displacement of NAPL at the trailing edge of blob movement (imbibition). The pressure difference required for blob mobilization is proportional to the difference in the drainage and imbibition capillary pressures. At constant pressure, blobs that are longer in the water flow direction are more readily mobilized than shorter blobs because a greater pressure difference can be established across the longer blobs.

There are limits, imposed by porous media permeability (k) and dh/dl on how high v_w can be in an actual field situation. As an example, consider gasoline trapped in a homogeneous sandy aquifer [$k = 10^{-10}$ m^2]. Using equation 16, it can be determined that a hydraulic gradient of 0.24 would be necessary to attain $(N_c)crit = 2 \times 10^{-5}$ and thus initiate blob mobilization. For complete NAPL removal ($N_c = 1.5 \times 10^{-3}$), a hydraulic gradient of 18 would be required. With less permeable media, even steeper, more unrealistic, gradients would be required for residual NAPL mobilization and recovery using Pump and Treat technology. It is evident that hydraulically induced removal of residually saturated NAPL is only going to be feasible in very permeable materials such as coarse sands and gravels. Acknowledging these limitations, partial hydraulic removal of residually saturated NAPL's may find some limited use as a first step in a contamination remediation scheme, utilizing other remediation tools, such as natural and enhanced chemical and biological transformation, to reduce the NAPL mass resistant to hydraulic removal.

Practically, γ_{wh} may be reduced by the addition of either water miscible solvents or surfactants that lower the surface tension of water. Many commercially available, biodegradable, surfactants, will lower γ_{wh} by an order of magnitude at best. However, interfacial tensions between water and crude oil have been lowered by four orders of magnitude with surfactant systems designed for enhanced crude oil recovery (32). Attainment of the ultra-low γ_{wh} values needed for bulk fluid NAPL mobilization is complicated by γ_{wh} dependence on surfactant average molecular weight and molecular weight distribution, and the electrolyte composition of the aqueous phase. In addition, surfactant systems acceptable in enhanced crude oil recovery may not be acceptable in remediation procedures due to residual contamination from the surfactant system itself.

Mixtures of water miscible organic solvents also do not yield large decreases in γ_{wh}. Methanol-water solutions of 40 and 80% methanol by volume have surface tensions of .046 and .026 N m^{-1} respectively at 30 C. So adding methanol to water to yield a 40% methanol-water solution will only decrease γ_{wh} by a factor of 1.5, not a large enough decrease to facilitate bulk NAPL mobilization in most instances.

Although surfactants and water miscible solvents may not be useful for bulk NAPL mobilization, they will increase NAPL water solubility. For hydrophobic organic compounds, an increase in the fraction of organic solvent present in the aqueous phase results in an exponential increase in compound solubility. Consider the above example with a 40% methanol-water solution. While this solution could not result in significant increases in bulk fluid mobilization, it would result in significant solubility increases for many NAPL. For example, hexane solubility in a 46% methanol-water solution is 42 times greater than in pure water (33).

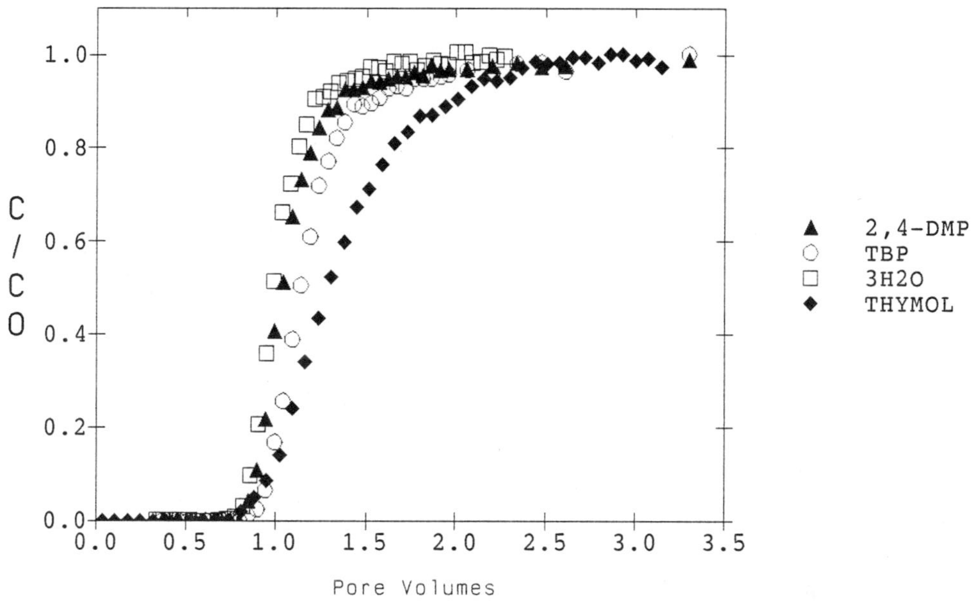

Fig. 4. Phenols' BTC's pre-heptane on Lula aquifer material

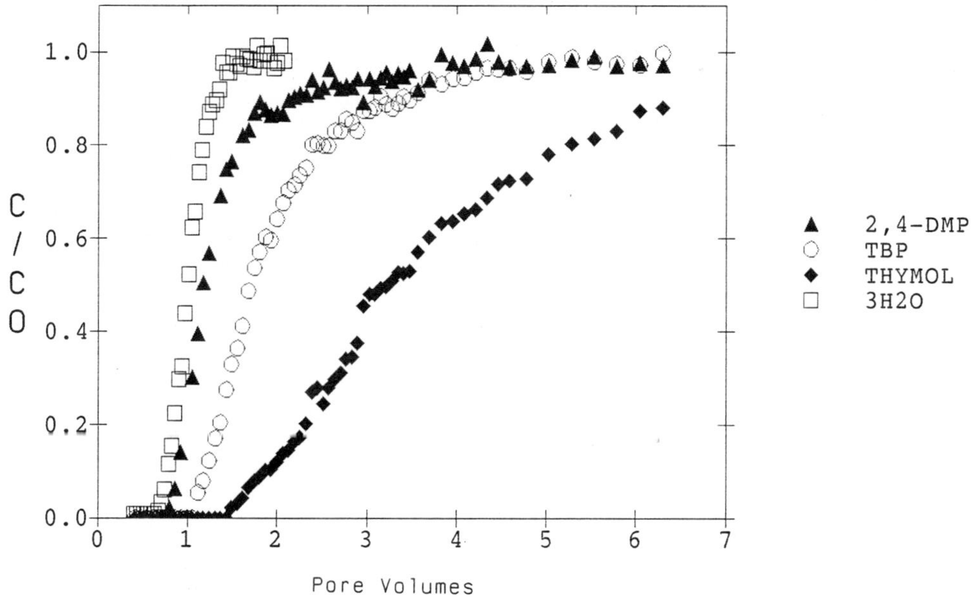

Fig. 5. Phenols' BTC's post-heptane on Lula aquifer material

Environmental Consequences of NAPL Residual Saturation

With the potential for 15 to 40% of the pore space in the saturated zone occupied by residual NAPL, it is clear that NAPL residual saturation represents a large and long term source of groundwater contamination that is recalcitrant to remediation. As we have seen, the natural organic carbon content of aquifer material is often too low to significantly sorb, and thus retard, the transport of many groundwater contaminants such BTX compounds, phenols, and common solvents. Contaminant transport from and through NAPL residual saturation will be governed by the rate of contaminant partitioning between the NAPL and the aqueous phase. Rates of Pump and Treat and bioremediation procedures will be controlled in part by the rate of contaminant release into the aqueous phase. Understanding contaminant transport from residual NAPL to the aqueous phase is requisite for accurate prediction of remediation time.

As discussed previously, the magnitude and rate of contaminant partitioning between a heterogeneous phase (soil organic carbon) and water was dependent on both contaminant solubility and geologic material organic carbon content. Consider again the heptane-water-aquifer material system under saturated conditions. In the natural state without the presence of any residually saturated NAPL, there was little retardation of a suite of alkylated phenols used as probe compounds (Fig. 4). The phenols used were 2,4-dimethyl phenol (DMP), p-tert-butyl phenol (TBP), and 2-isopropyl-5-methyl phenol (thymol). In the presence of 15% heptane residual saturation; however, the phenols retardation increased due to partitioning to the residual heptane (Fig. 5).

Retardation factors for the NOC's in the heptane-water-aquifer material system can be calculated using the following equation:

$$R_T = 1 + [(\rho_b / \theta_w \rho_w) K_{sw}] + [(\rho_h \theta_h / \theta_w \rho_w) K_{hw}] \qquad [19]$$

where ρ_b is the density of the bulk aquifer material [Mg m^{-3}], θ_w is the fraction of total column volume occupied by water [m^3 m^{-3}], ρ_w is the density of water (assume 1 Mg m^{-3}), K_{sw} is the distribution coefficient for contaminant partitioning between the geologic material and water, and ρ_h, θ_h, and K_{hw} are the corresponding values for heptane.

Generally, the phenols' R_T values calculated using equation [19] and R_T values determined experimentally using equation [2] were reasonably close. However, calculated R_T values were somewhat higher than the experimentally determined values (Table 4). Calculated R_T values utilized K_{hw} values that

Table 4. Retardation factors for phenols pre- and post-heptane on Lula aquifer material.

Phenols	Retardation factors		
	pre-heptane[1]	post-heptane[1] experimental	post-heptane[2] calculated
2,4-DMP	1.04	1.42	1.46
TBP	1.14	2.25	2.54
Thymol	1.26	4.75	6.02

[1]determined using eq. [2]
[2]determined using eq. [19]

were independently determined in shake-flask partitioning experiments where equilibrium in solute partitioning between the water and heptane phases was not limited by solute diffusion through the two phases and true equilibrium values could be accurately measured. However, in our column studies, where heptane was trapped as blobs in the aquifer material pore space, attainment of phenols' partitioning equilibria between the aqueous phase and heptane may have been limited by phenols' diffusion through the heptane blobs. This nonequilibrium partitioning was reflected in the asymmetric BTC for thymol transport through the residually saturated heptane (Fig. 5). Estimating the phenols' equilibrium partitioning values (K_{hw}) in the columns by extrapolation to $C = C_o$ introduced error into estimation of the R_T. This error may account for the lower R_T column values. Research with natural soil organic carbon also yielded column determined R_T values that tended to be slightly lower than batch equilibrium determined values (4).

Describing contaminant partitioning between the residual saturation of a simple NAPL, such as heptane, is relatively straightforward, and partitioning information collected in the laboratory should allow extrapolation to field scenarios. However, describing contaminant partitioning between a residually saturated complex NAPL, such as petroleum derived lubricants and fuels, and water is made difficult by the continuously changing composition of the NAPL. For complex NAPL, the rate of NAPL component partitioning into the aqueous phase will increase with component solubility. As a result, the residual NAPL will become more hydrophobic as the more hydrophilic components are weathered from the NAPL.

With this weathering process, contaminant solubilization kinetics will change and the NAPL will become more recalcitrant to removal. Opposing the movement toward increased hydrophobicity are chemical and biological oxidative processes acting to increase NAPL hydrophilicity. Which processes will predominate will depend on the relative rates of the oxidative and partitioning processes.

DISCLAIMER

Although the research described in this report has been funded by the U.S. Environmental Protection Agency through in-house efforts at the Robert S. Kerr Environmental Research Laboratory, it does not necessarily reflect the views of the Agency and no official endorsement should be inferred.

REFERENCES

1. Banerjee, P., M.D. Piwoni, and K. Ebeid. 1985. Sorption pf organic contaminants to a low carbon subsurface core. Chemosphere. 14:1057-1067.

2. Bouchard, D.C., and A.L. Wood. 1988. Pesticide sorption on geologic material of varying organic carbon content. Toxic. Indust. Health. 4:341-349.

3. Schwarzenbach, R.P., and J. Westall. 1981. Transport of nonpolar organic compounds from surface water to groundwater. Laboratory sorption studies. Environ. Sci. Technol. 15:1360-1367.

4. Bouchard, D.C., A.L. Wood, M.L. Campbell, P. Nkedi-Kizza, and P.S.C. Rao. 1988. Sorption nonequilibrium during solute transport. J. Contam. Hydrol. 2:209-223.

5. Bouchard, D.C., C.G. Enfield, and M.D. Piwoni. 1989. Transport processes involving organic chemicals. In Reactions and movement of organic chemicals in soils. Soil Sci. Soc. Am. Spec. Pub, Amer. Soc. Agron., Madison, WI.

6. Lapidus, L., and N.R. Amundson. 1952. Mathematics of adsorption in beds. VI. The effect of longitudinal diffusion in ion exchange and chromatographic columns. J. Phys. Chem. 56:984-988.

7. Hashimoto, I., K.B. Deshpande, and H.C. Thomas. 1964. Peclet numbers and retardation factors for ion exchange columns. Ind. Eng. Chem. Fund. 3:213-218.

8. Kay, B.D., and D.E. Elrick. 1967. Adsorption and movement of lindane in soils. Soil Sci. 104:314-322.

9. Nielsen, D.R., and J.W. Biggar. 1961. Miscible displacement in soils: Experimental information. Soil Sci. Soc. Am. Proc. 25:1-5.

10. Selim, H.M., J.M. Davidson, and R.S. Mansell. 1976. Evaluation of a two-site adsorption-desorption model for describing solute transport in soil. Proc. Summer Computer Simulation Conf., 12-14 July 1976. Washington, D.C., pp. 444-448.

11. Cameron, D.R., and A. Klute. 1977. Convective-dispersive solute transport with a combined equilibrium and kinetic adsorption model. Water Resour. Res. 13:183-188.

12. van Genuchten, M. Th. 1981. Non-equilibrium solute transport parameters from miscible displacement experiments. Res. Rep. 119, U.S. Salinity Laboratory and Dept. of Soil and Environ. Sciences, Univ. of California, Riverside.

13. Nkedi-Kizza, P., J.W. Biggar, H.M. Selim, M. Th. van Genuchten, P.J. Wierenga, J.M. Davidson, and D.R. Nielson. 1984. On the equivalence of two conceptual models for describing ion exchange during transport through an aggregated Oxisol. Water Resour. Res. 20:1123-1130.

14. Karickhoff, S.W., and K.R. Morris. 1985. Sorption dynamics of hydrophobic pollutants in sediment suspensions. Environ. Tox. Chem. 4:469-479.

15. Wu, Shian-chee, and P.M. Gschwend. 1986. Sorption kinetics of hydrophobic compounds to natural sediments and soils. Environ. Sci. Technol., 20:717-725.

16. Rao, P.S.C., A.G. Hornsby, D.P. Kilcrease, and P. Nkedi-Kizza. 1985. Sorption and transport of hydrophobic organic chemicals in aqueous and mixed solvent systems: Model development and preliminary evaluation. J. Environ. Qual. 14:376-383.

17. Enfield, C.G., and G. Bengtsson. 1988. Macromolecular transport of hydrophobic contaminants in aqueous environments. Ground Water 26:64-70.

18. Bear, J. 1979. Hydraulics of Groundwater. McGraw-Hill Inc., N.Y., N.Y.

19. Schwille, F. 1981. Groundwater pollution in porous media by fluids immiscible with water. In: Quality of Groundwater, Proceedings of an International Symposium. Elsevier Pub., Amsterdam, the Netherlands.

20. Collins, R.E. 1961. Flow of fluids through porous materials. Penwell Pub. Co., Tulsa, OK.

21. Chatzis, I., M.S. Kuntamukkula, and N.R. Morrow. 1984. Blob-size distribution as a function of capillary number in sandstones. Paper 13213, presented at SPE Annual Tech. Conf. and Exhib., Houston, Texas.

22. Chatzis, I., N.R. Morrow, and H.T. Lim. 1983. Magnitude and detailed structure of residual oil saturation. Soc. Petrol. Engin. J. 23:311-326.

23. Felsenthal, M. 1979. A statistical study of core waterflood parameters. J. Petrol. Tech. 31:1303-1304.

24. Conrad, S.H., E.F. Hagan, and J.L. Wilson. 1987. Why are residual saturations of organic liquids different above and below the water table? Proc. Petrol. Hydrocarbons Org. Chem. in Ground Water, NWWA.

25. Wilson, J.L., and S.H. Conrad. 1984. Is physical displacement of residual hydrocarbons a realistic possibility in aquifer restoration? Proc. Petrol. Hydrocarbons Org. Chem. in Ground Water, NWWA, Houston, TX, p.274-298.

26. Mohanty, K.K., H.T. Davis, and L.E. Scriven. 1980. Physics of oil entrapment in water-wet rock. Paper 9406, presented at SPE Annual Tech. Conf. and Exhib., Dallas, TX.

27. Pathak, P., H.T. Davis, and L.E. Scriven. 1982. Dependence of residual nonwetting liquid on pore topology. Paper 11016, presented at SPE Annual Tech. Conf. and Exhib., New Orleans, LA.

28. Morrow, N.R., and B. Songkran. 1981. Effect of viscous and buoyancy forces on nonwetting phase trapping in porous media. In: Surface Phenomena in Enhanced Oil Recovery, D.O. Shah, ed., Plenum Press, N.Y., N.Y.

29. Chatzis, I., and N. Morrow. 1981. Correlation of capillary number relationships for sandstones. Paper 10114, presented at SPE Annual Tech. Conf. and Exhib., San Antonio, TX.

30. Melrose, J.C., and C.F. Brandner. 1974. Role of capillary forces in determining microscopic displacement efficiency for oil recovery by waterflooding. J. Can. Petrol. Tech., 13:54-62.

31. Morrow, N.R. 1979. Interplay of capillary, viscous, and bouyancy forces in the mobilization of residual oil. J. Can. Petrol. Tech., 18:35-46

32. Foster, W.R. 1973. A low tension waterflooding process employing a petroleum sulfonate, inorganic salts, and a biopolymer. J. Petrol. Tech., 25:205-210.

33. Groves, F.R. 1988. Effects of cosolvents on the solubility of hydrocarbons in water. Environ. Sci. Technol. 22:282-286.

Key Words: sorption, partitioning, nonaqueous phase liquids

TRANSPORT AND DEGRADATION OF WATER-SOLUBLE CREOSOTE-DERIVED COMPOUNDS

Edward M. Godsy[1], Donald F. Goerlitz[1], and Dunja Grbić-Galić[2]

U.S. Geological Survey
Water Resources Division
Menlo Park, California 94025

Stanford University
Department of Civil Engineering
Stanford, California 94305

INTRODUCTION

Creosote is the most extensively used insecticide and industrial wood preservative today. It is estimated that there are more than 600 wood-preserving plants in the United States, and their collective use of creosote exceeds 4.5×10^6 kg/yr (von Rumker et al., 1975). Creosote is a complex mixture of more than 200 major individual organic compounds with differing molecular weights, polarities, and functionalities, along with dispersed solids and products of polymerization (Novotny et al., 1981). The major classes of compounds previously identified in creosote show that it consists of ~85% (w/w) polynuclear aromatic compounds (PAH), ~12% phenolic compounds, and ~3% heterocyclic nitrogen, sulfur, and oxygen containing compounds (NSO).

In 1983, an abandoned creosote wood-preserving plant (ACW) in Pensacola, Florida, was selected by the U.S. Geological Survey as one of the three national research demonstration areas to develop understanding of hazardous waste processes (Figure 1). The criteria governing the selection of this site included (1) the relatively simple mineralogy of the surfacial sand-and-gravel aquifer, (2) the apparently straightforward flow system within the aquifer, and (3) the availability of a preliminary data base on the flow system, extent of contamination, and the water chemistry.

Pine poles were treated with wood preservatives for nearly 80 years prior to the closing of the site in December 1981. Prior to 1950, creosote was used exclusively to treat poles and subsequent to that date, both creosote and pentachlorophenol (PCP) were used. Effluent from the treatment process consisted of water, cellular debris, diesel fuel, creosote, and PCP. The chemically complex, organic-rich mixture was discharged to shallow unlined waste disposal ponds and from there, large but unknown quantities of the waste infiltrated the soil beneath the ponds. Upon reaching the water table, the effluent mixed with the ground water, and two distinct phases resulted: (1) a denser than water hydrocarbon phase that moved vertically downward somewhat perpendicular to the direction of ground-water flow, and (2) an aqueous phase. The aqueous phase, or water-soluble fraction (WSF), has been enriched in phenolic compounds, single- and double-ring aromatic

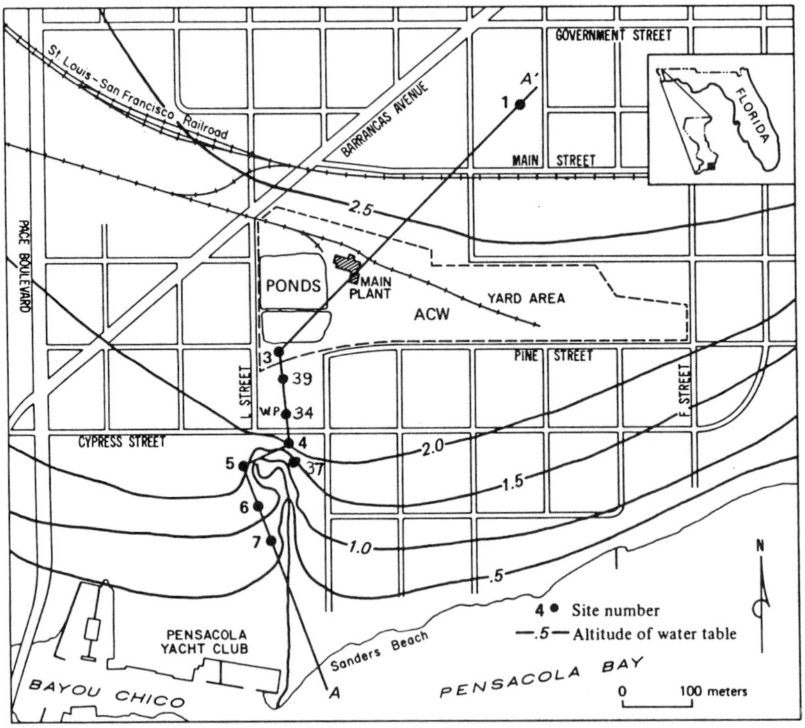

Figure 1. Shallow ground-water-level contours and well sites at the abandoned creosote works, Pensacola, Florida.

compounds, and single- and double-ring NSO compounds. As the contaminated water moves along with the ground-water flow, the dissolved contaminants are subject to physical, chemical, and biological processes that tend to retard the movement of these compounds relative to the ground water.

This paper describes the recent investigations into the cause(s) of the selective attenuation of the WSF compounds observed during downgradient movement in the aquifer, and the role of the resident methanogenic microbial consortium in that attenuation.

TRANSPORT PROCESSES

Background

Organic contaminants entering a ground-water system from the surface, will tend to move downward towards the water table. When the contaminated water reaches the water table, a contaminant will tend to move laterally in the direction of the ground-water flow. As the water moves away from the pollution source, the concentration of contaminants decreases due to dispersion and may decrease due to adsorption on aquifer particle surfaces, chemical, and/or biochemical reactions.

The mechanisms of advection and dispersion have an important control of the transport of organic pollutants in the subsurface environments. Total contaminant flow is composed of the portion that travels with the average ground-water flow (advection) and the portion that deviates from the average ground-water flow (dispersion). These result in a dilution of the contaminant concentration and a spreading of the contaminated area.

There are two types of dispersion: (1) dispersion that occurs at the pore scale (microdispersion) and (2) dispersion that occurs at the field scale due to aquifer heterogeneity (macrodispersion). Microdispersion is significant for transport in relatively slow-flowing ground water, and macrodispersion is important because of the heterogeneity of most aquifers (e.g. Sudicky et al., 1983). These phenomena make field and laboratory comparisons very difficult.

Quantitative relationships between sorption and the controlling factors have not been well established for aquifers containing less than 0.1% organic material, as was found at this site. Some adsorption of non-polar organic compounds was observed in columns filled with solids that do not contain organic matter, such as clean sand, limestone, and montmorillonite clay (Schwarzenbach and Westall, 1981). Goerlitz et al. (1985) demonstrated in laboratory columns containing sediments from the research site, that naphthalene was retarded during movement through the column and was presumably adsorbed onto the aquifer sediments, even though the organic content was less than 0.1%. This result suggests that some mechanism other than adsorption on the organic material may be responsible for the retardation of naphthalene in the column.

Polar organics appear to be more mobile than non-polar organics in aquifers with significant amounts of organic carbon, because they are poorly retained in the organic material in the aquifer (Roberts et al., 1982). Goerlitz et al. (1985) found that phenol, 3-methylphenol, and 3,5-dimethylphenol, all polar compounds, were not retarded during movement through laboratory columns containing aquifer material from the research site while pentachlorophenol, also a polar compound, was retarded. These results also suggest a mechanism other than adsorption onto the organic material in the aquifer.

Recent investigations by Westall et al. (1985) on the distribution ratio for the polar ionizable chlorinated phenols demonstrated that the major influence on the Kow value was the pH and ionic strength of the aqueous phase. Zachara et al. (1986) investigated the sorption isotherms for the hydrophobic ionizable organic base quinoline under differing pH and ionic strength aqueous solutions. It was also observed that the organic cation was strongly adsorbed when pH was below neutrality while the neutral quinoline molecule was only weakly adsorbed under the same conditions. They concluded that in subsurface materials with a low carbon content, quinoline sorption is controlled by pH, the nature and capacity of the exchange complex, and the ground water ion composition.

It is apparent that methods used to describe the movement of non-polar compounds in the subsurface are not adequate to describe the movement of hydrophobic ionizable organic compounds in low organic content aquifer material. The effects of pH and ionic strength are site-specific and must be determined independently.

The behavior of an ideal tracer (e.g. $CaCl_2$) which experiences neither adsorption nor chemical transformation during transport from storage in a one-dimensional solution-filled porous medium (Ogata and Banks, 1961), can be expressed as:

$$\epsilon \frac{\partial c}{\partial t} = D \frac{\partial^2 c}{\partial x^2} - q \frac{\partial c}{\partial x} \tag{1}$$

where ϵ = porosity (-)
t = time of transport (T)
c = solute concentration (M/L^3)
q = water flux (L^3/L^2-T)
D = effective dispersion coefficient (L^2/T)

x = spatial coordinate (L)

Linear sorption on the porous medium can be described as:

$$\bar{c} = kc \tag{2}$$

where: $\bar{c}$ = concentration of adsorbed solute (M/M)
k = partition coefficient (-)

Including the storage term on the solid in the transport equation, the result is:

$$\epsilon \frac{\partial c}{\partial t} + \rho \frac{\partial \bar{c}}{\partial t} = D \frac{\partial^2 c}{\partial x^2} - q \frac{\partial c}{\partial x} \tag{3}$$

where: ρ = bulk density (M/L^3)

Partial differentiation of Equation 2 yields:

$$\frac{\partial \bar{c}}{\partial t} = k \frac{\partial c}{\partial t} \tag{4}$$

then, combining Equation 3 with Equation 4 and simplifying, gives:

$$(\epsilon + \rho k) \frac{\partial c}{\partial t} = D \frac{\partial^2 c}{\partial x^2} - q \frac{\partial c}{\partial x} \tag{5}$$

The analytical solution of Equation 5 for constant concentration boundaries

$$C(x,0) = 0 \qquad x \geq 0$$
$$C(0,t) = C_0 \qquad t \geq 0$$
$$C(\infty,t) = 0 \qquad t \geq 0$$

after dividing through by $\epsilon + \rho k$ as given by Ogata and Banks (1961) is:

$$\frac{C}{C_0} = 0.5 \text{ erfc} \left[\frac{(\epsilon+\rho k)x-qt}{2\sqrt{D(\epsilon+\rho k)t}}\right] + 0.5 \exp\left[\frac{qx}{D}\right] \text{ erfc} \left[\frac{(\epsilon+\rho k)x+qt}{2\sqrt{D(\epsilon+\rho k)t}}\right] \tag{6}$$

where: x = column length (L)
C = output concentration (M/L^3)
C_0 = input concentration (M/L^3)
erfc = complimentary error function
exp = exponent, base e

The advancing front of a solute moves at a linear velocity that is less than the velocity of the solvent as follows:

$$R_t = 1 + \frac{\rho k}{\epsilon} = v_r$$

where: R_t = retardation factor of relative residence time
v_r = velocity of adsorbing solute divided by the velocity of a conservative tracer

The retardation factor is primarily influenced by the value of k because the values of ρ and ϵ vary to a lesser degree in natural environments.

Experimental Procedures

Sorption experiments were done using readily available HPLC equipment thus allowing automation of the elution and effluent monitoring (Goerlitz, 1984). The HPLC equipment consisted of two Isco Model 314, 500 mL syringe HPLC pumps; an Isco Model V^4 variable wavelength detector; a Waters Model

Table 1. Column elution experimental data.

porous media:	aquifer sand
column length:	35.0 cm
column diameter:	2.5 cm
column volume:	171.8 cm^3
bulk density:	1.53 g/cm^3
particle density:	2.47 g/cm^3
mean particle diameter:	0.038 cm
porosity:	0.381
pore volume:	65.44 cm^3
flow rate:	108.24 mL/day
flux rate:	22.05 cm^3/cm^2-day
pore velocity:	57.89 cm/day
Peclet number:	138
Reynolds number:	0.003
dispersion:	14.65 cm^2/day

Table 2. Adsorption data for the water soluble compounds in creosote.

Compound	Partition Coefficient, k	Retardation Factor, R_t	Log K_{ow}
$CaCl_2$	0	1.0	NA
phenol	0.0036	1.012	1.48
2-methylphenol	0.0072	1.026	1.93
3-methylphenol	0.0176	1.071	1.98
4-methylphenol	0.0257	1.103	2.00
3,5-dimethylphenol	0.0379	1.152	2.35
quinoline	ND	ND	2.03
isoquinoline	ND	ND	2.08
2(1H)-quinolinone	0.1370	1.550	1.26
1(2H)-isoquinolinone	0.1030	1.414	0.58
indole	0.0520	1.187	2.10
benzothiophene	0.1769	1.710	3.12
indene	0.1610	1.579	2.92

NA = Not Applicable
ND = Not Determined

R401 refractive index detector; an Isco fraction collector; and a computer controlled data collection and analysis system.

Initial adsorption experiments were done using 60/80 mesh glass beads to confirm that the dispersion due to the experimental apparatus was within theoretical limits using the same method described below.

Aquifer material was air dried, sieved through a 1.0 mm screen, remoistened to the point of aggregation and packed in small amounts into a 25 mm x 500 mm glass column. The packed material was then packed down firmly with a teflon tipped tamper after every 3 to 5 mm. After the column was packed, autoclaved, and the air replaced with sterile CO_2, a sterile solution of 0.010 M $CaCl_2$ was pumped through the column until no gas bubbles were visible.

Initially the 0.010 M $CaCl_2$ eluent was pumped until instrumental stability occurred. To start the experiment, the feed solution was abruptly changed for the test solution at the head of the column. The column characteristics were determined by changing the 0.010 M $CaCl_2$ eluent to a 0.015 M $CaCl_2$ solution. Soluble compounds were tested by adding the compound of interest to the eluent. Sparingly soluble compounds were introduced by directing the eluent through a "saturator" column. The saturator column was

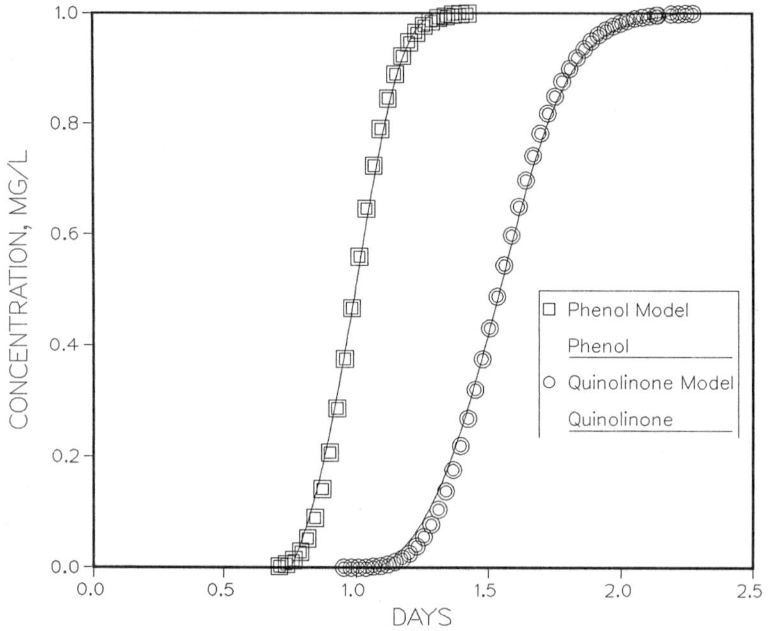

Figure 2. Elution profiles for phenol and 2(1H)-quinolinone from laboratory columns containing aquifer material from site 1.

prepared in a 4 mm x 300 mm commercially available empty stainless steel HPLC column. The column was packed with 60-80 mesh Chromosorb W coated with a 10% (w/w) of the selected compound. It has been shown that exposure of such a large area, 2-3 m^2/g for this substrate, to water brings about saturation with respect to the coating (May et al., 1978). By the use of two pumps, solutions at any desired concentration can be prepared as needed.

Results

Shown in Table 1 are the operating conditions and calculated parameters for the research site aquifer material studies. Table 2 contains the partition coefficients, k; the retardation factors, R_t; and the K_{ow} values for the above compounds. Figure 2 shows the elution profiles for phenol, and 2(1H)-quinolinone.

GROUND WATER ANALYSIS

Experimental Methods

Well Drilling Technique. Wells were drilled using the hollow-stem, continuous-flight auger method described by Schalf et al. (1981). This method allows boring into soils carrying the cuttings upward along the auger flights without the use of contaminating drilling fluids. To prevent sand and water intrusion into the hollow core while drilling, a knock-out plug was installed in the auger bit and O-rings were installed between the auger flights. Water samples were collected using a teflon bailer or a peristaltic pump directly from the auger flights.

Gas Chromatography-Mass Spectrometry-Data System. Organic solutes were identified using a Finnigan MAT GC/MS/DS. The GC/MS/DS consists of a Model 4510 quadrupole mass spectrometer interfaced to a Model 9610 gas chromatograph, both controlled by an Incos data system. Chromatography was performed by splitless injection on fused silica capillary columns and the oven temperature was programmed from 50°C at 10°C/min, to the maximum temperature limit of the column selected. Two columns (30 m x 0.25 mm i.d.), obtained from J & W Scientific, Inc., both having bonded liquid phases 0.25 μm thick, were used. One was a nonpolar DB-5 column and the other was a polar DB-WAX column with a bonded carbowax phase. The capillary columns were separately and directly connected to the ion source of the mass spectrometer which was operated in the electron impact mode.

Gas Chromatography. A Varian 6000 gas chromatograph equipped with an injection port splitter, fused silica capillary columns, and a flame ionization detector (FID) was used for the quantitative determinations of phenols, NSO, and PAH. The column oven was temperature-programmed from 75°C at 10°C/min to the maximum operating temperature for the selected column. The GC was equipped with columns identical to those described above. A Varian Vista 401 data system was used to record the chromatograms, peak retention times, and areas, and for computation of the concentration of the individual components in the sample extract.

Organic Analysis of Ground Water Samples. The water was centrifuged at 2000 x g for 10 min to remove particulates and the supernatant was divided into 5.00 mL subsamples. The bases, neutrals, and phenols were removed from the first subsample (pH 6.8) by CH_2Cl_2 extraction (all extractions were 1:1 v/v). The remaining organic acids in the aqueous isolate were determined directly by the HPLC method of Ehrlich et al. (1981) or acidified and the organic acids extracted into ether and determined by GC/MS/DS using the DB-WAX column.

Because of the complexity of the mixture of compounds occurring in creosote, a preliminary separation into three classes of compounds (neutrals, phenols, and bases) was necessary for the remaining subsample. This subsample was made acidic, pH 1.5, and the neutrals and phenols were extracted into CH_2Cl_2. The remaining aqueous layer was neutralized (pH 6.8) and the bases were extracted into CH_2Cl_2. The phenols were separated from the neutrals by shaking the mixture with 1.0 mL 5 M KOH solution. The neutrals remained in the CH_2Cl_2 solvent. The KOH solution was acidified to a pH of 1.5 and the phenols were extracted into CH_2Cl_2. Each of the isolate solutions were analyzed by GC/MS/DS and/or GC/FID.

Results

The results of the inorganic analyses for selected constituents are given in Table 3. The uncontaminated ground water is a Na-HCO_3-Cl type with dissolved solids generally less than 50 mg/L, which is typical for a shallow

Table 3. Inorganic analysis of water samples from 6.1 m deep wells. All results in mg/L except pH and temperature. Analyses by U.S.G.S.Quality of Water Service Unit, Ocala, Florida.

	Site Number					
	1	3	39	WP34	4	37
pH	6.9	5.5	5.9	6.5	6.5	5.8
temperature	23.5	23.0	23.0	23.5	23.5	23.0
DOC	8.4	357.0	192.0	49.2	20.4	6.2
alkalinity, as $CaCO_3$	45	176	136	237	142	75
ammonia-N	0.02	7.40	6.30	5.10	3.00	3.80
organic-N	1.20	8.60	4.90	2.80	1.20	2.90
nitrate-N	2.10	ND	ND	ND	ND	0.02
nitrite-N	0.01	0.10	0.06	0.04	0.07	0.02
phosphate-P	0.88	0.16	0.19	0.21	0.16	0.08
sulfate	6.1	58.0	0.1	0.3	0.1	0.1
magnesium	1.2	14.0	7.6	7.6	5.7	1.3
calcium	19.0	39.0	24.0	52.0	33.0	1.4
sodium	8.2	25.0	16.0	12.0	12.0	13.0
potassium	0.8	28.0	14.0	9.0	6.5	2.0
chloride	7.1	120.0	49.0	27.0	24.0	24.0
dissolved O_2	0.04	ND	ND	ND	ND	ND
sulfide	ND	2.9	4.0	1.9	1.1	0.9
methane	ND	13.0	10.0	11.0	13.0	11.2

ND = below detection limit, generally < 0.010 mg/L
Sulfide, dissolved oxygen, and methane analysis by M.J. Baedecker, U.S.G.S. Reston, Virginia

quartz sand-and-gravel aquifer. In the contaminated water at site 3, the concentrations of all constituents are higher and the dissolved solids are about 250 mg/L. Since these sediments do not contain carbonates, the higher alkalinities result from the degradation of the complex WSF organic compounds to volatile fatty acids.

The persistence of nitrogen and phosphorus compounds beyond the point at which the biodegradable organic compounds have disappeared, suggests that these compounds are not growth limiting. The lack of dissolved O_2, along with elevated concentrations of H_2S and CH_4, indicate anoxic conditions inthe area of contamination.

Results of organic analysis of water samples from sites 1, 3, 39, 40, 4, and 37 are summarized in Table 4 with the exception of the longer chain alkylphenols, which have been shown to resist methanogenic fermentation (Godsy et al., 1986). None of the major compounds in the WSF were detected in the water sample from the uncontaminated site 1. The listed compounds in Table 4 do not include similar compounds in amounts less than 0.01 mg/L each but on summation represent approximately 20% of the total.

The concentrations of phenol, 2-, 3-,4-methylphenol, quinoline, isoquinoline, 2(1H)-quinolinone, and 1(2H)-isoquinolinone decrease disproportionately downgradient when compared to chloride, the conservative tracer.

Table 4. Organic analyses of water samples from 6.1 m deep wells. All results in mg/L.

	Site Number					
	1	3	39	WP34	4	37
Organic Acids						
acetic	ND	45.14	18.30	5.89	5.32	3.12
formic	ND	0.13	0.85	0.73	0.97	ND
C_3-C_6 VFA	ND	44.41	ND	ND	ND	ND
benzoic	ND	27.49	ND	ND	ND	ND
PAH Compounds						
indene	ND	1.25	0.24	ND	ND	ND
naphthalene	ND	9.38	3.39	2.89	0.93	1.54
1-methylnaphthalene	ND	0.41	0.32	0.25	0.06	0.11
2-methylnaphthalene	ND	0.99	0.32	0.54	0.10	0.10
acenaphthene	ND	0.52	0.29	0.33	0.05	ND
Phenolic Compounds						
phenol	ND	26.01	6.90	0.04	0.08	ND
2-methylphenol	ND	13.27	4.60	0.45	0.03	ND
3-methylphenol	ND	26.65	9.20	0.29	0.07	ND
4-methylphenol	ND	11.97	4.13	0.13	0.03	ND
pentachlorophenol	ND	0.62	0.11	0.16	0.08	ND
NSO Compounds						
indole	ND	ND	ND	ND	ND	ND
quinoline	ND	11.20	0.01	ND	ND	ND
2-methylquinoline	ND	4.32	0.16	0.50	ND	0.03
isoquinoline	ND	3.61	0.01	ND	ND	0.01
2(1H)-quinolinone	ND	14.28	16.85	0.96	ND	1.69
1(2H)-isoquinolinone	ND	2.20	2.50	0.35	0.02	0.43
benzothiophene	ND	0.83	0.31	0.22	0.16	0.16
dibenzofuran	ND	0.30	0.04	0.16	ND	ND

ND = below detection limit, generally < 0.010 mg/L

The concentrations of the PAH, pentachlorophenol, benzothiophene, dibenzofuran, and 2-methylquinoline appear to decrease in proportion to the conservative tracer. This observation does not preclude that these compounds are anaerobically biodegradable, only that they do not disappear during downgradient travel in this section of the aquifer. At site 37, the increase in several of the organic compounds may be due to overland flow from the storage ponds during heavy rainfall events. Free creosote can be found from the water table (~1 m below land surface) to just below land surface at this site.

The C_3-C_6 volatile fatty acids and benzoic acid are removed during the first 53 m downgradient from the source ponds. Acetic and formic acids are known to be major intermediate products in the methanogenic fermentation of complex organic substrates, and should therefore persist during downgradient movement in the aquifer if the compounds are being biodegraded.

ANAEROBIC DEGRADATION

Background

Methane production from aromatic compounds was first observed 1906 by Soehngen (Evans, 1969). Tarvin and Buswell (1934) were subsequently able to show that a range of aromatic compounds could be metabolized to CH_4 and CO_2 with greater than 90% conversion of the substrates into these gases. The main emphasis in the work published after these two reports was on the elucidation of the biochemical mechanism of the breakdown (Evans, 1977). Recently, however, a body of literature has emerged describing the microbial degradation of aromatic compounds under anaerobic conditions. Benzoic acid has been used by most workers as a model aromatic compound and the methanogenic fermentation of benzoate has been demonstrated in sewage digestor sludge, rumen fluid, polluted river mud and freshwater lake mud.

Various pathways have been proposed for the anaerobic degradation of aromatic compounds depending on the nature of the substrate, but they are all similar in the major steps involved (Berry et al., 1987a): (1) the removal of substituent side groups by demethylation (Young and Rivera, 1985; Szewzyk et al., 1985), demethoxylation, dehydroxylation, and decarboxylation (Grbić-Galić and Young, 1985); (2) incorporation of oxygen from water (hydroxyl-group) into the ring structure of compounds that do not already have an oxygen substituent incorporated (Vogel and Grbić-Galić, 1986; Berry et al., 1987); (3) hydrogenation of the aromatic structure with the formation of an alicyclic intermediate; (4) fission of the alicyclic compounds to straight chain alcohols and acids; (5) conversion of these to acetate, H_2, and CO_2; and (6) under methanogenic conditions, conversion of these end-products to CH_4.

Chmielowski et al. (1965) screened 18 phenolic compounds to determine whether these were amenable to anaerobic fermentation by methanogenic cultures. Nine of these, including phenol and 4-methylphenol, were converted to CH_4 and CO_2. Healy and Young (1978) demonstrated that both phenol and catechol were stoichiometrically metabolized by a methanogenic consortium, and an anaerobic pathway was proposed for their metabolism and ring fission by Balba and Evans (1980). Recently, the anaerobic degradation of several substituted phenols (Boyd et al., 1983; Godsy and Goerlitz, 1986; Fedorak and Hrudey, 1986) and chlorinated phenols (Boyd and Shelton, 1984; Godsy et al., 1986) have been reported.

With the increased interest in the anaerobic degradation of the components of wastewaters from petroleum, coke operations, and coal conversion processes, the treatability of the NSO compounds is just beginning to be investigated. Wang et al. (1984) and Berry et al. (1987) reported on the methanogenic fermentation of indole. Although Wang et al. (1984) did not observe anaerobic degradation of quinoline and methylquinoline, Pereira et al. (1987) demonstrated microbial anaerobic oxidation of two-ring nitrogen heterocycles. However, their microbial systems did not catalyze further methanogenic fermentation of these compounds. Recently, Maka et al. (1986) observed anaerobic degradation of benzothiophene and dibenzothiophene by mixed cultures isolated from soil and sewage sludge in a complex growth medium, and Godsy and Grbić-Galić (1988) described the methanogenic degradation of benzothiophene in aquifer-derived microcosms. The authors are not aware of any reports on anaerobic degradation of oxygen-containing heterocycles.

Vogel and Grbić-Galić (1986) recently proved that the oxidation of homocyclic aromatic hydrocarbons was possible by demonstrating the hydroxylation of benzene and toluene to CH_4 and CO_2 by a mixed methanogenic culture. Mihelcic and Luthy (1988) showed that naphthalene and acenaphthene where anaerobically degradable in soil/water slurries under denitrifying conditions, but there have not been any reports of PAH undergoing degradation under methanogenic conditions.

Experimental Procedures

The large anaerobic microcosm for the WSF biodegradation studies was constructed in a 4 L glass bottle containing approximately 4 kg of aquifer material and 2.5 L water from site 3 at a depth of 6.1 m. The microcosm was constructed in an anaerobic glove box and flushed with O_2-free argon upon completion. The microcosm was fitted with both a liquid and a gas sampling port, and a U-tube manometer. One mL of amorphous FeS was added per liter as a reducing agent (Brock and O'Dea, 1977).

At 7-day intervals, the increase in gas volume was recorded using a water-wetted glass syringe and corrected for temperature and pressure. The gas composition was determined by GC. A Baseline 1010A gas chromatograph fitted with a thermal conductivity detector, was used. The GC was equipped with a 0.10 mL sample loop, and a 10-port switching valve fitted with a 1.5 m x 3 mm stainless steel 60/80 Chromosorb 102 column (1) and a 5 m x 3 mm stainless steel 60/80 Molecular Sieve 5A column (2). A sample was injected onto column 1 from the sample loop and held for 50 sec to allow the composite peak of Ar and CH_4 to pass onto column 2, while the CO_2 peak was retained. The valve was switched, reversing the column sequence allowing the CO_2 pass directly to the detector, followed by the Ar and CH_4 peaks that were separated in column 2. The operating conditions were as follows: column and detector temperature, 70°C; and helium carrier gas flow rate, 25 mL/min.

Organic acids were analyzed by HPLC as described previously. The remaining organic compounds were isolated from the aqueous phase by a 1:1 extraction into CH_2Cl_2, dried by passing through a Pasteur pipet containing oven-dried Na_2SO_4, and analyzed by GC-FID and confirmed by GC/MS/DS as outlined previously.

Microcosm biomass was determined by the acridine orange direct count (AODC) method described by Hobbie et al. (1977) and by Method 209 D of Standard Methods (American Public Health Association, 1985).

Results

Results indicate that, along with the phenolic compounds and organic acids previously reported to be anaerobically biodegradable (Godsy and Goerlitz, 1986), the nitrogen-containing compounds, quinoline, isoquinoline, 2(1H)-quinolinone, and 1(2H)-isoquinolinone are also anaerobically biodegradable to CH_4 and CO_2. These data (Figures 3, 4, and 5) also show that there is a three-step sequential degradation of these compounds occurring in the following order:

1. quinoline, isoquinoline, benzoic acid, and C_3-C_6 volatile fatty acids;

2. phenol;

3. 2-, 3-, 4-methylphenol, 2(1H)-quinolinone, and 1(2H)-isoquinolinone.

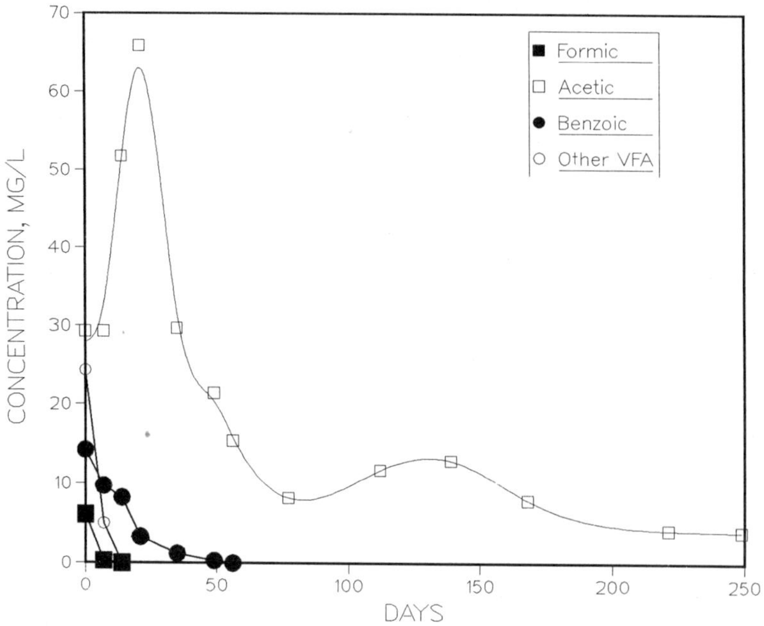

Figure 3. Concentration of organic acids in microcosms during methanogenesis of contaminated ground water from site 3.

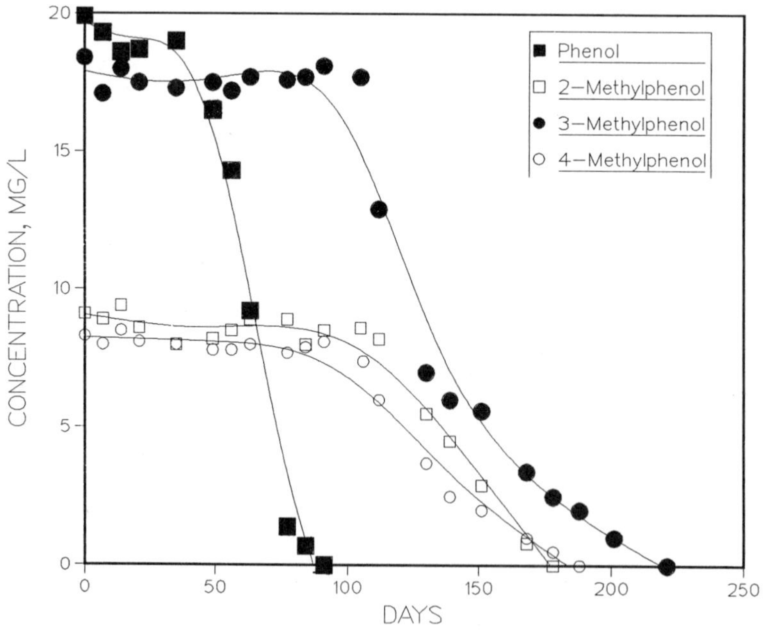

Figure 4. Concentration of phenolic compounds in microcosms during methanogenesis of contaminated ground water from site 3.

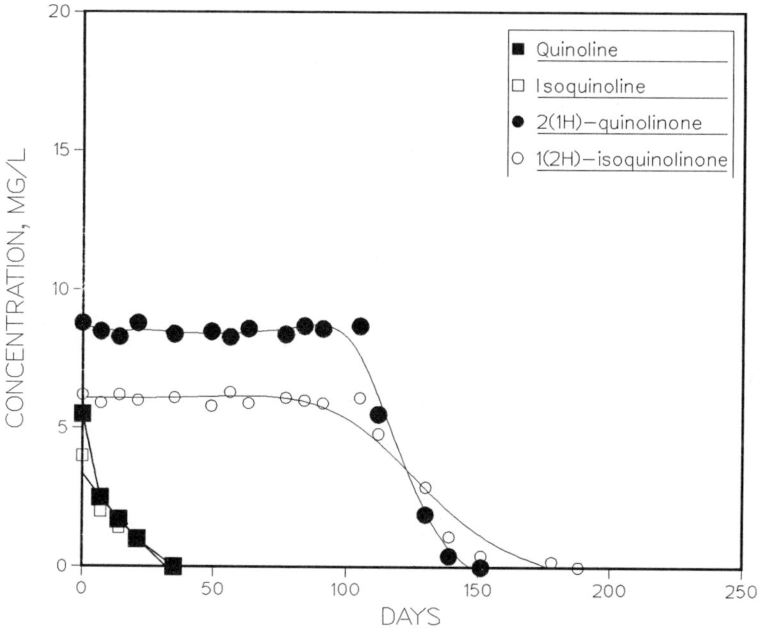

Figure 5. Concentration of nitrogen-conntaining heterocyclic compounds in microcosms during methanogenesis of contaminated ground water from site 3.

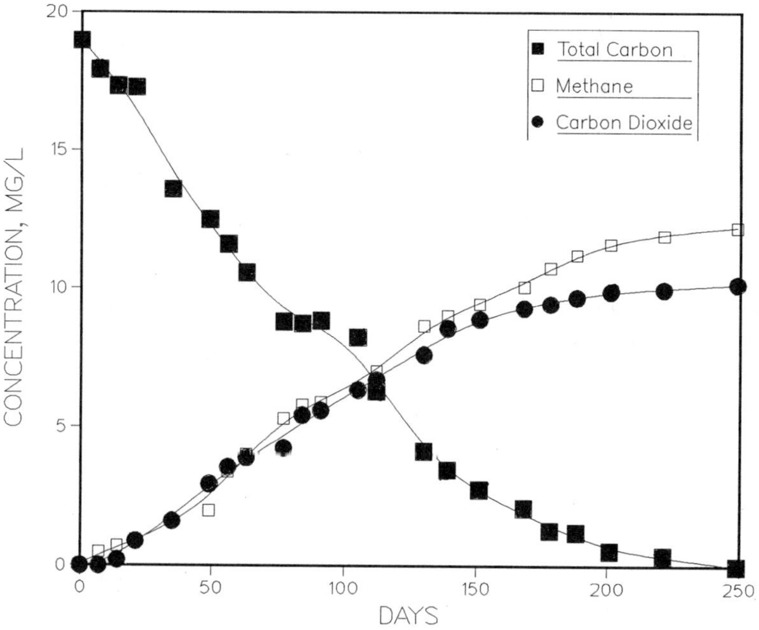

Figure 6. Carbon mass balance. All identified organic compounds in the contaminated ground water at site 3 were converted to mM carbon to facilitate the comparison to the evolved gases.

Acetic acid appears to be the major intermediate compound in the conversion of the biodegradable WSF compounds (Figure 3). This is very similar to conventional anaerobic domestic sewage sludge digestors where acetic acid is the major intermediate compound in the conversion of complex substrates to CH_4 and CO_2. The volatile fatty acids are rapidly converted to acetate in the microcosm along with those degradable compounds in step 1. An increase in acetate is also observed during each of the other steps of the sequential degradation.

A mass balance was performed on the anaerobic microcosm by first converting all identified components in the water to mmol of carbon and comparing this value with the mmol of CH_4 and CO_2 produced during the degradation. As shown in Figure 6, more CH_4 and CO_2 was produced than could be accounted for by the methanogenesis of the identified carbon compounds. The additional carbon apparently comes from the microbial degradation of organic compounds on the aquifer material. This observation was corroborated by volatile solids determination on the aquifer material which showed that the concentration of organic matter (6.6 mg/g dry weight) was several orders of magnitude higher than could be accounted for by the bacterial biomass alone. The nature of this organic matter is unknown at this time.

MODELING SUBSTRATE UTILIZATION

Background

The growth of a microbial population is a complex phenomenon composed of a number of simultaneously occurring events. Relative magnitudes of the respective rates determine the net effect upon the population. The primary events are the utilization of substrate and the concurrent growth of the organisms. These two events are closely related because it is only through the substrate utilization that energy and carbon are made available for cell growth. The cells must also use energy for maintenance and if no exogenous energy source is present, maintenance energy will be provided by energy reserves and the cell mass in the culture will decline due to death of the cells. From the macroscopic point of view, this leads to a decrease in the total mass of the culture.

Bacteria divide by binary fission and consequently the number of viable cells will increase in an exponential fashion. The reaction rate for bacterial growth can be expressed as a first-order equation:

$$\frac{\partial X}{\partial t} = \mu X \tag{7}$$

where: X = concentration of bacteria at time t (M/L^3)
μ = growth rate (T^{-1})

The growth yield, Y, is defined as the ratio of the rate of cell growth to the rate of substrate removal. This can be expressed mathematically as:

$$-\frac{\partial S}{\partial t} = \frac{\mu}{Y} X \tag{8}$$

Consequently it can be seen that the rate of substrate removal is first-order with respect to the concentration of viable cells.

The effect of the concentration of growth-limiting nutrient upon the growth rate (μ) is a function of the initial substrate concentration and increases as the substrate concentration is increased until it approaches a maximum value (μm).

The question of the best mathematical formula to express the relationship of the growth rate and substrate concentration has been a subject of much debate. No one yet knows enough about the mechanisms of microbial growth to propose a mechanistic equation which will characterize growth exactly. Instead, experimenters have observed the effects of various factors upon growth and have attempted to fit empirical mathematical equations to their observations. The equation with historical precedence and greatest acceptance is the one proposed by Monod (1949):

$$\mu = \frac{\mu m\ S}{Ks + S} \tag{9}$$

where: Ks = Monod saturation constant (M/L^3)

Ks determines how rapidly μ approaches μm and is defined as the substrate concentration at which μ is equal to half of μm.

The above relationships can be combined into an expression that relates substrate utilization to bacterial growth:

$$-\frac{\partial S}{\partial t} = \frac{\mu m\ X\ S}{Y(Ks + S)} \tag{10}$$

The Monod equation was developed from experiments using pure cultures of bacteria growing on single organic compounds. When growth of bacteria in environmental situations is considered, many complicating factors enter the picture. Two very important factors are: (1) the natural environment does not contain single organic compounds and (2) the natural environment contains complex bacterial communities rather than single species. These communities are usually in a continuous state of flux with constant changes in relative magnitudes of the species present. This can have a drastic effect upon observed kinetics so that the growth "constants" are seldom constant (Kompala et al., 1986).

Many investigators have studied the relationship between μ and S in mixed populations in order to ascertain whether it can be represented by the Monod equation (e.g. Lawrence and McCarty, 1970). It has generally been concluded that the Monod equation represents a reasonable model with which to describe a range of kinetic values describing bacterial growth in natural environments, and it is widely used.

On arithmetic axes, substrate-depletion curves may be sigmoidal (Monod), concave-up (first-order and Monod-no growth), concave down (logarithmic), or linear (zero-order) with time. In a natural system, the shape of the biodegradation curve is probably influenced by such factors as temperature, pH, dissolved oxygen, Eh, salinity, nutrients, toxicity, and the concentrations of microorganisms and compounds. The effects or interactions of such potentially important factors can make it difficult to predict the biodegradation kinetics of a particular substrate.

Substrate utilization by microorganisms in laboratory microcosms filled with the contaminated aquifer material generally can be treated in the same manner as for dispersed growth since all bacteria are exposed to the same concentrations of organic compounds that are present in the bulk liquid. This can be done with only the variables of S and X in one of the following expressions (Simkins and Alexander, 1984):

A. Monod Kinetics

$$-\frac{\partial S}{\partial t} = \frac{\mu m\ X\ S}{Y(Ks + S)}$$

B. Monod-No Growth

$$-\frac{\partial S}{\partial t} = \frac{\mu m\ Xo\ S}{Ks + S}$$

C. Zero Order in S

$$-\frac{\partial S}{\partial t} = \mu m\ Xo$$

D. First Order in S

$$-\frac{\partial S}{\partial t} = \frac{\mu m\ Xo}{Ks} S$$

E. Second Order in S

$$- \frac{\partial S}{\partial t} = \frac{\mu m}{Ks} S X$$

F. Logarithmic

$$- \frac{\partial S}{\partial t} = \mu m X$$

The preceding mathematical expressions generally concern a single population of microorganisms growing on a readily degradable substrate. Under these conditions, the Monod equation generally can describe substrate utilization by bacteria. These equations also can apply to such complex mixtures of microorganisms and substrates as domestic sewage but commonly do not apply to complex mixtures of difficult to degrade, slightly toxic compounds such as those in the WSF of creosote.

The other equations presented are simplifications of the Monod equation that can be used under certain specified conditions. Because the relationships of Ks to So and of Xo to So are not known, the other equations were fit to the data. The zero-order equation can be used when Xo >> So and So >> Ks; the first-order approximation can be used when Xo >> So and So << Ks; the second-order equation can be used when So << Ks; the logarithmic equation can be used when So >> Ks; and the Monod-no growth equation can be used when Xo >> So.

Figure 7 shows the best mathematical fits of the substrate depletion models to the experimental data for the degradation of organic carbon present in the WSF. Not shown are the linear zero-order equation and the Monod equation, which could not be fit to this data using realistic kinetic constants. The second-order curve (not shown) coincided closely with the logarithmic curve and is omitted for clarity.

DISCUSSION AND CONCLUSIONS

The transport processes of concern in ground water contamination include advection, dispersion, adsorption, and biodegradation. Before one can begin to asses the importance and rates of biodegradation during transport, the physical processes occurring during transport must be determined.

The Peclet Number is a dimensionless group of numbers that measures the dispersion tendency or the ratio of the rate of transport by advection to the rate of transport by axial dispersion ($Pe = vL/D$) where v is the interstitial velocity (q/ϵ). Ogata and Banks (1961) show that for $Pe > 100$, dispersion can be practically neglected, whereas for $Pe < 5$ the flow regime approaches complete mixing. The Pe for these experiments (138) shows that there is very little dispersion due to the experimental apparatus and virtually no adsorption of organic compounds onto the experimental apparatus. The low value for the Re (0.003), a dimensionless group relating the ratio of inertial forces to viscous forces in the solvent ($Re = vd/\nu$) where d is the average diameter of the porous medium and ν is the kinematic viscosity of the solvent, demonstrates that the flow through column is laminar.

Modeling the effects of adsorption on solute transport assumes that the solute and the aquifer material react in an instant equilibrium, i.e., no kinetic effects, that the ratio of the adsorbed solute to the solute dissolved in water is constant, i.e., linear isotherm, and that adsorption and desorption is a reversible process.

The retardation factors (R_t) obtained in this study (Table 2) correlate with the octanol:water partition coefficient (K_{ow}) values obtained from the literature - the greater the K_{ow} value, the greater the R_t. The exceptions were 2(1H)-quinolinone and 1(2H)-isoquinolinone, which were retained to a greater extent than the K_{ow} value would predict. Zachara et al. (1986) demonstrated that for ionic heterocyclic compounds, sorption may be controlled by pH, the nature and capacity of the exchange complex, and the ground

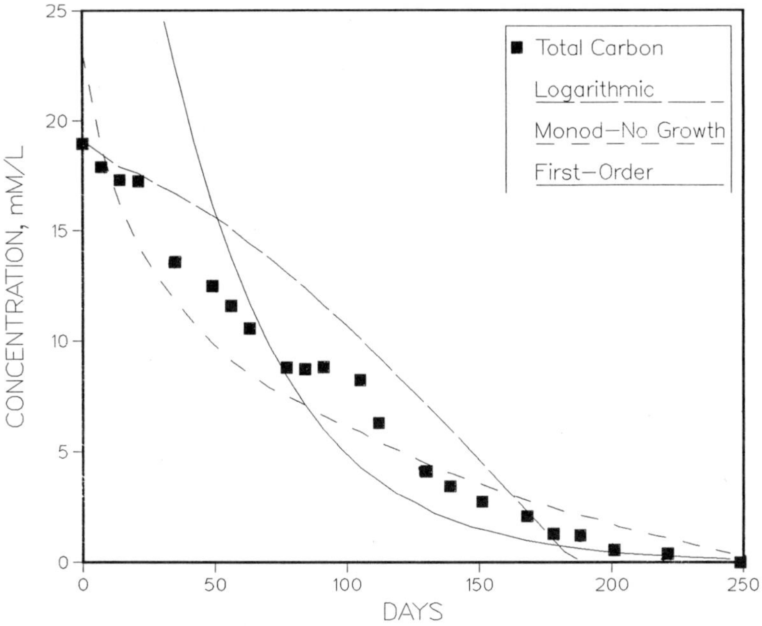

Figure 7. Comparison of best-fit mathematical models to calculated carbon substrate-disappearance curve.

water ion composition. These possibilities are presently under investigation.

It is apparent that the WSF compounds are only slightly adsorbed to the aquifer material, and that the observed selective attenuation of these compounds during downgradient movement cannot be solely attributed to sorption.

Evidence for the anaerobic degradation of aromatic compounds in general abound in the literature, but very few studies have been performed on the environmental fate of these compounds in the subsurface. Even fewer studies have been done that integrate the field analyses with laboratory studies that are carried out under conditions that closely simulate the field conditions.

Aquifer materials typically have a large surface area to which bacteria can attach, which results in high concentrations of attached bacteria relative to the concentration of free cells in the surrounding pore water. In the contaminated aquifer at the wood-preserving plant, >90%/unit volume of the total bacteria, or ~5.0×10^5 bacteria/g dry weight, were found to be attached to the sediment. The concentration of bacteria is a function of the substrate available to support their growth. Because the plant was in operation for nearly 80 years, the organisms surrounding the aquifer material have reached an equilibrium state and can be most likely described by a steady-state approach, whereby microbial growth equals the death rate of the microorganisms.

Inorganic analyses of water samples from the contaminated area clearly demonstrate that anoxic conditions prevail. The contaminated ground water is devoid of dissolved O_2, is approximately 60-70% saturated with respect to

CH_4, and contains a relatively high concentration of H_2S. Sufficient nitrogen and phosphorus are present to allow for microbial degradation of susceptible organic compounds.

The degradation of the major components in the WSF in laboratory microcosms revealed that there is a three step-degradation. The first step consists of the degradation of quinoline, isoquinoline, benzoic acid, and conversion of C_3-C_6 volatile fatty acids to acetate. Figure 6 shows that methanogenesis starts concomitantly with the degradation of the WSF compounds. The second step is the degradation of phenol followed by the simultaneous degradation of 2-, 3-, 4-methylphenol, 2(1H)-quinolinone, and 1(2H)-isoquinolinone. The mass balance of the identified compounds with the CH_4 and CO_2 produced showed that the total gas produced was 107% of the identified carbon after 250 days. This larger value undoubtedly arises from the unidentified organics on the contaminated aquifer material that are slowly released into the aqueous phase. Detailed analyses of these materials is presently underway.

It is interesting to note that the degradation of the compounds in the WSF takes place at a pH that is less than 6. Domestic anaerobic sewage sludge digestors operate optimally within a pH range of 6.6 to 7.6. Below a pH of 6.6, reactor failure often occurs presumably due to the increased concentration of undissociated fatty acids (ten Brummeler et al., 1985). However, Willams and Crawford (1984) have reported active methanogenesis in peat bogs where the pH ranged from 3.8 to 4.3. They noted that methane production in this system resulted from the reduction of CO_2 and not from acetate. Methanogenesis at the research site may continue at this low pH because of the absence of fatty acids other than acetate and traces of formate once fermentation is underway.

The ground water velocity in the area of the research site has been determined to range from 0.3 to 1.2 m/day. Flow velocities at the 6.1 m sampling depth have been determined to be on the order of 1 m/day. This value allows for easy comparison of laboratory and field data - distance travelled downgradient approximately equals the residence time in the microcosms.

Figures 8, 9, 10, and 11 show the concentrations of the organic acids, phenolic compounds, NSO, and PAH compounds, respectively, at the downgradient sites relative to their concentration at site 3 and normalized to the conservative tracer. Figure 8 shows that acetic acid increases in concentration during downgradient movement in the aquifer. This observation is consistent with the results from laboratory microcosms and with the concept that acetic acid is a major intermediate in methanogenesis of complex organics.

The PAH compounds (Figure 11) were not monitored in the laboratory microcosms because of the propensity for these compounds to be adsorbed onto the rubber stoppers of the microcosm. It is interesting to note that in the ground-water system, indene disappears faster than can be accounted for by dilution and dispersion alone. This observation is presently under investigation in microcosms modified for the study of the PAH's.

The approximate downgradient distances of the contaminated sites from the source ponds are as follows: site 3, 6 m; site 39, 53 m; WP 34, 99 m; and site 4, 122 m. Comparison of the composition of the major WSF compounds in the ground water samples to the composition in microcosm after a comparable time of residence facilitates recognition of the disappearance pattern observed in the ground water samples. During the first 50 days of residence in the microcosms, or travel from site 4 to site 39, the C_3-C_6 volatile fatty acids are rapidly converted to acetic acid and ultimately to

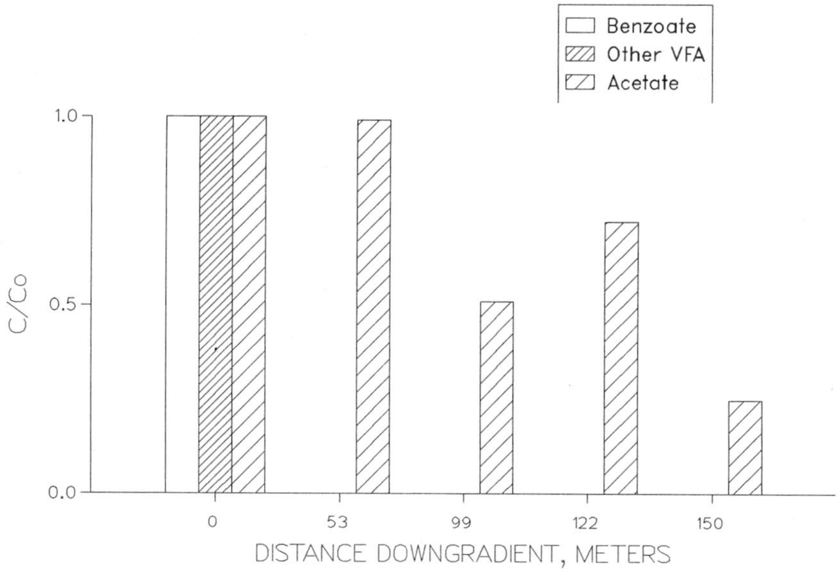

Figure 8. Concentration of organic acids relative to the concentration at site 3 normalized to the concentration of the conservative tracer.

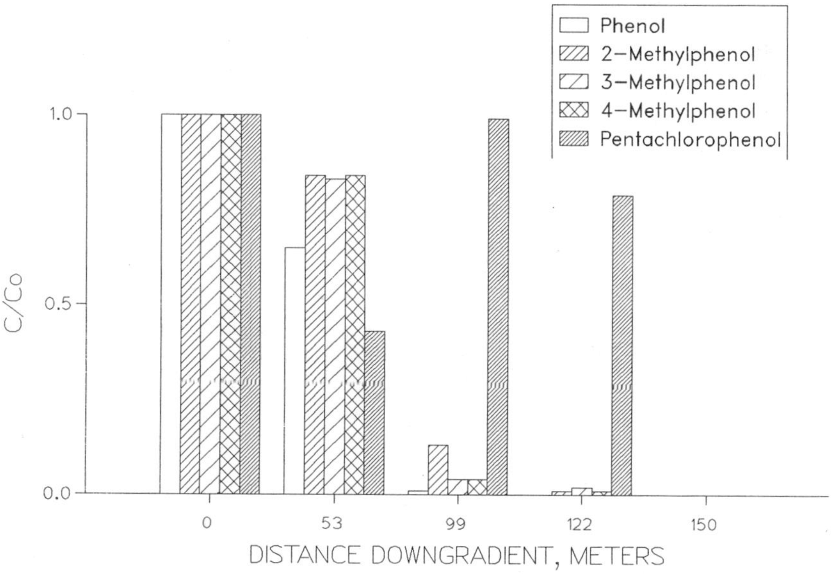

Figure 9. Concentration of phenolic compounds relative to the concentration at site 3 normalized to the concentration of the consevative tracer.

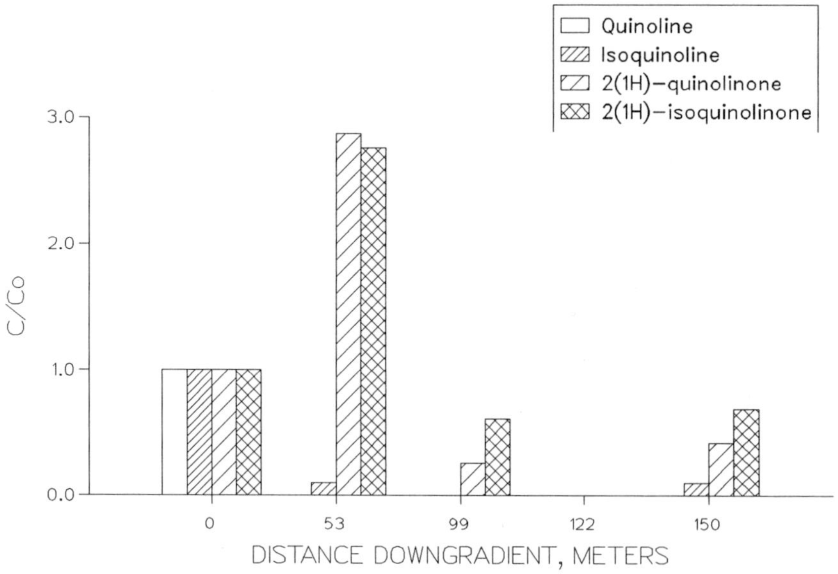

Figure 10. Concentration of nitrogen-containing heterocyclic compounds relative to the concentration of the conservative tracer.

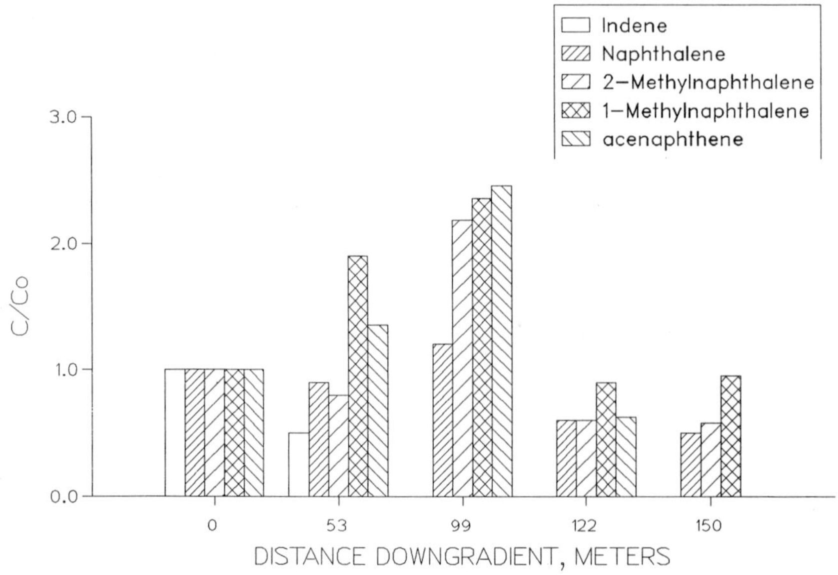

Figure 11.Concentration of polynuclear aromatic compounds relative to the concentration at site 3 normalized to the concentration of the conservative tracer.

CH_4 and CO_2. Benzoic acid, quinoline, and isoquinoline are also biodegraded. Phenol degradation occurs between days 50 and 99 in the microcosms and phenol also disappears from the ground water during transit from site 39 to WP 34. After 100 days in the microcosms, 2-, 3-, 4-methylphenol, 2(1H)-quinolinone, and 1(2H)-isoquinolinone are biodegraded and are removed from the system after about 200 days. A similar pattern of disappearance is observed during travel from WP 34 to site 4 for 2-, 3-, and 4-methylphenol; however, the biodegradation of 1(2H)-isoquinolinone is delayed somewhat compared to the microcosms.

2(1H)-quinolinone has been shown to be an intermediate in the aerobic degradation of quinoline (Bennett et al., 1985) and along with 1(2H)-isoquinolinone, have been shown to be anaerobic transformation products of their respective reduced counterparts (Pereira et al., 1987). Since quinoline and isoquinoline are readily biodegraded earlier in the microcosms and closer to the source in the ground water, it is expected that the relative concentration of the oxidized compounds would increase until they undergo biodegradation themselves.

The substrate disappearance curves generated from best-fit models of the total carbon biodegradation data indicate that these equations do not adequately describe the anaerobic microbial utilization of this complex multi-component substrate. As ground-water solute-transport models become more sophisticated and better capable of describing the movement of organics, the models that incorporate bacterial degradation may require specific data for each compound or, at least, for groups of related compounds.

Integration of both laboratory results and field observations was required to determine the ultimate fate of the subsurface contaminants at ACW. Results of this study demonstrate that the disproportionate decrease of selected organic compounds observed during downgradient movement in the aquifer could be attributed to microbial degradation of selected compounds. The anoxic conditions in the contaminated area, high concentrations of CH_4, the increased numbers of methanogenic bacteria, and lack of significant adsorption, suggest that methanogenic fermentation was the predominant microbiological process. This observation was confirmed in the laboratory microcosms.

REFERENCES

American Public Health Association, 1985, "Standard methods for the examination of water and wastewater," Sixteenth Edition.

Balba, M.T., and Evans, W.C., 1980, The anaerobic dissimilation of benzoate by Pseudomonas aeruginosa coupled with Desulfovibrio valgaris, with sulphate as a terminal electron acceptor, Biochem. Soc. Trans., 8: 624.

Bennett, J.L., Updegraff, D.M., Pereira, W.E. and Rostad, C.E., 1985, Isolation and identification of four species of quinoline-degrading pseudomonads from a creosote-contaminated site at Pensacola, Florida, Microbios Letters, 29:147.

Berry, D.F., Madsen, E.L., and Bollag, J.M., 1987, Conversion of indole to oxindole under methanogenic conditions, Appl. Environ. Microbiol., 53: 180.

Berry, D.F., Francis, A.J., and Bollag, J.M., 1987a, Microbial metabolism of homocyclic and heterocyclic aromatic compounds under anaerobic conditions, Microbiol. Rev., 51:43.

Boyd, S.A., and Shelton, D.R., 1984, Anaerobic biodegradation of chlorophenols in fresh and acclimated sludge, Appl. Environ. Microbiol., 47:272.

Boyd, S.A., Shelton, D.R., Berry, D., and Tiedje, J.M., 1983, Anaerobic biodegradation of phenolic compounds in digested sludge, Appl. Environ. Microbiol., 46:50.

Brock, T.D., and O'Dea, K., 1977, Amorphous ferrous sulfide as a reducing agent for culture of anaerobes, Appl. Environ. Microbiol., 33:254.

Chmielowski, J., Grossman, A., and Labuzek, S., 1965, Biochemical degradation of some phenols during methane fermentation, Zesz. nauk. Politech. Slask., Inz. Sanit., 8:97.

Ehrlich, G.E., Goerlitz, D.F., Bourell, J.H., Eisen, G.V., and Godsy, E.M., 1981, Liquid chromatographic procedure for fermentation product analysis in the identification of anaerobic bacteria, Appl. Environ. Microbiol., 42:878.

Evans, W.C., 1969, Microbial transformation of aromatic compounds, in: "Fermentation Advances," D. Perlman, ed. ,Academic Press, New York.

Godsy, E.M., and Goerlitz, D.F., 1986, Anaerobic microbial transformations of phenolic and other selected compounds in contaminated ground water at a creosote works, Pensacola, Florida, in: "Movement and Fate of Creosote Waste in Ground Water, Pensacola, Florida: U.S. Geological Survey Toxic Waste--Ground-Water Contamination Program, U.S. Geological Survey Water-Supply Paper 2285," H.C. Mattraw, Jr., and Franks, B.J., eds., Tallahassee.

Godsy, E.M., Goerlitz, D.F., and Ehrlich, G.G., 1986, Effects of pentachlorophenol on methanogenic fermentation of phenol, Bull. Environ. Contam. Toxicol., 36:271.

Godsy, E.M., and Grbić-Galić, D., 1988, Biodegradation parhways for benzothiophene in methanogenic microcosms, in: "U.S. Geological Survey Toxic Waste--Ground-Water Contamination Program," U.S. Geological Survey Open-File Report (in press).

Goerlitz, D.F, 1984, A column technique for determining sorption of organic solutes on the lithologic structure of aquifers, Bull. Environ. Contam. Toxicol., 32:37.

Goerlitz, D.F., Troutman, D.E., Godsy, E.M., and Franks, B.J., 1985, Migration of wood-preserving chemicals in contaminated groundwater in a sand aquifer at Pensacola, Florida, Environ. Sci. Technol., 19:955.

Grbić-Galić, D. and Young, L.Y., 1985, Methane fermentation of ferulate and benzoate: anaerobic degradation pathways, Appl. Environ. Microbiol., 50:292.

Hobbie, J.E., Daley, R.J., and Jasper, S., 1977, Use of nuclepore filters for counting bacteria by fluorescence microscopy, Appl. Environ. Microbiol., 33:1225.

Kompala, D.S., Ramkrishna, D., Jansen, N.B., and Tsao, G.T., 1986, Investigation of bacterial growth on mixed substrates: experimental evaluation of cybernetic models, Biotechnol. Bioeng., 28:1044.

Lawrence, A.W., and McCarty, P.L., 1970, Unified basis for biological treatment design and operation, J. Sanitary Engrg. Div. ASCE., 96:775.

Maka, A., McKinley, V.L., Conrad, J.R., and Fannin, K.F., 1987, Degradation of benzothiophene and dibenzothiophene under anaerobic conditions by mixed cultures, Absts. Annu. Meet. Amer. Soc. Microbiol., O 54:194.

May, W.E., Wasik, S.P., and Freeman, D.H., 1978, Determination of the aqueous solubility of polynuclear aromatic hydrocarbons by a coupled column liquid chromatographic technique, Anal. Chem., 50:175.

Mihelcic, J.R., and Luthy, R.G., 1988, Microbial degradation of acenaphthene and naphthalene under denitrification conditions in soil-water systems, Appl. Environ. Microbiol., 54:1188.

Monod, J., 1949, The growth of bacterial cultures, Ann. Review Microbiol., 3:371.

Novotny, M., Strand, J.W., Smith, S.L., Wiesler, D., and Schwende, F.J., 1981, Compositional studies of coal tar by capillary gas chromatography-mass spectrometry, Fuel, 60:213.

Pereira, W.C., Rostad, C.E., Updegraff, D.M., and Bennett, J.L., 1987, Fate and movement of azaarenes and their anaerobic biotransformation products in an aquifer contaminated by wood-treatment chemicals. J. Environ. Tox. Chem., 6:163.

Ogata, A., and Banks, R.B., 1961, A solution of the differential equation of longitudinal dispersion in porous media. "U.S. Geological Survey Professional Paper 411-A," Washington, D.C.

Roberts, P.V., Schreiner, J., and Hopkins, G.D., 1982, Field study of organic water quality changes during groundwater recharge in the Palo Alto Baylands, Water Res., 16:1025.

Schalf, M.R., McNabb, J.F., Dunlap, W.J., Crosby, R.L., and Fryberger, J.S., 1981, "Manual of ground-water sampling procedures," NWWA/EPA Series, National Water Well Association, Worthington, Ohio.

Schwarzenbach, R.P., and Westall, J., 1981, Transport of nonpolar organic pollutants from surface water to groundwater: laboratory sorption studies, Environ. Sci. Technol., 15:1350.

Simkins, S., and Alexander, M., 1984, Models for mineralization kinetics with the variables of substrate concentration and population density, Appl. Environ. Microbial., 47:1299.

Syewzyk, U., Syewzyk, R., and Schink, B., 1985, Methanogenic degradation of hydroquinone and catechol via reductive dehydroxylation to phenol. FEMS Microbiol. Ecol., 31:79.

Sudicky, E.A., Cherry, J.A., and Frind, E.O., 1983, Migration of contaminants in groundwater at a landfill: a case study, 4. A natural-gradient dispersion test, J. Hydrol., 63:81.

Tarvin, D., and Buswell, A.M., 1934, The methane fermentation of organic acids and carbohydrates, J. Amer. Chem. Soc., 56:1751.

ten Brummeler, E., Huslshoff Pol, L.H., Dolfing, J., Lettinga, G., and Zehnder, A.J.B., 1985, Methanogenesis in an upflow anaerobic sludge blanket reactor at pH 6 on an acetate-propionate mixture, Appl. Environ. Microbiol., 49:1472.

Vogel, T.M., and Grbić-Galić, D., 1986, Incorporation of oxygen from water into toluene and benzene during anaerobic fermentative transformation, Appl. Environ. Microbiol., 52:200.

von Rumker, R., Lawless, E.W., and Meiners, A.F., 1975, Production, distribution, use and environmental impact potential of selected pesticides, U.S. Environmental Protection Agency, EPA 540/1-74-001.

Wang, Y., Suidan, M.T., and Peffer, J.T., 1984, Anaerobic biodegradation of indole to methane, Appl. Environ. Microbiol., 48:1058.

Westall, J.C., Leuenberger, C., and Schwarzenbach, R.P., 1985, Influence of pH and ionic strength on the aqueous-nonaqueous distribution of chlorinated phenols, Environ. Sci. Technol., 19:193.

Williams, R.T., and Crawford, R.L., 1984, Methane production in Minnesota peatlands, Appl. Environ. Microbiol., 47:1266.

Young, L.Y., and Rivera, M.D., 1985, Methanogenic degradation of four phenolic compounds, Water Res., 19:1325.

Zachara, J.M., Ainsworth, C.C., Felice, L.J., and Resch, C.T., 1986, Quinoline sorption to subsurface materials: role of pH and retention of the organic cation, Environ. Sci. Technol., 20:620.

SUBSURFACE PROCESSES OF NONAQUEOUS PHASE CONTAMINANTS

Danny D. Reible* and Tissa H. Illangasekare**

* Department of Chemical Egr.
Louisiana State University
Baton Rouge, LA 70803

** Department of Civil Egr.
University of Colorado
Boulder, CO 80309

ABSTRACT

The processes affecting the fate and transport of nonaqueous phase liquids in the subsurface are reviewed. The processes examined include bulk phase infiltration through the vadose zone, the mobility of a discontinuous nonaqueous phase residual and partitioning between these phases and the subsurface water. Experiments on the infiltration of nonwetting immiscible hydrocarbon phases in water-wetted sands and sandy soils suggest that the residual water excludes the nonaqueous phase from the fine pores of the medium and reduces capillary suction. As a result, a sharp front model of nonaqueous phase infiltration is shown to adequately predict infiltration rates using the water conductivity in sand and the correlation of Brooks and Corey (1964) to predict the phase conductivity and capillary rise heights to parameterize the effective capillary suction. Estimates of the rate of mass transfer between the infiltrating nonaqueous phase and the residual soil water suggest that local equilibrium between phases is approximately satisfied for hydrophilic compounds. Estimates of the size of discontinuous residual ganglia of nonaqueous phase liquid indicated, however, that significant mass transfer resistances exist between the nonaqueous residual and infiltrating recharge water. A preliminary model of the approach to equilibrium between the nonaqueous phase ganglia and recharge water is postulated.

INTRODUCTION

Subsurface contamination by organic and other pollutants pose a serious risk to the quality of ground-water supplies. Because of the concern for ground-water quality, analysis of the problem has often been limited to dilute aqueous phase contamination. Over the past several years, however, there has been increasing recognition that subsurface contamination by the generally low solubility organics is often in the form of a concentrated, separate phase. And as pointed out by Reible (1987) and Hunt et al. (1988), it is unlikely that effective and efficient remediation schemes to remove the ground-water contamination risk can be implemented without direct consideration of the nonaqueous phase.

Pumping and treatment of ground water, for example, is a common method of treating subsurface contamination. In this remediation technology, contaminated ground water is removed from the subsurface by pumping and treated above ground before being pumped back below the surface. If undissolved contaminant exists in the subsurface, however, the separated or nonaqueous phase material can act as an essentially continuous source of pollutants to ground water. Consider, for example, a gasoline spill that contains 1% of the soluble organic benzene. A minimum of approximately 350 L of water must be for every L of gasoline spilled and if remediation is delayed until the average water tion of benzene is 1 ppm due to convection and dispersion away from the spill site, of water that must be removed to volume of the initial gasoline spilled rises to [illegible],000. These ratios can rise by an additional several orders of magnitude for less soluble organics.

Thus treatment of the aqueous phase is unlikely to provide a timely or cost effective solution to subsurface contamination if a separated phase is present. Direct analysis of the nonaqueous phase is required to define the extent and magnitude of contamination and to design effective remediation schemes. Unfortunately this analysis is hindered by the sampling protocol for subsurface contamination, which often does not differentiate between nonaqueous and aqueous phase contamination. As will be indicated in this paper, not only is differentiation between aqueous and nonaqueous phase contamination required, but the physical form of any separated phase contamination must be identified to adequately design remediation schemes.

The purpose of this paper is to review the subsurface processes that affect nonaqueous phase contamination. The processes and parameters that define the dynamics of a continuous bulk phase, a discontinuous residual phase and partitioning constituents from these phases will be identified. The emphasis will be on organic contaminants lighter than water and in the unsaturated or vadose zone where such contaminants can normally be found as a separated phase. In addition to reviewing the processes and identifying the key parameters, recent attempts to model these processes will be described. The discussion will focus on results generated in our laboratories which has emphasized one and two dimensional experiments in which an immiscible nonaqueous phase (oil or gasoline surrogate) is applied to laboratory sands or a field soil. The experiments are primarily designed to track the dynamics of the bulk nonaqueous phase using visualization with dyes and gamma ray attenuation to identify the dynamic spatial distribution of liquid saturation. The experiments have been conducted in both homogeneous media and more realistic layered media. Some preliminary experimentation and modeling has also been conducted on partitioning of miscible components of the nonaqueous phase. The preliminary experiments and modeling of partitioning have been designed to guide subsequent inquiry on partitioning between aqueous and nonaqueous subsurface phases. In particular, the potential for nonequilibrium partitioning limited by diffusional resistances has been investigated.

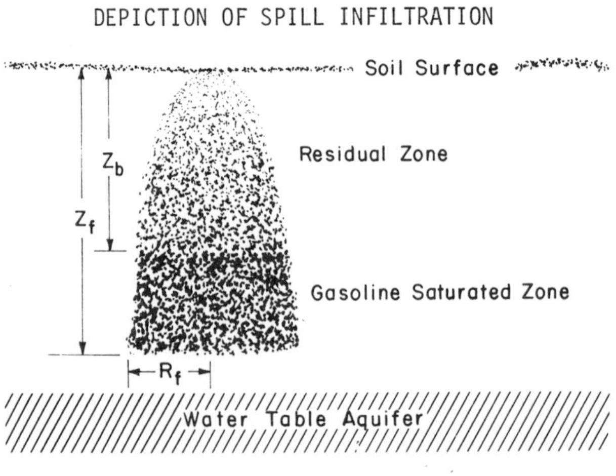

Figure 1 - Conceptual Model of Spill Infiltration

The subsurface processes that influence nonaqueous phase contaminants can best be introduced through description of a contamination scenario. The processes introduced will then be examined in more detail in the subsequent text. Consider a sparingly soluble organic phase, e.g. gasoline, that is spilled or otherwise applied to the soil surface. Let us first consider gasoline as a single component organic contaminant. The gasoline will infiltrate through the unsaturated zone primarily due to the influence of gravity and capillary forces. For a sufficiently small spill, all of the gasoline will eventually infiltrate and the upper soil layers will begin to drain. This process is depicted in Figure 1. Behind an infiltrating, essentially saturated plug of gasoline, a partially drained, residually contaminated zone will appear. The depth and lateral spread of the contaminated zone will be a function of the magnitude of the gravity and capillary forces and the properties of the gasoline and the media. If sufficient gasoline is applied to the surface, the free nonaqueous phase can penetrate to the water table and spreading of the gasoline on the capillary fringe of the water table aquifer will occur. As the aquifer is recharged with rain and surface waters much of the initial residual of gasoline remaining after the infiltration can be displaced and transported further into the vadose zone and to the water table. Water table height fluctuations can also incorporate some of the floating gasoline into the water table.

These processes become even more complicated when the gasoline is recognized to be a complex mixture of organic components. Partitioning can occur between the gasoline phase and the air, residual water and soil of the vadose zone. In the residually saturated gasoline zone that remains after passage of the gasoline infiltration front, vaporization and dissolution can be especially important. A significant air-filled porosity exists that allows transport of the vapors away from the nonaqueous liquid phase and recharge water passing through this zone can transport contaminants further into the soil. Similar partitioning processes can occur at the water table, ultimately leading to the drinking-water contamination that drives environmental concerns and regulatory action.

BULK PHASE INFILTRATION

Processes and Important Parameters

Consider first the infiltration of the bulk nonaqueous phase liquid in the vadose zone. The infiltration is controlled by the interactions of four phases, air, residual water, soil and the bulk nonaqueous phase. Air often has a negligible influence on liquid infiltration in the vadose zone and is generally not considered. The other three phases influence the infiltration by controlling the conductivity and capillary suction. Chemical properties that influence the phase mobility include the kinematic viscosity, dielectric constant and interfacial tension with water. Soil properties that are important include the heterogeneity of the medium, intrinsic permeability, residual water content, pore size and shape and clay content.

Let us examine first those factors that most influence conductivity, or the response of the system to a given head gradient. The intrinsic permeability (κ) is a function only of the geometry of the medium in an inert medium such as glass beads. The conductivity of a fluid in a medium ($K = \kappa g/\nu$, where g is the gravitational constant and ν is the fluid's kinematic viscosity), however, is dependent on the fluid properties and in an inert medium, the conductivity of the nonaqueous phase liquid (napl) is related to the water conductivity by

$$K_{napl} = K_{water} \frac{\nu_{water}}{\nu_{napl}} \tag{1}$$

Here, the subscript napl refers to the nonaqueous phase liquid. This relationship is a consequence of Darcy's Law and it assumes that the infiltrating chemical essentially fills the pore spaces. The conductivity as defined in Equation (1) gives a direct indication of the rate of infiltration of a nonaqueous phase. A saturated plug driven only by gravity (negligible capillary suction) will infiltrate at a velocity equal to this conductivity because the head gradient that appears in Darcy's Law is unity under such conditions.

When only part of the pore space is available for the flow of the chemical, the conductivity of the nonaqueous phase liquid must be modified by a relative permeability. Relative permeability is defined as the ratio of the phase permeability (or conductivity) at the actual volumetric content to the permeability (or conductivity) under saturated (i.e. pore filled) conditions. Reible et al. (1989a) has observed that the correlations for relative permeability presented by Brooks and Corey (1964) adequately predict the relative permeability of a nonaqueous phase in an initially air and residual water filled vadose zone. In terms of fluid volumetric content (θ- volume fluid per total volume of pores plus soil), this relationship can be written for the infiltrating phase (which is assumed wetting with respect to air)

$$\frac{\kappa(\theta)}{\kappa} = \kappa_r = \left[\frac{\theta - \theta_{ir}}{\varepsilon - \theta_{ir}}\right]^{(2+3b)/b} \quad (2)$$

Here, b is a grain size parameter that varies from about 2.8 in sands to about 10 in clay soils (Crosby et al., 1984), θ is the volumetric content and ε is the soil porosity. The subscript ir refers to irreducible and θ_{ir} defines the volumetric content below which no flow is observed. In sands wetted by a residual water film this may be low or zero for a nonaqueous phase liquid while in clays it may be a significant fraction of the total porosity.

The dielectric constant and clay content of the medium modify this picture in that low dielectric constant fluids such as hydrocarbons infiltrate much more quickly in clay soils than high dielectric constant fluids such as water. This has been associated with the double layer that neutralizes the excess negative charge often observed on clay particles (Schramm et al., 1985). Low dielectric constant fluids in the pore space are not as polarized by the excess surface charge as high dielectric constant and thus high electric susceptability fluids. Schramm et al. (1985) observed as much as a two orders of magnitude increase in conductivity over that predicted by Equation (1) for low dielectric constant fluids in a clay soil. In clays that are wetted by a residual water film, this effect is minimized in that much of the excess surface charge can be neutralized by the water. Green et al. (1981), for example, noted that the permeability of low dielectric constant fluids can increase compared to that of water when the clay is initially partially water-wetted. These conflicting results make it impossible to predict with any certainty the conductivity of different dielectric constant fluids in clay soils. As indicated by Nielsen et al. (1986) the thickness of the double layer and thus its importance in bulk phase infiltration can also be modified significantly by the type and concentration of ions in the pore fluid.

The residual water in the vadose zone also influences the mobility of the nonaqueous phase. If the infiltrating nonaqueous phase exhibits a low interfacial tension with water, it will tend to displace the residual water film during infiltration. The build up of a displaced water layer in front of the infiltrating layer results in modification of the phase infiltration rate and increased retention of the nonaqueous phase by the originally water-wetted soil. During infiltration of low miscibility, high water-interfacial tension fluids (automatic transmission fluid and iso-octane), Reible et al. (1989a) observed little or no displacement of the residual water and that soil particle surfaces and the pendular regions at particle contact points were largely unavailable to the nonaqueous phase fluid. The net result of the lack of residual water displacement is reduction of the pore space available for the nonaqueous phase flow and the need for a relative permeability as defined in Equation 2. Reible et al. (1989a) observed that the relative permeability of automatic transmission fluid and iso-octane infiltrating sands was between 0.29 and 0.38 while Equation 2 predicts a relative permeability of about 0.34 with θ_{ir}=0 and b=2.8 (Crosby et al., 1984).

The residual water in the pore spaces prior to infiltration of a nonaqueous phase also influences the net capillary forces acting during the infiltration. Since water is more wetting, it tends to fill the fine pores and limit the infiltrating organic to the larger pore spaces, thus effectively reducing the capillary suction. This can be seen in Figure 2, in which the capillary rise of isooctane is shown by volumetric content (as measured by a gamma ray attenuation system) as a function of height above a saturated layer in a medium sand. Compared to the capillary rise curve in dry sand, the presence of the residual water both reduces the magnitude of the capillary suction effects (i.e. reduces the capillary rise) and eliminates the tailing associated with capillary rise into the fine pores of the medium.

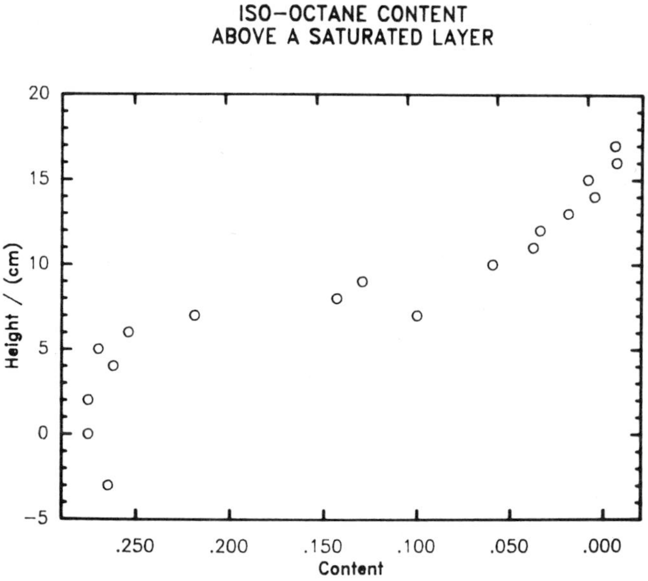

Figure 2 - Capillary Rise of i-Octane above a Saturated Layer in Sand

The capillary pressure versus saturation curve exhibits hysteresis, that is the capillary pressure along wetting and draining saturation paths differ. Reible et al. (1989a) observed a capillary rise height of 5 cm of isooctane in a sand residually saturated with respect to water but a height of about 8 cm in the sand residually saturated with isooctane. Due to the exclusion of the fine pores of the medium by residual water, both wetting and draining capillary rise curves were sharp. The sharpness of the capillary rise or capillary pressure versus content curves makes it easy to define an air entry pressure in the system which quantifies the magnitude of capillary suction effects.

Because the important fluid and soil properties outlined above differ in different media, the most important factor controlling the mobility of a bulk nonaqueous phase is medium heterogeneity. As with water infiltration, a chemical phase will tend to collect on low permeability layers. High permeability zones such as fractures or macropores in the subsurface provide effective conduits for nonaqueous phase motion. Even if a soil is laterally homogeneous, it tends to be layered vertically with zones of high and low permeability material. The effect of such layers on the subsurface flow is especially complicated in the vadose zone. In Figure 1, for example, gasoline is depicted as infiltrating as an essentially saturated plug. In layered media, an essentially saturated gasoline front would probably exist in a low permeability layer and in a higher permeability layer above. A higher permeability layer below, however, would drain more quickly than the low permeability layer could provide infiltrating gasoline, resulting in complex saturation profiles that influence the relative permeability and the capillary suction. There have been very few attempts to address this issue quantitatively in models of nonaqueous phase subsurface movement. The complex multiphase flow parameter functions (e.g. relative permeability and capillary pressure versus saturation) make dealing with realistic soil heterogeneities in conventional models of nonaqueous phase flow a difficult task.

Modeling Bulk Phase Infiltration

There exists no predictive model of nonaqueous phase infiltration that incorporates all of the complexities discussed qualitatively above. Typically, existing models focus on bulk phase movement in an essentially homogeneous, incompressible media without soil-liquid interactions such as the double layer phenomena.

The movement of each phase in a porous medium is modeled by multiplying the gradient in total potential by an appropriate conductivity to define fluid flux. The conductivities are functions of the fluid and media properties as described above. The total potential is the sum of a pressure potential and the gravitational potential and two less understood potentials, the solute potential and the electrochemical potential (Nielsen et al., 1986). In the unsaturated zone, the pressure potential is generally negative and gives rise to the capillary suction. Each of the potentials can be converted to the familiar units of head by dividing by the gravitational constant.

In most models of either water or nonaqueous phase fluid infiltration, the solute and electrochemical potentials are assumed invariant and therefore provide no driving force for flow. Generally, the medium is assumed rigid and Darcy's law is assumed to apply to the unsaturated flow although it was originally developed for saturated flow. If the air flow is neglected, the movement in one-dimension (z) is governed by the set of partial differential equations

$$\frac{\partial \theta_{water}}{\partial t} = \frac{\partial}{\partial z}\left\{ K_{water}(\theta_{water},\theta_{napl}) \left[\frac{\partial h_{water}}{\partial z} - 1 \right] \right\}$$

$$\frac{\partial \theta_{napl}}{\partial t} = \frac{\partial}{\partial z}\left\{ K_{napl}(\theta_{napl},\theta_{water}) \left[\frac{\partial h_{napl}}{\partial z} - 1 \right] \right\} \qquad (3)$$

Here K_{water} and K_{napl} represent the relative conductivities and θ_{water} and θ_{napl} the volumetric content of the aqueous and nonaqueous phases, respectively. Generally, the θ derivative on the left hand side is written

$$\frac{\partial \theta}{\partial t} = \left(\frac{\partial \theta}{\partial h}\right)\left(\frac{\partial h}{\partial t}\right) \qquad (4)$$

where $\partial\theta/\partial h$ refers explicitly to the capillary pressure versus saturation relationships that are implied in Equation (3). Air appears only implicitly in Equation (3) in that

$$\varepsilon = \theta_{water} + \theta_{napl} + \theta_{air} \qquad (5)$$

Although the air content appears only implicitly, the three phase relative permeability and capillary pressure relationships are generally different from their two phase counterparts. Because of the difficulty in measuring three phase relative permeabilities, however, two phase data is often used to predict three phase relationships. Faust (1985) has presented a relationship for this purpose based on the work of Stone (1973). Kool and Parker (1987) and Parker et al. (1987) have extended this approach to hysteretic relative permeability and capillary pressure relationships.

Equations (3) describe general two phase infiltration in the unsaturated zone. If the nonaqueous phase contains a partitioning component, these equations must be supplemented with a mass balance equation on the component (e.g. as described in Pinder and Abriola, 1986). Let us instead attempt to simplify Equations (3) by considering only the bulk phase and examining assumptions that may reduce the complexity of the model.

Figure 3 shows the volumetric content of an infiltrating liquid as a function of depth as observed in a one-dimensional column experiment in sand. This figure displays the same behavior indicated in Figure 1, that of a sharply-defined saturated plug of infiltrating liquid preceding a residually saturated draining layer. The saturation profile was measured by gamma ray attenuation measurements during the infiltration of water, automatic transmission fluid, isooctane and gasoline into sand and a sandy surface soil. In every case, the shape of the saturation profile at the leading and trailing fronts remained sharp and essentially independent of time until the saturated plug began to fade as the result of mass lost to the trailing residual zone.

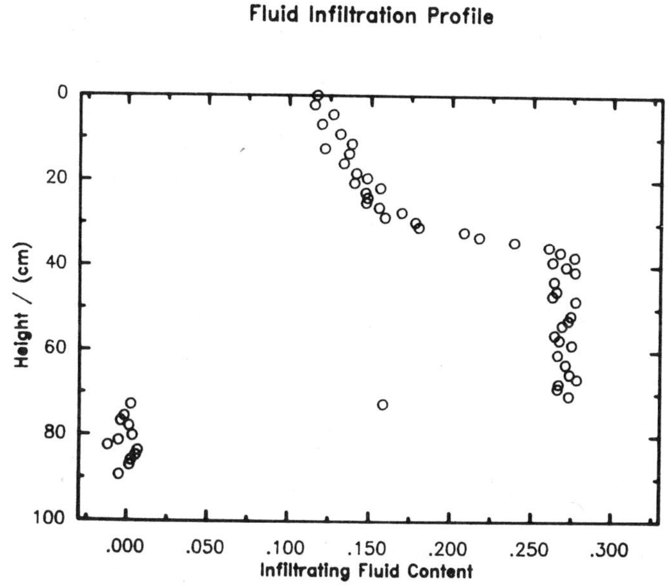

Figure 3 - Volumetric content of infiltrating liquid as a function of height

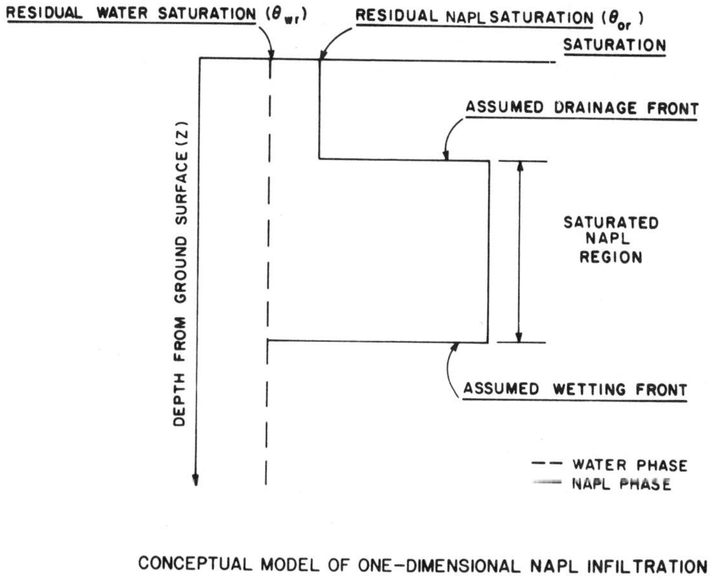

Figure 4 - Conceptual model of bulk phase infiltration

In all experiments, the sand was initially residually saturated with respect to water. As indicated previously, little or no displacement of the residual water was observed during the infiltration of the organic phase. This is especially significant in that a porous media drained to field capacity or residual saturation with respect to water is a common initial condition prior to a nonaqueous phase infiltration. Thus in many situations of interest, it may be possible to reduce the three flowing phases of air, water and chemical to the single flowing chemical phase.

This view of the infiltration process gives rise to the conceptual model shown in Figure 4. The residual water content, shown with a broken line, is unaffected by the infiltrating nonaqueous phase. The nonaqueous phase wetting front is approximated as sharp followed by an essentially saturated layer. Finally, the saturated layer is followed by a drainage front and a residual organic layer that extends to the surface. The organic content in the residual layer was about 10% (saturation of about 25%) in the sand experiments. This residual does not represent an irreducible residual in that slow drainage of the organic was observed to continue. At least half of this material can be displaced by subsequent water infiltration.

The observed sharpness of the infiltration front is consistent with a relatively narrow pore size distribution. At least part of this apparent uniformity in pore size is due to the presence of the residual water which excludes the nonwetting organic phase from the fine pores as indicated earlier. The sharpness of the wetting and drainage fronts of the nonaqueous phase liquid suggest that a reasonable model of the sandy media is a bundle of uniform diameter capillary tubes. One such tube is shown in Figure 5. Note that since the residual water is immobile, the nonaqueous phase fluid is the wetting phase (compared to the displaced air). Figure 5 also shows the variation in pressure head in a vertical capillary tube. The driving force for the infiltration by this model is the ponded depth at the surface minus the negative of the effective capillary pressure head at the wetting front of the infiltrating liquid. A small air head gradient is shown below the infiltrating front which is assumed negligible. The effective capillary suction head is approximately constant since the infiltration experiments indicated that the wetting saturation profile is approximately independent of time. By the parallel capillary model, the effective capillary suction head should be approximately equal to the capillary rise height of the nonaqueous phase fluid in the saturation environment of the infiltration. That is, the effective capillary suction at the leading front should be equal to the capillary rise height of the nonaqueous phase in a residually water saturated porous medium. The effective capillary suction at the drainage front should be the capillary rise height of the nonaqueous phase in a medium residually saturated with respect to both water and chemical.

Subject to the assumptions outlined above, the multiphase model of the nonaqueous phase infiltration given by Equation (3) can be reduced to a single phase flow equation equivalent to Richard's equation. That is, the equation governing the chemical phase reduces to

$$\frac{\partial \theta_{napl}}{\partial t} + \frac{\partial}{\partial z} K(\theta_{napl}) = \nabla^2 h_{napl} \tag{6}$$

Since the infiltration fronts are assumed sharp, the left hand side of Equation (6) is zero in the saturated plug region behind the wetting front. Thus the model can be reduced still further to a moving boundary problem with Laplace's equation satisfied behind the boundary and a boundary condition of head equal to the negative of the capillary suction (h_f) at the infiltrating front ($\mathbf{z}_f$).

$$\nabla^2 h_{napl} = 0 \qquad \text{subject to } h_{napl} = -h_f \text{ at } \mathbf{z} = \mathbf{z}_f \tag{7}$$

Although this model has been developed for application in the vadose zone, the model would also be applicable to the water-table, nonaqueous phase interactions as long as the saturation profiles between phases can be assumed constant, allowing the use of effective capillary suction heads. A numerical model based on this approach using the boundary element method (e.g. Brebbia et al., 1984) is under development. The boundary element method is especially convenient for this problem in that discretization of only the moving boundary is required.

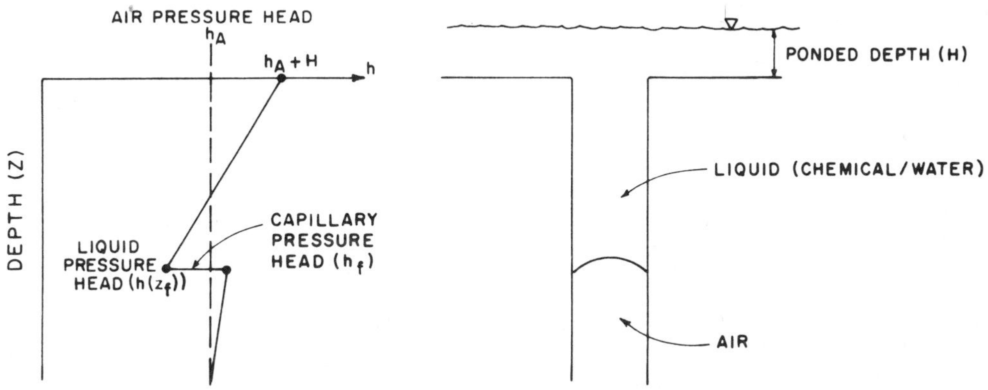

Figure 5 - Uniform Capillary Tube Model of Water-Wetted Medium

Figure's 6 and 7 display a comparison of the simplified model predictions to one-dimensional experimental infiltration data in sand with automatic transmission fluid and isooctane, respectively. In one-dimension in a homogeneous medium, the simplified model is analytical in form (Reible et al., 1989a). For these comparisons, capillary rise measurements were used to estimate capillary suction heads. The effective conductivity in the simple model was calculated from the water conductivity by Equation (1) multiplied by a relative permeability defined by Equation (2). The size grain parameter in Equation (2) was estimated from the statistics presented by Crosby et al. (1984) as approximately 2.8. The irreducible residual of the nonaqueous phase in the sand was taken to be zero, although the predictions of the model in a sandy medium are relatively insensitive to the irreducible residual level. Thus the model curves shown in Figure's 6 and 7 do not represent fits to the experimentally observed organic infiltration rate but are predictions based on measured or estimated properties of the medium and the organic liquid.

RESIDUAL NONAQUEOUS PHASE MOBILITY

After infiltration and drainage of a nonaqueous phase liquid, a residual of the nonaqueous phase liquid remains, immobilized by capillary forces. Since the nonaqueous phase liquid does not typically wet the soil medium, the residual is in the form of discontinuous ganglia. Some of the residual ganglia can be displaced by recharge water and spread deeper into the vadose zone but the mobility of the form is limited due to the strength of capillary forces and the difficulty in transmitting inertial forces through the discontinuous phase. The volume of the nonaqueous phase residual can be as much as the volume of the residual water or more. We have observed an initial residual of about 10% of the soil volume after drainage of isooctane in a variety of sands and in a sandy Louisiana surface soil. Much of this initial residual could be displaced by infiltrating recharge water, but a stable residual amounting to about 5% of the soil volume was not displaceable under a natural hydraulic gradient.

The mobility of the discontinuous ganglia has been the subject of extensive research in the petroleum industry (e.g. Ng et al., 1978; Larsen et al., 1981; Payatakes, 1982; and Chatzis et al., 1983). This literature was applied to the displacement of a residual nonaqueous

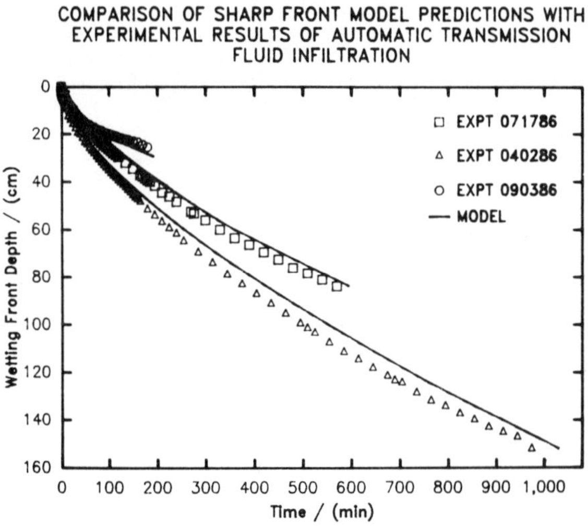

Figure 6 - Simplified infiltration model predictions compared to automatic transmission fluid data

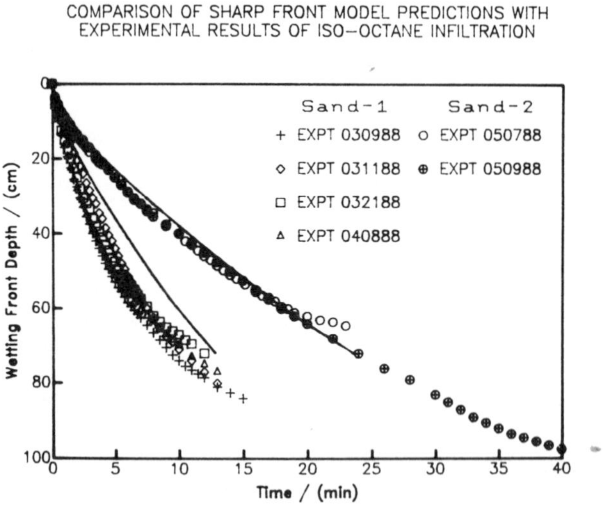

Figure 7 - Simplified infiltration model predictions compared to i-octane data

contaminant by Wilson and Conrad (1984). Recently, Hunt et al. (1988) applied this background to the subject of steam extraction of contaminants from a discontinuous residual. Despite these efforts, the prediction of the size and distribution of the ganglia that make up the discontinuous residual is semi-quantitative, at best. It is possible, however, to make a crude estimate of the size and exposed area of ganglia that may be stable to displacement by recharge water. These size and exposed area estimates will also be used below to estimate the kinetics of partitioning between the nonaqueous and the aqueous recharge phases.

The pressure difference between the nonaqueous phase liquid and another fluid in an assumed cylindrical pore is given by

$$\Delta P = \frac{2\sigma_i}{r_t} \tag{8}$$

Here, σ_i is the interfacial tension and r_t is the pore throat diameter. For a nonaqueous ganglion to be stable in the absence of the flow of a displacing fluid, this pressure must balance the hydrostatic pressure in the ganglion. For a nonaqueous phase ganglion to be stable in the presence of flow of a displacing fluid, this pressure must balance the pressure drop across the ganglion as a result of the displacing fluid flow. Considering recharge water infiltrating under the force of gravity as the potential displacing fluid, the pressure gradient across the ganglion is simply the hydrostatic pressure of the water phase surrounding the ganglion. That is, the head gradient in an infiltrating fluid phase is approximately unity if capillary suction effects are neglected. Since the density of most organics is less than that of water (with an important exception being the chlorinated organics), the maximum displacing pressure drop across a nonaqueous phase ganglion is given by the water hydrostatic pressure,

$$P_h = \rho_{water}\, g\, L \tag{9}$$

where ρ_{water} is the density of water, g the gravitational constant and L the vertical length of the ganglion. Equating these two pressures, the maximum vertical length of a stable ganglion is given by

$$L = \frac{2\sigma_i}{r_t\, \rho_{water}\, g} \tag{10}$$

This equation gives the maximum length of the ganglion. It should provide, however, a reasonable estimate of the actual length since the ganglion are formed from the draining of a zone essentially saturated with the nonaqueous phase liquid. That is, the ganglion are formed from a liquid body initially larger than the stable length that drains until the stable length is reached. In reality, a large distribution of ganglion sizes exist (Payatakes, 1982) and estimates such as those made here are normally limited to estimating the maximum size of the ganglion (e.g. Hunt et al. 1988). Note that the larger ganglion contain the largest volume of residual fluid and may therefore be the most important with respect to defining the risk associated with the contamination.

Let us assume that all or most of the volume of nonaqueous phase residual is to be found in ganglion of a size described above, that the porous medium is adequately modeled by a square array of spherical soil particles and that the nonaqueous liquid will drain until an interface at the maximum pore diameter is stable. This model is depicted in Figure 8.
The maximum hydraulic radius of a pore in a square array of spheres is located at the contact point of two adjacent square arrays and is given by

$$r_t \approx \frac{2\ \text{Area}}{\text{Wetted Perimeter}} = \frac{2\, d_p^2}{4\, d_p} = \frac{d_p}{2} \tag{11}$$

For an example sand grain diameter of 0.3 mm, this suggests a maximum effective pore throat radius of 0.15 mm. A typical immiscible hydrocarbon-water interfacial tension is about 35-50 dyne/cm and the density of water is 1 gr/cm^3. These values suggest a maximum ganglion length of about 5 cm to be stable to water displacement during infiltration of recharge water in this hypothetical medium. Ng et al. (1978) report that a nonwetting residual oil "blob" typically exhibits an aspect ratio (ratio of length to diameter) of between 1 and 8. This suggests that the diameter of our largest ganglia are of the order of 1 cm (or r_g=0.5 cm).

Thus in the hypothetical medium examined above, the discontinuous ganglia might be as large as 5 cm long and 1 cm in diameter. The medium is equivalent to the sands employed by Reible et al. (1989a) in which an initial isooctane content of about 10% and a final residual isooctane content (after water displacement) of about 5% was observed. This is consistent with the difference in maximum ganglion length between a ganglion supporting its own weight and a ganglion surrounded by water flowing with unit hydraulic gradient.

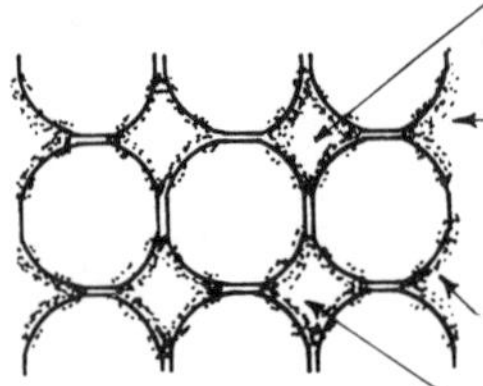

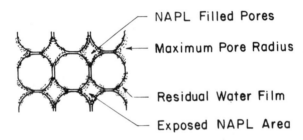

Figure 8 - Conceptual model of grain arrangement and pore structure

CHEMICAL PARTITIONING BETWEEN LIQUID PHASES

The above discussion defines some of the key processes affecting the dynamics of the bulk nonaqueous phase. In general, however, one or more of the components of the nonaqueous phase dominate the risk associated with that phase. To examine the mobility of those contaminants of interest, the dynamics of the partitioning between the aqueous and nonaqueous phase must be examined. Reible et al. (1989b) examined the equilibrium partitioning of soluble aromatic constituents of gasoline between the nonaqueous phase and subsurface water. Let us consider here the kinetics of the partitioning between the nonaqueous phase and residual water during vadose zone infiltration and between the nonaqueous phase and infiltrating recharge water.

The nonaqueous phase liquid will be assumed to be composed predominantly of an essentially immiscible component and one or more minor components partially miscible in water. Gasoline, for example, might be idealized as a mixture of largely immiscible alkanes and a small fraction of partially miscible aromatic compounds. For the purposes of the preliminary analysis of the transport and partitioning processes in this paper, let us consider a model gasoline composed of 94% isooctane, 5% toluene and 1% benzene. Let us examine the partitioning of the more soluble aromatic components of the model gasoline into 1) the residual water in the vadose zone during infiltration of the model gasoline, and, 2) the recharge water that infiltrates subsequent to drainage of the vadose zone soil to a residual nonaqueous phase liquid saturation.

Partitioning during Bulk Phase Infiltration

Prior to contamination by a nonaqueous liquid, the soil vadose zone is composed of soil grains, air and residual water. The upper layers of the vadose zone drain quickly to a residual water content that might be as much as 10% of the soil volume (perhaps 25% of the pore space). Because water tends to wet the vadose zone soil grains, this water is held in a thin film on the surface of the soil grains and in pendular regions at the contact points between them. Loss processes such as evaporation and diffusion or uptake by plant roots are slow processes compared to the hydraulic draining process, resulting in a long period of essentially constant water content. If a nonaqueous phase liquid contaminant is applied to the soil during this period, the separate phase can infiltrate through the vadose zone by displacing the air in the pore spaces. As indicated previously, Reible et al. (1989a) observed that immiscible organic phases did not displace the residual water due to the high interfacial tension between phases and the preferential wetting of the medium by water.

This phenomena also has implications for partitioning of contaminants from the nonaqueous to residual aqueous phases. This was demonstrated in experiments conducted by the author in which a strongly sorbing but essentially water-insoluble dye (Oil Red O) was added to isooctane and allowed to infiltrate through a dry and water-wetted sandy surface soil. During infiltration into the water-wetted soil, essentially no sorption of the dye onto the soil media was observed while during infiltration into dry soils, essentially complete and instantaneous sorption of the dye was observed. Thus mass transfer is assumed to occur between the aqueous and nonaqueous phases and between the aqueous and soil phases but not directly between the nonaqueous and soil phases, as shown below.

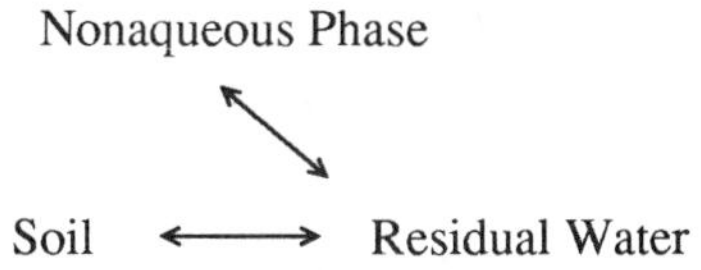

The equilibrium between the two liquid phases can be quantified by conventional fugacity relationships that depend on the activity of the partitioning component in the aqueous and nonaqueous phases. Typically, the nonaqueous phase is assumed ideal. For the 1% benzene in isooctane example, however, the activity coefficient in the isooctane is about 1.5 by either Scatchard-Hildebrand Theory (Prausnitz, 1969) or UNIFAC (Fredunsland et al., 1975), suggesting an error of 50% in equilibrium concentration by assuming ideality. The activity coefficient of benzene in the aqueous phase is quite large (2430) and approximately equal to the inverse of the aqueous solubility in mole fraction units. On a weight basis, the partition coefficient between the isooctane and the residual water is about 0.0039 grams benzene/ gram water per gram benzene/ gram isooctane. For a 1% by weight benzene in isooctane solution, the equilibrium concentration of benzene in the residual water is about 39 parts per million (ppm). A similar calculation for isooctane containing 5% toluene suggests that the residual water would contain 89 ppm toluene at equilibrium. Details of the equilibrium calculation procedures can be found in Reible (1989b). Note that the calculated equilibrium concentrations are sufficiently low that cosolvent effects would not be significant.

The aqueous phase concentrations in this example are also sufficiently low to suggest that the partitioning from the residual water to the soil grains can be described by linear partitioning. That is, the aqueous concentrations are very much less than the aqueous solubility of pure benzene or toluene and linear sorption isotherms would be expected (Walters and Luthy, 1984). For medium and high organic carbon content soils such as might be expected in the near surface vadose zone, the soil-water partition coefficient is generally taken as the product of the fraction organic carbon and the organic carbon-water partition coefficient. The organic carbon-water partition coefficient can be correlated with the octanol-water partition coefficient or the aqueous solubility of the constituent (Lyman et al., 1983).

While the above calculations illustrate the determination of the equilibrium between the infiltrating organic phase and the residual water in the vadose zone, they cannot describe the dynamics of the mass transfer process. The potential exists for nonequilibrium partitioning with residual water because of the diffusional resistances in the water film and interior to the soil particles. Let us examine the potential for diffusional resistances by considering a hydrophobic contaminant for which the dynamics would be controlled by diffusional resistances in the water film. For spherical soil grain particles the diffusion path through the water film is given by

$$\lambda = \frac{d_p}{6} \frac{\theta_{water}}{1-\varepsilon} \tag{12}$$

where d_p is the grain diameter, θ_w the volumetric residual water content and ε the porosity of the medium. For sand grains of 0.3 mm diameter with a porosity of 40% and a residual water content of 10% (the conditions of the experiments of Reible et al., 1989a), this suggests a diffusion path of about 0.0083 mm. With a water diffusivity of about 10^{-5} cm^2/sec, this corresponds to a characteristic diffusion time of about 0.1 sec. Clearly diffusional resistances in the residual aqueous film are small for typical sandy grain soils. Soil particles that are large and are themselves porous may lead to diffusional mass transfer resistances, but based on the above calculation, it is expected that local equilibrium modeling should normally describe partitioning into a residual water film from an infiltrating nonaqueous phase liquid. Note, however, that the capacity of the soil phase may be very large for hydrophobic contaminants and many characteristic diffusion times may be required to achieve equilibrium for such compounds.

Partitioning from a Residual Nonaqueous Phase

The kinetics of partitioning between a discontinuous residual of nonaqueous phase liquid and recharge water infiltrating through the vadose zone is largely determined by the size and spatial distribution of the discontinuous ganglia of the nonaqueous phase. The approximate size of the ganglia was estimated earlier based on a balance of the capillary holding pressure and the displacing pressure drop in the water phase. Continuing with the approximation presented there of representing the porous media as a square array of spherical soil grains, the exposed side area of the ganglia shown in Figure 8 is 21% of the area of an elementary side of the array of spheres.

$$A_{pore} = \left(d_p^2 - \pi \frac{d_p^2}{4} \right) = 0.21\, d_p^2 \tag{13}$$

This suggests that the exposed area of the ganglion is 21% of the surface area of the equivalent cylinder. The exposed area to volume ratio of the ganglion is thus $0.41/r_g$=0.82 cm^{-1}.

This information can be tied to a model of the dynamics of partitioning between the infiltrating recharge water and the nonaqueous phase liquid ganglion. A mass balance on the recharge water passing along the exposed surface of the nonaqueous phase liquid can be written

$$U \frac{d}{dz}\left(C_{water} V\right) = k_{eff} A \left(\frac{MW_{napl}}{MW_{water}} \frac{\rho_{water}}{\rho_{napl}} K_{24} C_{napl} - C_{water}\right) \quad (14)$$

This is essentially identical to a relationship posed by Hunt et al. (1988) to describe nonequilibrium partitioning between a nonaqueous phase residual and extracting steam. Here U is the velocity of the infiltrating recharge water, MW is the molecular weight and k_{eff} is an effective mass transfer coefficient that accounts for resistances in each of the fluid and solid phases. K_{24} is the partition coefficient between the aqueous and nonaqueous phases in mole fraction units. It is given by the ratio of the activity coefficient of the component in the nonaqueous phase divided by the activity coefficient in the aqueous phase. In the absence of capillary forces, an infiltrating liquid is driven only by gravity and the gradient in pressure head is unity. Thus U can be approximated through Darcy's Law as the water conductivity of the vadose zone soil divided by the porosity of the soil ($U = K/\varepsilon$). Using this approximation and using the exposed area per volume ratio of the ganglion, the above equation can be written

$$\frac{d}{dz}\left(C_{water}\right) = \frac{0.41\, k_{eff}\, \varepsilon}{r_g K} \left(\frac{MW_{napl}\, \rho_{water}}{MW_{water} \rho_{napl}} K_{24} C_{napl} - C_{water}\right) \quad (15)$$

This equation can be integrated, for example with constant parameters to give

$$C_{water}(z) = \frac{MW_{napl}\, \rho_{water}}{MW_{water} \rho_{napl}} K_{24} C_{napl} \left[1 - \exp\left\{ -\frac{0.41\, k_{eff}\, \varepsilon}{r_g K} z \right\} \right] \quad (16)$$

The assumption that K_{24} is constant implies that the activity coefficients are independent of concentration. This model indicates that the aqueous concentration approaches the equilibrium concentration at a rate related to the effective mass transfer coefficient. Practical use of this model now requires definition of the effective mass transfer coefficient.

The effective mass transfer coefficient is in general attenuated by the resistances in each sorbing phase. If k_{water} represents the mass transfer coefficient in the aqueous phase and k_{napl} represents the mass transfer coefficient in the nonaqueous phase, the overall coefficient based on the concentration driving force in the aqueous phase is given by

$$\frac{1}{k_{eff}} = \frac{1}{k_{water}} + \frac{K_{24}\, \rho_{water}\, MW_{napl}}{k_{napl}\, \rho_{napl}\, MW_{water}} \quad (17)$$

Hunt et al. (1988) employed the correlation of Wilson and Geankopolis (1966) to estimate the mass transfer resistances in the flowing aqueous phase.

$$k_{water} = 1.09 \frac{U_s}{\varepsilon} \left(\frac{U_s d_p}{D}\right)^{-2/3} \quad (18)$$

This correlation suggests that the mass transfer coefficient in the water phase is about 3.3 X 10^{-3} cm/sec using K=1 cm/min to estimate the Darcy velocity, U_s, $\varepsilon=0.4$, $d_p=0.3$ mm and $D=10^{-5}$ cm^2/sec. Reible (1987) observed that the problem of defining the aqueous phase resistance is largely that of defining the dispersion in the medium, normally taken as proportional to the pore velocity. Let us assume for the purposes of this proposal, however, that the above is a reasonable estimate for the aqueous phase mass transfer coefficient.

Now consider the mass transfer coefficient in the nonaqueous phase ganglion. Due to the dissolution of a miscible contaminant of the nonaqueous phase, the interface between the aqueous and nonaqueous phases can be depleted of the partitioning species. The depleted contaminant can be replenished by desorption from the residual water and the soil grains and by transport from adjacent pores. Let us conservatively approximate the diffusion path as one pore diameter recognizing that if the nonaqueous phase is limiting the rate of mass transfer that this diffusion path will grow as the surface of the ganglion becomes depleted. With this approximation, the mass transfer coefficient in the nonaqueous phase liquid becomes,

$$k_{napl} \approx \frac{D_{eff}}{2\, r_t} \tag{19}$$

Using an effective diffusion coefficient in the pore space of about 10^{-6} cm^2/sec and the pore throat radius of 0.15 mm, this equation provides an estimate of the mass transfer coefficient of about 6.7 X 10^{-5} cm/sec.

The overall coefficient is then approximately 2.6 X 10^{-3} cm/sec. With a porosity of 40% and a water conductivity of 1 cm/min, the model predicts that recharge water in intimate contact with a 6 cm long ganglion of nonaqueous phase residual would attain a contaminant concentration 26% of the equilibrium value. Contact with a nonaqueous phase ganglion for 45 cm would be required to achieve aqueous concentrations within 90% of equilibrium. Because of the possibility of additional resistances within the soil particles (if not surface sorbing), the thin film of residual water and the conservative assumptions of diffusion path in the nonaqueous phase residual ganglia, the actual aqueous concentrations could exhibit even larger deviations from equilibrium.

SUMMARY

In this paper, some of the complexities of the movement of contaminants in a subsurface nonaqueous phase have been identified. The mobility of these contaminants have been shown to be a strong function of the physical form of the nonaqueous phase liquid as well as the media and chemical properties.

Despite the complexity of the fate and transport processes affecting a nonaqueous phase, some physically meaningful approximations can be made to understand the qualitative behavior of the system and, in some cases, to make practical quantitative estimates of subsurface behavior of the nonaqueous phase. The presence of a wetting residual water phase in the media was shown to limit capillary suction influences on the nonaqueous phase and direct sorption of nonaqueous contaminants onto the soil particles. The exclusion of water-filled fine pores was shown to result in sharp wetting and drainage fronts for an infiltrating nonaqueous phase liquid. As a result of these effects, a comparatively simple model of nonaqueous phase movement in the vadose zone can be formulated.

It is also possible to develop a simplified model of the dynamics of a residual nonaqueous phase liquid and estimates were made of the size of the stable discontinuous ganglia and the rate of mass transfer between these stable ganglia and infiltrating recharge water. It was shown that diffusion resistances may limit the attainment of equilibrium between the aqueous and nonaqueous phases during infiltration.

The paper has also identified areas where further research is required. Media/ fluid interactions such as electrical interactions are clearly important but are not yet sufficiently well understood to make quantitative estimates of their effect. The difficulties of heterogeneous media are difficult to address through either laboratory or numerical modeling. Finally, some of the simplified models presented here for describing the dynamics of partitioning between the aqueous and nonaqueous phases have been inadequately tested in the laboratory. These and other studies must be continued to better define the fate and transport processes affecting a nonaqueous phase in the subsurface so that a better assessment of the risk associated with the land application of hazardous substances can be made.

ACKNOWLEDGEMENTS

This work was partially supported by the LSU Hazardous Waste Research Center and the US Environmental Protection Agency and their support is gratefully acknowledged. The encouragement and suggestions provided by Louis Thibodeaux are especially appreciated. The authors would also like to thank those students who have participated in aspects of the research including Dharmesh Doshi, Ibrahim Ayoub and Mark Malhiet.

REFERENCES

Abriola, L.M. and G.F. Pinder, A multiphase approach to the modeling of porous media contamination by organic compounds, 1, Equation development, *Water Resources Research, 21* (1) 11, 1985.

Brooks, R.H. and A.T. Corey, Hydraulic properties of porous media, Colorado State University, *Hydrology Paper No. 3*, March 1964.

Chatzis, I., N.R. Morrow and H.T. Lim, Magnitude and detailed structure of residual oil saturation, *Society of Petroleum Engineering Journal, 23*, 311, 1983.

Crosby, B.J., G.H. Hornberger, R.B. Clapp and T.R. Ginn, A statistical exploration of the relationships of soil moisture characteristics to the physical properties of soils, *Water Resources Research, 20*, 682, 1984.

Faust, C.R., Transport of immiscible fluids within and below the unsaturated zone: A numerical model, *Water Resources Research, 21* (4) 587, 1985.

Fredunsland, A., R.L. Jones and J.M. Prausnitz, Group contribution estimation of activity coefficients in nonideal liquid mixtures, *AIChE Journal, 21*, 1086, 1975.

Green, W.J., G.E. Lee and R.A. Jones, Clay-soils permeability and hazardous waste storage, *Journal of the Water Pollution Control Federation, 53*, 1263, 1981.

Hunt, J.R., N. Sitar and K.S. Udell, Nonaqueous phase liquid transport and cleanup, 1, Analysis of mechanisms, *Water Resources Research, 24* (8) 1247, 1988.

Kappusamy, T., J. Sheng, J.C. Parker and R.J. Lenhard, Finite-element analysis of multiphase immiscible flow through soils, *Water Resources Research, 23* (4) 625, 1987.

Kool, J.B. and J.C. Parker, Development and evaluation of closed-form expressions for hysteretic soil hydraulic properties, *Water Resources Research, 23* (1) 105, 1987.

Larsen, R.G., H.T. Davis and L.E. Scriven, Displacement of residual nonwetting fluid from porous media, *Chemical Engineering Science, 36* (1) 75, 1981.

Lyman, W.J., W.F. Reehal and D.H. Rosenblatt, *Handbook of Chemical Property Estimation Methods,* McGraw-Hill, New York, 1983.

Nielsen, D.R., M.Th. van Genuchten and J.W. Biggar, Water flow and solute transport processes in the unsaturated zone, *Water Resources Research, 22* (9) 89, 1986.

Ng, K.M., H.T. Davis and L.E. Scriven, Visualization of blob mechanics in flow through porous media, *Chemical Engineering Science, 33* (8) 1009, 1978.

Parker, J.C., R.J. Lenhard and T. Kappusamy, A parametric model for constitutive properties governing multiphase flow in porous media, *Water Resources Research, 23* (4) 618, 1987.

Pathak, P., H.T. Davis and L.E. Scriven, Dependence of residual nonwetting liquid on pore topology, Society of Petroleum Engineers 57th Annual Fall Technical Conference, 1982.

Payatakes, A.C., Dynamics of oil ganglia during immiscible displacement in water-wet porous media, *Annual Review of Fluid Mechanics, 14*, 365, 1982.

Phannkuch, H.O., Determination of the contaminant source strength from mass exchange processes at the petroleum/ground water interface in shallow aquifer systems, *Proceedings of the NWWA Conference on Petroleum Hydrocarbons and Organic Chemicals in Ground Water*, 111, 1984.

Pinder, G.F. and L.M. Abriola, On the simulation of nonaqueous phase organic compounds in the subsurface, *Water Resources Research, 22* (9) 109, 1986.

Prausnitz, J.M. *Molecular Thermodynamics of Fluid-Phase Equilibria,* Prentice-Hall, New Jersey, 1969.

Reible, D.D. Subsurface contamination by multiphase processes - Research and policy implications for the EPA, AAAS Environmental Science and Engineering Fellow Report, 1987.

Reible, D.D., T.H. Illangasekare, D.V. Doshi and M.E. Malhiet, Infiltration of immiscible contaminants in the unsaturated zone, submitted to *Ground Water,* 1989a.

Reible, D.D., M.E. Malhiet and T.H. Illangasekare, Modeling gasoline fate and transport in the unsaturated zone, in press, *Journal of Hazardous Materials,* 1989b.

Schramm, M., A.W. Warrick and W.H. Fuller, Permeability of soils to four organic liquids and water, *Hazardous Waste and Hazardous Materials, 3* (1) 21, 1986.

Stone, H.L., Estimation of three-phase relative permeability and residual oil data, *Journal of Canadian Petroleum Technology, 12* (4) 53, 1973.

Walters, R.W. and R.G. Luthy, Equilibrium sorption of polycyclic aromatic hydrocarbons from water onto activated carbon, *Environmental Science and Technology, 18* (6) 1984.

Wilson, E.J. and C.J. Geankopolis, Liquid mass transfer at very low Reynolds numbers in packed beds, *Industrial Engineering Chemistry - Fundamentals, 5* (1) 9, 1966.

Wilson, J.L. and S.H. Conrad, Is physical displacement of residual hydrocarbons a realistic possibility in aquifer restoration?, *Proceedings of Petroleum Hydrocarbons and Organic Chemicals in Ground Water- Prevention, Detection and Restoration,* National Water Well Association, Houston, Tx, 1984.

MULTIMEDIA TRANSPORT OF POLLUTANTS

MULTIMEDIA PARTITIONING OF DIOXIN

Curtis C. Travis and Holly A. Hattemer-Frey

Office of Risk Analysis
Oak Ridge National Laboratory
Oak Ridge, Tennessee 37831

INTRODUCTION

Because of their extreme toxicity, much concern and debate has arisen about the nature and extent of human exposure to dioxins. Since municipal solid waste (MSW) incinerators are known to emit polychlorinated dibenzo-p-dioxins (PCDDs) and polychlorinated dibenzofurans (PCDFs) (Rappe et al., 1987a), many people who live near MSW incinerators feel that they will be exposed to high levels of dioxin and subsequently develop cancer. What is often overlooked in this debate, however, is the fact that the general population is continuously being exposed to trace amounts of dioxin as exemplified by the fact that virtually all human adipose tissue samples contain dioxin levels of three parts per trillion (ppt) or greater (Patterson et al., 1986 and Ryan et al., 1983). The purpose of this study is to investigate how 2,3,7,8,-tetrachlorodibenzo-p-dioxin (TCDD) is partitioned in the environment and to identify the major pathways of human exposure.

Background levels of TCDD have been measured in air, soil, sediment, suspended sediment, fish, cow milk, human adipose tissue, and human breast milk samples (Beck et al., 1987; Crummett, 1987; Czuczwa and Hites, 1985; Eitzer and Hites, 1987; O'Keefe et al., 1983; Patterson et al., 1986; Rappe et al., 1987b, 1988; van den Berg et al., 1986). A steady-state mathematical model patterned after Mackay et al.'s (1985a-b) Level-III Fugacity Model is used to evaluate the concentration of dioxin in various environmental media. The model can estimate (a) unknown (nondetectable) environmental concentrations from the physicochemical properties of TCDD and (b) known (detectable) concentrations of TCDD in various environmental media, thus providing a coherent account of concentrations in all media. These concentrations are then used to predict the amount of dioxin entering the food chain and to estimate the average daily intake of dioxin by humans.

THE FUGACITY FOOD CHAIN MODEL

The Fugacity Food Chain model, a non-equilibrium, multimedia transport model, was used to study the environmental fate of TCDD. The model divides the environment into six compartments: air, water, soil, sediment, suspended sediment (in water), and biota (in water). The air, water, and soil compartments are interactively connected, while the sediment, suspended sediment, and biota compartments are connected only with the water compartment (Mackay et al., 1985a-b). The model was modified to account for uptake of TCDD through the terrestrial food chain. While this model is not an exact replica of the environment, fugacity models are considered acceptable for predicting the equilibrium partitioning of non-particulate chemicals (Cohen and Ryan, 1985; Mackay and Paterson, 1982; Mackay et al., 1985a-b).

The input parameters required to predict the cross-media partitioning of a chemical include: (1) the physicochemical and biochemical properties of the compound; (2) estimates of emission rates for the compound into air, water, and soil; and (3) estimates of degradation rate coefficients for reactions that remove the compound from the system.

Physicochemical and Biochemical Properties

The distribution, decomposition, and accumulation of any chemical in the environment depends largely upon its physical and chemical properties. Table 1 lists the physical and chemical properties of TCDD.

Table 1. Physical and chemical properties of TCDD.

1. Log K_{ow}*	6.85	Crosby, 1985
2. Water solubility	6.0×10^{-8} mol/m^3	Marple et al., 1986
3. Vapor pressure	4.5×10^{-6} Pa	Rordorf, 1985
4. Molecular weight	322 g/mol	Schroy et al., 1985

* K_{ow} = octanol-water partition coefficient

Assessing the environmental fate of organics depends largely on being able to predict their bioaccumulation in living organisms, including fish, cattle, and humans. Biotransfer factors (BTFs) (equilibrium concentration of TCDD in tissue (ng/kg) divided by the daily intake of TCDD (ng/kg)) and bioconcentration factors (BCFs) (equilibrium concentration of TCDD in tissue (ng/kg) divided by the concentration of TCDD (ng/kg) in water (for aquatic organisms) and soil (for vegetation)) depict a chemical's tendency to bioaccumulate in living organisms. The BTFs for cow adipose tissue and milk and the BCFs for fish and vegetation are listed in Table 2.

Table 2. Biotransfer and bioconcentration factors for TCDD.

1. Daily intake-to-cow whole tissue	0.22	Jensen et al., 1981
2. Daily intake-to-dairy cow whole milk	0.058	Travis and Arms, 1988
3. Water-to-fish	49,000	Mehrle et al., 1987
4. Soil-to-vegetation	0.003	Travis and Arms, 1988

While most of the TCDD that ends up in water sorbs onto suspended sediments or equilibrates with biota, photolysis of TCDD in near-surface waters is an important degradative pathway (Mill, 1985), with a half-life of about 40 hours (Dulin et al., 1986; Podoll et al., 1986).

THE ENVIRONMENTAL FATE OF TCDD

The low solubility, low vapor pressure, and high octanol-water partition coefficient of TCDD result in its partitioning mainly between soil (85.3%) and sediment (14.3%) with less than 1.0% partitioning into air, water, and suspended sediment. Table 3 gives the observed background and predicted environmental concentrations for TCDD.

Table 3. Background and predicted environmental concentrations for TCDD.

Phase	Predicted (ng/kg)	Background (ng/kg)	Reference(s)
Air	$5.3x10^{-6}$	$2.7-6.9x10^{-6}$	Eitzer and Hites, 1986
Water	$4.5x10^{-6}$	<1.0	Czuczwa and Hites, 1985
Soil	2.2	<1-9.0	Crummett, 1987; US EPA, 1986
Sediment	0.79	<1.0-26.0	Czuczwa et al., 1984
Suspended sediment	0.79	0.10-19.0	Mackay et al., 1985a
			Stalling et al., 1983

Predicted concentrations are within the range of background values cited above.

ESTIMATING UPTAKE OF TCDD THROUGH THE FOOD CHAIN

The food chain is the primary pathway of human exposure to a large class of lipophilic compounds like TCDD, DDT, and most pesticides (Travis and Arms, 1987). A food chain model (Travis and Hattemer-Frey, 1987; Hattemer-Frey and Travis, 1988) was used to estimate human exposure to TCDD from ingestion of agricultural produce (fruits, grains, and vegetables), beef, milk, and fish.

Vegetation

Accumulation of TCDD in vegetation involves root uptake and atmospheric deposition. Root uptake has been correlated with K_{ow} (Baes, 1982; Briggs et al., 1982) and can be quantified in terms of B_v, the soil-to-vegetation BCF (see Table 2). The concentration of TCDD in vegetation due to root uptake (CVR) is estimated from the following equation:

$$\text{CVR (mol/kg)} = C_s * B_v * 0.001/1.5 \qquad (1),$$

where C_s equals the predicted concentration of TCDD in soil ($6.85x10^{-9}$ mol/m^3), and $1.5g/cm^3$ is the density of soil. The concentration of TCDD in vegetation due to root uptake is estimated to be 0.004 ng/kg.

Deposition or direct application of TCDD onto outer plant surfaces contributes substantially to vegetative contamination, since most of the TCDD released into the atmosphere is eventually deposited onto plant, soil, and water surfaces (Czuczwa et al., 1984). The concentration of TCDD on vegetation due to atmospheric deposition (CVD) can be estimated using the following equation (Travis et al., 1986).

$$\text{CVD (mol/kg)} = \frac{(C_a)\ (V_d)\ (r/Y)}{\lambda} \qquad (2),$$

where C_a = the air concentration of TCDD
= $2.2x10^{-16}$ mol/m^3 (Eitzer and Hites, 1986);

V_d = atmospheric deposition velocity
= 0.0023 m/s (Radian Corporation, 1987);

r/Y = interception fraction to productivity ratio for TCDD
= 0.4 m2/kg for vegetation consumed by Cows (Baes et al., 1984)
= and 0.04 m^2/kg for vegetation consumed by humans (Baes et al., 1984):

and λ = the weathering constant of TCDD deposited onto the plant
= 5.7×10^{-7} sec^{-1} (Baes et al., 1984; Highland, 1986).

Vegetation-specific differences in the interception fraction necessitate different estimates of the concentration of TCDD on vegetation that cows (CVCOWS) and humans (CVHUM) consume. CVCOWS equals 3.5×10^{-13} mol TCDD/kg plant or 0.11 ng/kg. Thus, the total concentration of TCDD on vegetation for cows equals CVR + CVCOWS = 0.12 ng/kg. These data show that foliar deposition accounts for 97% of the TCDD contamination of vegetation consumed by cattle. CVHUM equals 3.5×10^{-14} mol/kg or 0.01 ng/kg, and the total concentration of TCDD on exposed produce and vegetables equals CVR + CVHUM = 0.02 ng/kg. In this case, foliar deposition accounts for 73% of the TCDD contamination of exposed produce and vegetables consumed by humans.

Beef and Milk

Because of its low water solubility and high lipophilicity, TCDD readily accumulates in living organisms. Ingestion of contaminated plants and soil represents a major exposure pathway for terrestrial organisms. The predicted daily intake of TCDD by feed lot cattle can be estimated from the following equations:

$$\text{Intake from water (mol/d)} = C_w * QW * 0.001 \qquad (3),$$

$$\text{Intake from soil (mol/d)} = C_s * QS * 0.001/1.5 \qquad (4),$$

$$\text{Intake from vegetation (mol/d)} = CVR + CVCOWS * QV \qquad (5),$$

where C_w is the concentration in water (mol/m^3) and QW, QS, and QV are the intake of water, soil, and vegetation (kg/d) by cattle. Feed lot cattle destined for slaughter consume about 5.0 kg of exposed forage and 3.0 kg grains per day (Baes et al., 1984), 37.6 kg of water per day (Spector, 1959), and 0.1 kg soil per day (Fries, 1986). Assuming that grains contain no TCDD (Stevens and Gerbec, 1987), the predicted daily intake of TCDD by cattle is given in Table 4.

Table 4. Predicted daily intake of TCDD by feed lot cattle.

Source	Daily Intake (ng/day)	Percentage of the Total Intake
1. Water	0.0002	<.01%
2. Soil	0.15	20%
3. Forage	0.60	80%

The vegetation ingestion pathway accounts for most of the TCDD intake by feed lot cattle.

The concentrations of TCDD in cow adipose tissue (CB) and cow milk (CM) were calculated according to the following equations:

$$CB\ (mol/kg) = (C_s * QS * B_f * 0.001/1.5) + (C_w * QW * B_f * 0.001) + (CV * QV * B_f) \quad (6)$$

and

$$CM\ (mol/kg) = (C_s * QS * B_m * 0.001/1.5) + (C_w * QW * B_m * 0.001) + (CV * QV * B_m) \quad (7)$$

where C_s = the predicted concentration of TCDD in soil (mol/m^3);
C_w = the predicted concentration of TCDD in water (mol/m^3);
QS = the amount of soil ingested per cow per day (kg/d);
QW = the amount of water ingested per cow per day (L/d);
QV = the amount of vegetation ingested per cow per day (kg/d);
B_f = the BTF for TCDD in cow adipose tissue (d/kg);
B_m = the BTF for TCDD in cow milk fat (d/kg).
and CV = concentration of TCDD in vegetation (CVR+CVCOWS) (mol/kg);

Fish

Fish can bioaccumulate TCDD residues directly from the water and from ingesting contaminated food items (Isensee and Jones, 1975). Bioconcentration factors of 45 to 29,000 have been found in laboratory experiments that studied the accumulation of TCDD from water by fish (Branson et al., 1985; Isensee and Jones, 1975; Matsumura and Benezet, 1973; Mehrle et al., 1987). These reported BCFs, however, were determined through short-term studies in which the concentration of TCDD in fish did not reach equilibrium. Using a computer model, Mehrle et al. (1987) estimated that the average BCF for equilibrium concentrations of TCDD in fish is 49,000.

TCDD concentrations in fish (CF) can be estimated using the following equation:

$$CF\ (mol/kg) = C_w * BCF * 0.001 \quad (8),$$

Using a BCF of 49,000, the predicted concentration of TCDD in fish is 0.2 ng/kg.

ESTIMATING THE AVERAGE DAILY INTAKE OF TCDD FOR HUMANS

Table 5 gives predicted average daily intake of TCDD by the general population of the US from inhalation and ingestion of contaminated food items and drinking water (Travis and Hattemer-Frey, 1987).

Table 5. Predicted average daily intake of TCDD by humans.

Source	Daily Intake (ng/day)	Percent of the Total Daily Intake
1. Air	0.001	2%
2. Water	6.5×10^{-6}	<0.01%
3. Food (total)	0.046	98%
a. Vegetables	0.005	11%
b. Milk	0.013	27%
c. Meat	0.023	50%
d. Fish	0.005	10%

These data show that the food chain, especially meat and dairy products, account for 98% of human exposure to TCDD. Consumption of contaminated fish and shellfish does not comprise a major source of human exposure to TCDD contrary to suggestions by O'Keefe et al. (1983) and Stalling et al. (1985). The model estimated the average daily human intake of TCDD to be 0.05 ng/d.

EXPOSURE TO MSW INCINERATOR EMISSIONS

It is widely believed that MSW incineration is the major source of human exposure to dioxin. To assess the extent of human exposure to facility-emitted PCDDs and PCDFs, Travis and Hattemer-Frey (1988) evaluated risk assessment documents prepared for 11 proposed MSW incinerators designed to use modern, efficient pollution control equipment. The geometric mean of the data from those 11 documents was used to represent exposure to a typical, modern MSW incinerator. (Hence, the following results may not apply to older incinerators that do not use efficient pollution control equipment.) Results suggest that MSW incineration is not a major source of human exposure to dioxins and furans for the following reasons.

First, exposure to background levels of PCDDs and PCDFs account for more than 99% of the total daily intake by the maximally exposed individual living near a typical, modern MSW incinerator. Table 6 shows that the predicted daily intake of PCDDs/PCDFs by the maximally exposed individual is 80 times less than exposure to background levels.

Table 6. Total daily intake of PCDDs/PCDFs by maximally exposed individual living near a typical, modern MSW incinerator.

Source	Daily Intake (pg TEDFs/day)	Percentage of the Total Daily Intake	Reference
Background*	87.0	99.3%	Beck et al., 1988a Ono et al., 1987
Incinerator	0.6	0.7%	Travis & Hattemer-Frey, 1988
Total	87.6	100%	

*Geometric mean of estimates reported by Beck et al., 1988a and Ono et al., 1987 for a 70 kg individual.

These data indicate that human exposure to PCDDs and PCDFs emitted from a typical, state-of-the-art MSW incinerator is not excessive relative to exposure to background levels, since the individual lifetime cancer risk associated with exposure to facility-emitted PCDDs and PCDFs is one in a million ($1x10^{-6}$), while the cancer risk associated with exposure to background environmental contamination is two in ten thousand ($2x10^{-4}$).

SOURCES OF PCDD/PCDF INPUT

If MSW incineration is not the major source of human exposure to PCDDs and PCDFs, what is? Although the magnitude of PCDD/PCDF input into the environment remains unknown, principal sources of PCDDs and PCDFs are suspected to be: (1) high temperature industrial processing facilities, such as metal processing/treatment plants and copper smelting plants (Marklund et al., 1986; Rappe et al., 1987a); (2) motor vehicles (Ballschmitter et al., 1986; Bumb et al., 1980; Marklund et al., 1987; Rappe et al., 1988); (3) chemical manufacturing processes (Hutzinger et al., 1985); (4) chemical/industrial waste incineration (Hutzinger et al., 1985; Weerasinghe and Gross, 1985); (5) MSW incinerators (Rappe et al., 1987a; Travis and Hattemer-Frey, 1988; Marklund et al., 1986; Hutzinger et al., 1985); and (6) pulp and paper mills (Beck et al., 1988b and Rappe et al., 1988). The magnitude of emissions from sources for which empirical data are available are discussed below.

Industrial Sources

Rappe et al. (1987a) and Marklund et al. (1986) reported that emissions of PCDDs and PCDFs from high-temperature industrial processing plants seem to be of the same magnitude as emissions from MSW incinerators. The Swedish EPA (1987) estimated that emissions of PCDDs and PCDFs from these types of plants operating in Sweden range from 50 to 150 grams (expressed in Toxic Equivalents Dioxins and Furans (TEDFs)) per year, while normal-sized, modern MSW incinerators (50 to 200,000 tons of waste per year capacity) emit about 1 to 50 grams TEDFs per year (Rappe et al., 1987a; Travis and Hattemer-Frey, 1988). (When PCDDs and PCDFs concentrations are expressed in TEDFs, the toxicity of all isomers is weighted relative to the known toxicity of 2,3,7,8-TCDD. Thus, a release of one gram of TEDFs is equivalent to a release of one gram of TCDD.)

Rappe et al. (1988) found that levels of PCDDs and PCDFs in air collected from a West German industrial area located 600 meters from a copper smelting plant (3.6×10^{-2} ng/m^3) were about 12 times higher than measured background levels in suburban air collected about 13 km from Hamburg, West Germany (3.0×10^{-3} ng/m^3) and about three times higher than levels measured around a MSW incinerator operating near Hamburg (1.3×10^{-2} ng/m^3).

Rappe et al. (1987a), moreover, concluded that because there are many more industrial sources, their contribution to total PCDD/PCDF emissions could be much greater than the contribution from MSW incinerators. Marklund et al. (1986) observed that "total emissions from industrial incinerators could be of the same magnitude or even higher than emissions from MSW incinerators," while Nakano et al. (1987) reported that "PCDDs and PCDFs in the urban air are surmised to be derived from domestic and industrial waste incinerators." Thus, various researchers agree that industrial sources may contribute equal or even larger amounts of PCDDs and PCDFs into the environment than MSW incinerators.

Motor Vehicles

Studies have shown that motor vehicle emissions may be a larger source of PCDD/PCDF input than emissions from MSW incinerators. Ballschmiter et al. (1986) argued that a more prevalent source of PCDD/PCDF input than MSW incineration must exist to account for the widespread, background contamination of PCDDs and PCDFs and suggested that the ubiquitous, non-point character of motor vehicle emissions "strongly recommends this source for consideration as a major environmental input."

Marklund et al. (1987) found that total emissions of PCDDs and PCDFs from cars in Sweden using unleaded gasoline were 10 to 100 grams TEDFs per year, which is equivalent to the amount of PCDDs and PCDFs emitted from 2 to 20 MSW incinerators of normal size and technology. Thus, in Sweden, motor vehicles and MSW incinerators emit about the same amount of PCDDs and PCDFs. Jones et al. (1987) reached the same conclusion for the United States and contends that due to the widespread source of motor vehicle emissions, roadside exposures are equal to or greater than exposures from the elevated stacks of MSW incinerators.

Rappe et al. (1988) also measured PCDD/PCDF levels in ambient air within a traffic tunnel. PCDD/PCDF levels in the tunnel (2.8×10^{-2} ng/m^3) were two times higher than levels downwind from the MSW incinerator (1.3×10^{-2} ng/m^3) and nine times higher than the suburban air concentration (3.0×10^{-3} ng/m^3). Rappe et al. (1988) concluded that "measurements made in the traffic tunnel clearly indicate that motor vehicles are a source of PCDDs/PCDFs in the ambient air." Hence, several researchers confirm that motor vehicle emissions may also be a significant source of PCDD/PCDF in the environment.

Pulp and Paper Mills

The US EPA (1987) first reported that pulp and paper mills using a chlorine bleaching process may be another source of PCDD/PCDF input into the environment. Swanson et al. (1988) analyzed the effluents from one mill producing bleached pulp in Sweden and found that PCDD/PCDF emissions ranged from 2 to 5.8 grams TEDFs per year. Since the mill they sampled used a less efficient processing method than most operating mills in Sweden, they estimated that Swedish pulp and paper mills as a whole discharge about 5 to 15 grams of TEDFs per year (Swanson et al., 1988). These preliminary findings suggest that the amount of PCDDs and PCDFs formed in pulp and paper mills in Sweden is small relative to other sources.

Thus, these data indicate that environmental concentrations cannot be linked to any one combustion source. Combustion processes in general (not just MSW incinerators) are the dominant source of PCDDs and PCDFs in the environment. It is premature to conclude that MSW incineration is the major source. The magnitude of PCDD/PCDF emissions from and PCDD/PCDF concentrations in the ambient air and other environmental media (e.g., soil and cow's milk) around other operating combustion sources known to emit dioxins and furans (e.g., copper smelting plants, steels mills, motor vehicles, and pulp and paper mills) are needed before definitive statements about the major source(s) of PCDD/PCDF input can be made.

CONCLUSIONS

Organic compounds end up in the media in which they are most soluble. TCDD, a highly lipophilic, persistent compound, sequesters almost completely in soil and sediment and bioconcentrates in aquatic and terrestrial organisms, including man. The food chain, especially beef and dairy products, accounts for 98% of human exposure to TCDD. Inhalation and water consumption are not major sources of human contamination. Using measured (and predicted) concentrations of TCDD in various environmental media yields an intake estimate of 0.05 ng per day.

MSW incinerators are one source of PCDD/PCDF input into the environment. Empirical evidence demonstrates, however, that well-operated, modern MSW incinerators may not be the major source of human exposure to PCDDs and PCDFs, since exposure to background levels overwhelms exposure to facility-emitted contaminants. The relatively small contribution by incinerators to total daily intake suggests that some, as of yet unidentified, source(s) (possibly automobiles or industrial sources) are substantially contributing to background levels of dioxins and furans. We recommend that future research efforts focus on characterizing the source(s) of background levels of PCDDs and PCDFs, because they may pose far greater threats to human health than PCDD/PCDF emissions from modern MSW incinerators.

ACKNOWLEDGEMENTS

Operated by Martin Marietta Energy Systems, Inc. for the US Department of Energy under Contract No. DE-AC0584OR21400.

REFERENCES

Baes, C.F., III. 1982. Prediction of radionuclide Kd values from soil-plant concentration ratios. Trans. Amer. Nuclear Soc. 41:53-54.

Baes, C.F., III, R.D. Shart, A. Sjoreen and R. Shor. 1984. A Review and Analysis of Parameters for Assessing Transport of Environmentally Released Radionuclides Through Agriculture. U.S. Department of Energy Oak Ridge National Laboratory, ORNL-5786.

Ballschmiter, K., H. Buchert, R. Niemczyk, A. Munder, and M. Swerv. 1986. Automobile exhausts versus municipal waste incineration as sources of the polychloro-dibenzodioxins (PCDD) and - furans (PCDF) found in the environment. Chemosphere 15:901-915.

Beck, H., K. Eckart, W. Mathar, and R. Wittowski. 1988a. PCDD and PCDF Body Burden from Food Intake in the Federal Republic of Germany. Presented at the Seventh International Symposium on Chlorinated Dioxins and Related Compounds, Las Vegas, Nevada, October 4-9, 1987, Chemosphere (submitted).

Beck, H., K. Eckart, W. Mathar, and R. Wittowski. 1988b. Occurrence of PCDD and PCDF in different kinds of paper. Chemosphere 17(1):51-57.

Beck, H., K. Eckart, M. Kellert, W. Mathar, Ch-S. Ruhl, and R. Wittowski, 1987. Levels of PCDFs and PCDDs in samples of human origin and food in the Federal Republic of Germany. Chemosphere 16(8/9):1977-1982.

Branson D.R., I.T. Takahashi, W.M. Parker, and G.E. Blau. 1985. Bioconcentration kinetics of 2,3,7,8-tetrachlorodibenzo-p-dioxin in rainbow-trout. Environ. Toxicol. Chem. 4(6):779-788.

Briggs, G.G., R.H. Bromilow, and A.A. Evans. 1982. Relationships between lipophilicity and root uptake and translocation of non-ionized chemicals by Barley. Toxicol. Environ. Chem. 7:173-189.

Bumb, R.R., W.B. Crummett, S.S. Cutie, J.R. Gledhill, R.H. Hummel, R.O. Kagel, L.L. Lamparski, E.V. Luoma, D.L. Miller, T.J. Nestrick, L.A. Shadoff, R.H. Stehl, and J.S. Woods. 1980. Trace chemistries of fire: A source of chlorinated dioxins. Science. 210:385-390.

Cohen, Y., and P.A. Ryan. 1985. Multimedia modeling of environmental transport: Trichloroethylene test case. Environ. Sci. Technol. 19:412-417.

Crosby, D.G. 1985. The degradation and disposal of chlorinated dioxins. In Dioxins in the Environment, M.A. Kamrin and P.W. Rodgers, Eds., Hemisphere Press Corp., Washington, D.C., pp. 195-204.

Crummett, W.B. 1987. Dow Chemical Company. Personal correspondence.

Czuczwa, J.M., and R.A. Hites. 1985. Dioxins and dibenzofurans in air, soil and water. In Dioxins in the Environment, M.A. Kamrin and P.W. Rodgers, Eds., Hemisphere Press Corp., Washington, D.C., pp. 85-99.

Czuczwa, J.M., B.D. McVeety, and R.A. Hites. 1984. Polychlorinated dibenzo-p-dioxins in sediments from Siskiwit Lake, Isle Royale. Science 226:568-569.

Dulin, D., H. Drossman, and T. Mill. 1986. Products and quantitative yields for photolysis of chloroaromatics in water. Environ. Sci. Technol. 20(1):72-77.

Eitzer, B.D., and R.A. Hites. 1987. Dioxins and furans in the ambient atmosphere: A baseline study. (Abstract) Presented at the Seventh

International Symposium on Chlorinated Dioxins and Related Compounds, Las Vegas, Nevada, October 4-9, 1987. Chemosphere (submitted).

Eitzer, B.D., and R.A. Hites. 1986. Concentrations of dioxins and dibenzofurans in the atmosphere. Inter. J. Environ. Anal. Chem. 27:215-230.

Fries, G.F. 1986. Assessment of Potential Residues in Food Derived from Animals Exposed to TCDD Contaminated Soil. Presented at "Dioxin '86" Conference, Fukuoka, Japan.

Hattemer-Frey, H.A., and C. C. Travis, 1988. Pentachlorophenol: Environmental partitioning and human exposure. Archives of Environmental Contamination and Toxicology. (In press.)

Highland, J. 1986. Generic Risk Assessment Vol. 2. Prepared for the Ontario Waste Management Corporation. Environ Corporation, Washington, D.C.

Hutzinger, O., M.J. Blumich, M. van den Berg, and K. Olie. 1985. Sources and fate of PCDDs and PCDFs: An overview. Chemosphere 14(6/7): 581-600.

Isensee, A.R., and G.E. Jones. 1975. Distribution of 2,3,7,8-Tetrachlorodibenzo-p-dioxin (TCDD) in aquatic model ecosystem. Environ. Sci. Technol. 9(7):668-672.

Jensen, D.J., R.A. Hummel, N.H. Mahle, C.W. Kocher, and H.S. Higgins. 1981. A residue study on beef cattle consuming 2,3,7,8-tetrachlorodibenzo-p-dioxin. J. Agric. Food Chem. 29:265-268.

Jones, K.H., J. Walsh, and D. Alston. 1987. The statistical properties of available worldwide MSW combustion dioxin/furan emissions data as they apply to the donduct of risk assessments. Chemosphere 16(8/9):2183-2186.

Mackay, D., S. Paterson, and B. Cheung. 1985a. Evaluating the environmental fate of dhemicals: The fugacity-level III approach as applied to 2,3,7,8-TCDD. Chemosphere 14(6/7):859-863.

Mackay, D., S. Paterson, B. Cheung, and W.B. Neely. 1985b. Evaluating the environmental behavior of chemicals with a fugacity level III model. Chemosphere 14(3/4):335-374.

Mackay, D., and S. Paterson. 1982. Calculating gugacity. Environ. Sci. Technol. 15(9):1006-1014.

Marklund, S., C. Rappe, M. Tysklind, and K-E Egeback. 1987. Identification of polychlorinated dibenzofurans and dioxins in exhausts from cars run on leaded gasoline. Chemosphere 16(1):29-36.

Marklund, S., L-O. Kjeller, M. Hansson, M. Tysklind, C. Rappe, C. Ryan, H. Collazo, and R. Dougherty. 1986. Determination of PCDDs and PCDFs in incineration samples and pyrolytic products. In Chlorinated Dioxins and Dibenzofurans in Perspective, C. Rappe, G. Choudhary, and L. Keith, Eds., Lewis Publishers, Chelsea, Michigan.

Marple, L., R. Brunck, and L. Throop. 1986. Water solubility of 2,3,7,8-tetrachlorodibenzo-p-dioxin. Environ. Sci. Technol. 20:180-182.

Matsumura, F., and H.J. Benezet. 1973. Studies on the bioaccumulation and microbial degradation of 2,3,7,8-tetrachlorodibenzo-p-dioxin. Environ. Health Perspec. 5:253-258.

Mehrle, P., D.R. Buckler, E.E. Little, L.M. Smith, J.D. Petty, P.H. Peterman, D.L. Stalling, G.M. DeGraeve, J.J. Kyle, and W.J. Adams. 1987. Toxicity and bioconcentration of 2,3,7,8-tetrachlorodibenzodioxin and 2,3,7,8-tetrachlorodibenzofuran in rainbow trout. Environ. Toxicol. Chem. (Submitted).

Mill, T. 1985. Prediction of the environmental fate of tetrachlorodibenzodioxin. In Dioxins in the Environment, M.A. Kamrin and P.W. Rodgers, Eds., Hemisphere Press Corp., Washington, D.C., pp. 173-193.

Nakano, T., M. Tsuji, and T. Okuno. 1987. Level of chlorinated organic compound in the atmosphere. Chemosphere 16(8/9):1781-1786.

O'Keefe, P., C. Meyer, D. Hilker, K. Aldous, B. Jelus-Tyror, K. Dillon, R. Donnelly, E. Horn, and R. Sloan. 1983. Analysis of 2,3,7,8 tetrachlorodibenzo-p-dioxin in Great Lakes fishes. Chemosphere 12(3):325-332.

Ono, M., Y. Kashima, T. Wakimoto, and R. Tatsukawa, 1987. Daily intake of PCDDs and PCDFs by Japanese through food. Chemosphere 16(8/9):1823-1828.

Patterson, D.G., J.S. Holler, C.R. Lapeza, Jr., L.R. Alexander, D.F. Groce, R.C. O'Connor, S.J. Smith, J.A. Liddle, and L.L. Needham. 1986. High-Resolution Gas Chromatographic/High-Resolution Mass Spectrometric Analysis of Human Adipose Tissue for 2,3,7,8-Tetrachlorodibenzo-p-dioxin. Anal. Chem. 58:705-713.

Podoll, R.T., H.M. Jaber, and T. Mill. 1986. Tetrachlorodibenzodioxin: rates of volatilization and photolysis in the environment. Environ. Sci. Technol. 20(5):490-492.

Radian Corporation. 1987. Final Report, Deposition Modeling Analysis for Particulate Emissions from the Proposed Los Angeles City Energy Recovery LANCER Facility.

Rappe, C., L-O. Kjeller, P. Bruckmann, and K-H. Hackhe. 1988. Identification and quantification of PCDDs and PCDFs in urban air. Chemosphere 17(1):3-20.

Rappe, C., R. Andersson, P-A. Bergquist, C. Brohede, M. Hansson, L-O. Kjeller, G. Lindstrom, S. Marklund, M. Nygren, S.E. Swanson, M. Tysklind, and K. Wiberg. 1987a. Overview on environmental fate of chlorinated dioxins and dibenzofurans. Sources, levels and isomeric pattern in various matrices. Chemosphere 16(8/9):1603-1618.

Rappe, C., M. Nygren, G. Lindstrom, H.R. Buser, O. Blaser, and C. Wuthrich. 1987b. Polychlorinated dibenzofurans, dibenzo-p-dioxins and other chlorinated contaminants in cow milk from various locations in Switzerland. Environ. Sci. Technol. 21(10):964-970.

Rordorf, B.F. 1985. Thermodynamic and thermal properties of polychlorinated compounds: The vapor pressures and flow tube kinetics of ten dibenzo-para-dioxins. Chemosphere 14(6/7):885-892.

Ryan, J.J., J.C. Pilon, H.B.S. Conacher, and D. Firestone. 1983. Inter-laboratory study on determination of 2,3,7,8-tetrachlorodibenzo-p-dioxin in fish. J. Assoc. Off. Anal. Chem. 66(3):700-707.

Schroy, J.M., F.D. Hileman, and S.C. Cheng. 1985. Physical/chemical properties of 2,3,7,8-TCDD. Chemosphere. 14(6/7):877-880.

Spector, W.S. 1959. Handbook of Biological Data. W.B. Saunders Company, Philadelphia, p. 196.

Stalling, D.L., J.D. Petty, L.M. Smith, and W.J. Dunn, III. 1985. Dioxins and furans in the environment: A problem for chemometrics. In Dioxins in the Environment, M.A. Kamrin and P.W. Rodgers, Eds., Hemisphere Publishing Company, Washington, D.C., pp. 101-126.

Stalling, D.L., L.M. Smith, J.D. Petty, J.W. Hogan, J.L. Johnson, C. Rappe, and H.R. Buser. 1983. Residues of polychlorinated dibenzo-p-dioxins and dibenzofurans in laurentian Great Lakes fish. In Human and Environmental Risks of Chlorinated Dioxins and Related Compounds, R.E. Tucker, A.L. Young, and A.P. Gray, Eds., Plenum Press, New York, pp. 221-240.

Stevens, J.B., and E. Gerbec. 1987. Dioxin in the Food Chain: A Model for Calculating Health Risk from RDF Incinerators. Prepared for the Northern States Power Company, Minneapolis, Minnesota. School of Public Health, University of Minnesota, Minneapolis.

Swanson, S.E., C. Rappe, J. Malmstrom, and K.P. Kringstad. 1988. Emissions of PCDDs and PCDFs from the pulp industry. Chemosphere 17(4):681-691.

Swedish Environmental Protection Agency. Dioxin. May, 1987.

Travis, C.C., and A.D. Arms. 1987. The food chain as a source of toxic chemical exposure. In Toxic Chemicals, Health and the Environment. L.B. Lave and A.C. Upton, Eds., The Johns Hopkins University Press, Baltimore, MD; Chapter 5, pp. 95-113.

Travis, C.C., and A.D. Arms. 1988. Bioconcentration of organics in beef, milk, and vegetation. Environ. Sci. Technol. 22(3):271-274.

Travis, C.C., and H.A. Hattemer-Frey. 1987. Human exposure to 2,3,7,8-TCDD. Chemosphere 16(10-12):2331-2342.

Travis, C.C., and H.A. Hattemer-Frey. 1988. A perspective on human exposure to dioxin from municipal solid waste incinerators. J. Air. Poll. Control Assoc. (submitted).

Travis, C.C., G.A. Holton, E.L. Etnier, C. Cook, F.R. O'Donnell, D.M. Hetrick, and E. Dixon. 1986. Assessment of inhalation and ingestion population exposures from incinerated hazardous wastes. Environ. Inter. 12:553-540.

U.S. Environmental Protection Agency. (US EPA) 1986. The National Dioxin Study Tiers 3,5,6, and 7 (Draft Report), Washington, D.C.

U.S. Environmental Protection Agency (US EPA). 1987. The National Dioxin Study: Tiers 3,4,5, and 7. Office of Water Regulation and Standards, EPA-440/87/003, Washington, DC.

van den Berg, M., F.W.M. van der Wielen, K. Olie, and C.J. van Boxtel. 1986. The presence of PCDDs and PCDFs in human breast milk from the Netherlands. Chemosphere 15(6):693-706.

Weerasinghe, N.C.A., and M.L. Gross. 1985. Origins of polychlorodibenzo-p-dioxins (PCDD) and polychlorodibenzofurans (PCDF) in the environment. In Dioxins in the Environment, M.A. Kamrin and P.W. Rodgers, Eds., Hemisphere Publishing Company, Washington, D.C., pp. 133-151.

Key Words

1. Food chain
2. Background dioxin levels
3. Dioxin
4. Human exposure
5. Municipal waste incineration

MODELING THE UPTAKE AND DISTRIBUTION OF ORGANIC CHEMICALS IN PLANTS

Sally Paterson and Donald Mackay

Institute for Environmental Studies
University of Toronto
Toronto, Canada M5S 1A4

ABSTRACT

Increasing attention is being devoted to experimental measurement and modeling of the uptake, release and partitioning of toxic and persistent organic chemicals in plants. A primary reason for this is that contaminants present in air and soil may accumulate in plants which are subsequently eaten by humans directly or consumed indirectly by the plant - domestic animal - meat and dairy product route. It is suggested that development of plant toxicokinetic models, which is in its infancy, may follow the same route as pharmacokinetic fish-water models which have been developed to a greater level of advancement. An approach to modeling plant partitioning is described based on the fugacity concept. Avenues by which the model may be improved are discussed.

INTRODUCTION

It has become increasingly apparent that vegetation provides a significant source for human and animal intake of organic chemicals. Organochlorines and other persistent organic chemicals, including PCBs and dioxins, as well as pesticides and herbicides have been detected globally in a variety of grains, fruits, vegetables, foliage and lichens[1-7]. These chemicals thus become directly available to humans and herbivores through the consumption of vegetation, and indirectly through the consumption of meat, poultry and dairy products.

Many of these contaminants are very slowly degraded and are highly hydrophobic. In the environment, they tend to become sorbed to organic matter in soils and sediments which subsequently become reservoirs for plant uptake. Hydrophobic chemicals tend to accumulate in the roots and in leaf waxy cuticles. Water soluble chemicals may also enter the root system and may subsequently be transported by means of the transpiration stream through the xylem to the shoot system.

Pathways of chemicals to vegetation include:- root uptake and possible subsequent translocation; uptake from vapor in surrounding air; external contamination of foliage or stem by soil or particulate matter, followed by possible penetration of the cuticle; and uptake through oil channels, for oil-containing plants. Plants lose chemicals by transpiration and metabolism as well as direct volatilization from leaf surfaces.

Experimental work by Briggs and co-workers[1,8], Bacci et al.[9] and Kerler and Schonher[10,11] provides evidence that strong relationships exist between physical chemical properties of chemicals and their partitioning from air, water and soil into various plant tissues, such as roots, stem and foliage. Correlations have been observed between octanol/water partition coefficient (K_{OW}), vapor pressure and water solubility and accumulation in, and transport to, various parts of the plant.

It is noteworthy that present efforts to develop partitioning correlations for chemicals between soil, air and plants are similar to the situation which existed in the early 1970's when similar correlations were sought between concentrations of contaminants in fish and water. In these studies, correlations were developed using K_{OW} as the primary descriptor of chemical properties. The most successful early studies were of the steady state relationships which existed after prolonged exposure, i.e. when near-equilibrium was established. Whole fish concentrations were used in these correlations. Later, more detailed and valuable correlations were developed when it was determined that lipids were the primary site of accumulation. This enabled fish-to-fish variations to be accounted for by considering varying lipid contents. Only when the basic equilibrium partitioning behavior had been elucidated were the kinetics of uptake successfully addressed. Chemical-specific uptake rates were measured and correlated and the roles of water and food as sources of chemical determined. Various kinetic approaches were tried including expressions employing rate constants and mass transfer resistances. The broad basic relationships are now fairly clear and current work is focussing on detailed correlations of transfer rates as a function of chemical properties, fish size, species and condition. Pharmacokinetic models are now emerging describing the transport of chemical into and within fish.

Perhaps chemical models for plants will evolve similarly. If so, the first models will be simple partitioning models into whole plants, or plant parts, treating near-equilibrium situations. It is this early model development stage which is treated in this paper. The key information required is a description of the partitioning properties of whole plants or parts of plants for a variety of chemicals. It is likely that the appropriate chemical properties will be Henry's Law constants H (or air/water partition coefficients K_{AW}) and K_{OW}. In fugacity terms the aim is to establish correlations for fugacity capacities or Z values, as a function of chemical and plant tissue properties.

Later, when these partition coefficients or Z values are known with some reliability, it should be possible to propose expressions describing the kinetics of chemical uptake and transport, and ultimately assemble the plant equivalent of pharmacokinetic models. Such models will elucidate the relative uptake rates of chemical from air (both by absorption of gaseous chemical and deposition of aerosol particles) and from soil by water uptake through the root system.

A considerable literature already exists concerning the uptake and translocation of chemicals of agricultural importance, notably herbicides and plant growth regulating substances. These chemicals are usually fairly water-soluble and have log K_{OW} values in the range 0 to 3. Often they are weakly dissociating acids and they (or their metabolites) may undergo specific interactions with plant tissues. Their half lives are generally of the order of weeks. They thus have a quite different set of properties from the more hydrophobic (log $K_{OW} > 4$), nonpolar, stable chemicals which are of primary concern in human food sources. Correlations which have been developed for agrochemicals must thus be applied with extreme caution to hydrophobic chemicals.

Accordingly, this paper describes a preliminary attempt to quantify partitioning of organic contaminants from the environmental media of soil and air into plant tissues and organs. We develop and discuss a novel, equilibrium, fugacity-based model of a hypothetical plant which enables concentrations of chemicals in plant parts to be estimated from air and/or soil concentrations.

Fugacity

The fugacity concept, which we outline briefly below, has been used extensively to model the fate of organic chemicals in the environmental media of air, water, soil and sediment[12,13]. Fugacity can be considered a surrogate for concentration which facilitates manipulation of algebraic equations and interpretation of dynamic processes which describe the fate and distribution of environmental contaminants.

Fugacity f with units of pressure (Pa) is related to concentration C (mol/m^3) through a fugacity capacity Z (mol/m^3.Pa) where

$$C = fZ$$

Fugacity is an equilibrium criterion, since when two adjacent phases, with concentrations C_1 and C_2 respectively, are in equilibrium, their fugacities are equal and partitioning can be described in terms of their Z values, that is

$$C_1/C_2 = fZ_1/fZ_2 = Z_1/Z_2 = K_{12}$$

where K_{12} is a dimensionless partition coefficient.

A unique Z value exists for each chemical in each environmental medium and is dependent on the physical chemical properties of the chemical, such as aqueous solubility, vapor pressure and octanol/water partition coefficient (K_{OW}), and the properties of the phase including temperature. Derivation of these Z values has been extensively described, a summary being given in Table 1.

PLANT MODEL

In an attempt to describe the equilibrium distribution of organic chemicals in plants, we have developed a novel, fugacity-based plant model which is illustrated schematically in Figure 1. Two bulk compartments, air and soil, surround the plant. The air compartment is considered to consist of subcompartments of pure air and aerosols, designated A_1 and A_2 and the soil consists of subcompartments of air, water, organic and mineral matter, designated S_1 to S_4 respectively.

The plant is assumed to consist of seven compartments of defined dimensions and properties, namely roots, stem, xylem, phloem, inner leaf, cuticle, fruit and seeds, designated P_1 to P_7 respectively.

The dimensions assumed for the compartments of the plant are hypothetical but are similar to those of a soybean at an intermediate stage of growth.

The roots are assumed to occupy a volume of 100 cm^3 or 0.0001 m^3. The stem length is set at 20 cm and combined with a diameter of 1 cm to give a total stem volume of 15.7 cm^3. The xylem and phloem are each considered to have a diameter equal to one-tenth that of the stem, or 0.1 cm, and a total length of 50 cm in the plant, resulting in a capacity for each of approximately 0.392 cm^3. The average leaf area was assumed to be 100 cm^2 with a thickness of 1 mm. A total of 20 leaves per plant results in a total foliage volume of 200 cm^3. The leaf is subdivided into cuticle and inner leaf with cuticle having a thickness equal to 1/100 that of the total leaf, i.e. 10 μm. The resulting volumes for inner leaf and cuticle are 198 cm^3 and 2 cm^3 respectively, expressed in the model as volume fractions $\emptyset_{P5}$ and $\emptyset_{P6}$ of total leaf.

Table 1. Definition of Z values.

Compartment		
Air (Z_{A1})	$1/RT$	$R = 8.314$ (Pa.m^3/mol K) T = absolute temperature K
Aerosols (Z_{A2})	$6x10^6/(P_L^S RT)$	P_L^S = liquid vapor pressure (Pa)
Soil Air (Z_{S1})	$1/RT$	
Soil Water (Z_{S2})	$1/H$ or C^S/P^S	H = Henry's Law constant (Pam^3/mol) C^S = aqueous solubility (mol/m^3) P^S = vapor pressure (Pa)
Soil Organics (Z_{S3})	$K_{OC} \cdot D_{S3}/H$	K_{OC} = organic carbon partition coefficient $= 0.41\ K_{OW}$ D_{S3} = soil mineral density (kg/L)
Soil Mineral (Z_{S4})	not defined	
Plant roots (Z_{P1})	$RCF.Z_{S2} \cdot D_{P1}$	$RCF = 0.82 + 0.014\ K_{OW}$
Plant stem (ZP_2)	$SXCF.Z_{S2} \cdot D_{P2}$	$SXCF = 0.82 + 0.0065\ K_{OW}$
Plant xylem contents (Z_{P3})	$1/H$	
Plant phloem contents (Z_{P4})	$1/H$	
Plant inner leaf (Z_{P5})	$1/H$	
Plant cuticle (Z_{P5})	K_{OW}/H	
Plant fruit, seeds (Z_{P7})	not defined	
Plant bulk leaf	$\emptyset_{P5} \cdot Z_{P5} + \emptyset_{P6} \cdot Z_{P6}$	

Table 2. Densities and Volumes of Model Compartments

Compartment	Subscript	Density (kg/m^3)	Volume (m^3)
Air	A1	1.19	-
Aerosols	A2	1500	-
Soil air	S1	1.19	-
Soil water	S2	1000	-
Soil organics	S3	1500	-
Soil mineral	S4	-	-
Plant roots	P1	900	0.0001
Plant stem	P2	900	1.57×10^{-5}
Plant xylem	P3	1000	3.92×10^{-7}
Plant phloem	P4	1000	3.92×10^{-7}
Plant inner leaf	P5	506	1.98×10^{-4}
Plant cuticle	P6	1000	2.0×10^{-6}
Plant fruit and seed	P7	-	-
Bulk leaf	BL	1000	0.0002

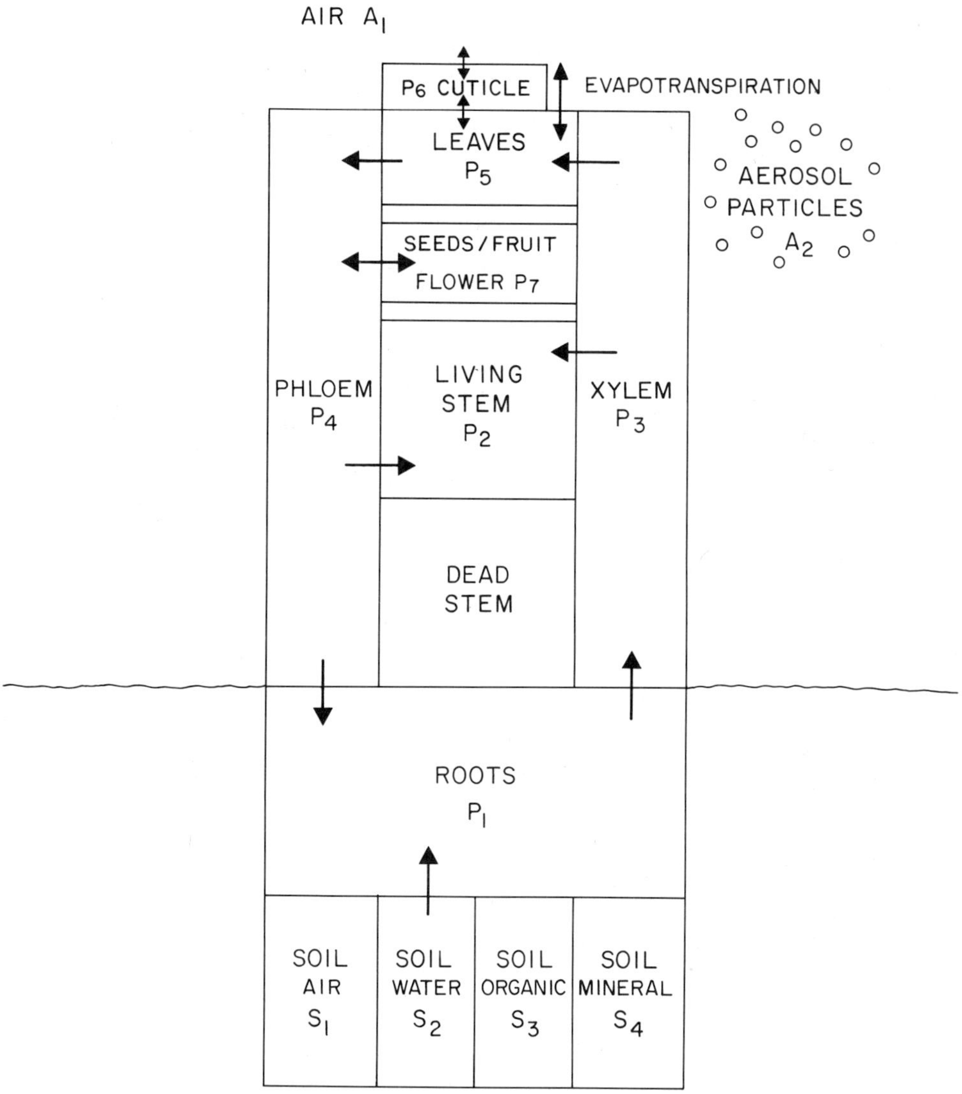

Figure 1. Schematic Plant Model

Volumes of air and soil are not defined, because only concentrations, and not amounts, are required by the model for these phases.

The roots and stem are each considered to have a density D_{P1} and D_{P2} of 900 kg/m^3[(14)]. The xylem contents or transpiration stream and the phloem contents which transports nourishment to plant organs and tissues are assumed to have the properties of water and therefore have densities D_{P3} and D_{P4} of 1000 kg/m^3.

For purposes of this model, the leaf is assumed to consist of two subcompartments:- i) an inner space, occupied by air, water and lipid material with inner leaf volume fractions of 0.5, 0.48 and 0.02 respectively, and ii) the outer waxy cuticle which covers the surface and has properties similar to octanol. The resulting densities of total inner leaf, D_{P5}, and cuticle, D_{P6} are 506 and 1000 kg/m^3 respectively.

The plant compartments and their dimensions are given in Table 2 along with density and organic content of air, soil and plant compartments where applicable.

Fugacity Capacities

Fugacity capacities or Z values for air and soil and their subcompartments are calculated by conventional methods. Other Z values are calculated from them using partition coefficients as described below.

Roots and Stem. In the development of Z values for root and stem of the plant we have modified correlations developed by Briggs and coworkers[1,8] for concentration factors in root and stem.

Briefly, Briggs et al[1] developed a root concentration factor RCF where

$$RCF = \frac{\text{Concentration in roots (fresh wt)}}{\text{concentration in water}}$$

and

$$\log (RCF - 0.82) = 0.77 \log K_{OW} - 1.52$$

or

$$RCF = 0.82 + 0.03 K_{OW}^{0.77}$$

The value of the power 0.77 is critically important if this correlation is to be applied to more hydrophobic chemicals. Noting that RCF is C_R/C_W and K_{OW} is C_O/C_W, subscripts R, W and O referring to root, water and octanol, we can rearrange the equation to give

$$C_R = 0.82 C_W + 0.03(C_O/C_W)^{0.77}.C_W$$

The term 0.82 C_W presumably represents the capacity of the water contained in the root for chemical, i.e. the root behaves as if it was 82% water. The second term presumably represents the capacity of the root tissues which equals the root water capacity when K_{OW} is 73. For more hydrophobic chemicals, which are of primary interest here, this second term dominates. It is difficult to justify the 0.77 power on thermodynamic partitioning grounds, a power of 1.0 being more likely. Such a power would imply that the plant tissues behave like octanol, i.e. they enable the chemical to form a near-ideal solution. It is possible that in the

48 hour duration of the experiment the more hydrophobic chemicals did not achieve equilibrium, thus RCF is underestimated. Hypothesizing that the power is 1.0, results in the correlation

$$RCF = 0.82 + 0.14\ K_{OW}$$

A plot of this second curve illustrated in Figure 2 fits the data well up to a value of log K_{OW} approximately equal to 2.5. Beyond this point, the experimental points lie below the correlated curve. This is expected if, due to their low water solubility, these chemicals had not achieved equilibrium in the roots.

Measuring partitioning between macerated stems and external solution, and assuming the contribution of the aqueous phase in the stems to uptake of chemical from the xylem transpiration stream is similar to that in roots, Briggs et al[8] developed the following correlation

$$\log (K_{stem/xylem\ sap} - 0.82) = 0.95 \log K_{OW} - 2.05$$

Applying the above assumption of non-equilibrium partitioning (in this case into the stem) for more hydrophobic chemicals and linearizing the correlation results in the relationship for a stem/xylem sap concentration factor SXCF where

$$SXCF = 0.82 + 0.0065\ K_{OW}$$

Since RCF and SXCF describe partitioning from roots and stem to water and xylem sap respectively, they can be considered to be the ratio of their respective Z values

$$RCF = Z_{P1}D_{S2}/(Z_{S2}.D_{P1})$$

and $$SXCF = Z_{P2}D_{P3}/(Z_{S2}.D_{P2})$$

It is necessary to correct for density because the Briggs correlations express concentration on a mass fraction (g/g) basis whereas partitioning in terms of Z values is described as the ratio between concentrations in mass/volume units (eg. g/m^3 or mol/m^3).

A similar approach to the calculation of these Z values using the Briggs correlation in their original form has been discussed by Calamari et al[14].

Therefore since K_{OW} describes equilibrium partitioning between octanol and water, Z values for roots and stem respectively can be described as Z_{P1} and Z_{P2}

$$\begin{aligned} Z_{P1} &= RCF.Z_{S2}.D_{P1}/D_{S2} \\ &= (0.82\ Z_W + 0.014\ Z_O)D_{P1}/D_{S2} \end{aligned}$$

$$\begin{aligned} Z_{P2} &= SXCF.Z_{S2}.D_{P2} \\ &= (0.82\ Z_W + 0.0065\ Z_O)D_{P2}/D_{P3} \end{aligned}$$

where Z_W and Z_O are the fugacity capacities of water and octanol respectively. For purposes of this model, the contents of the xylem and phloem are assumed to have a fugacity capacity equal to that of water and

$Z_{P3} = 1/H$ xylem contents
$Z_{P4} = 1/H$ phloem contents

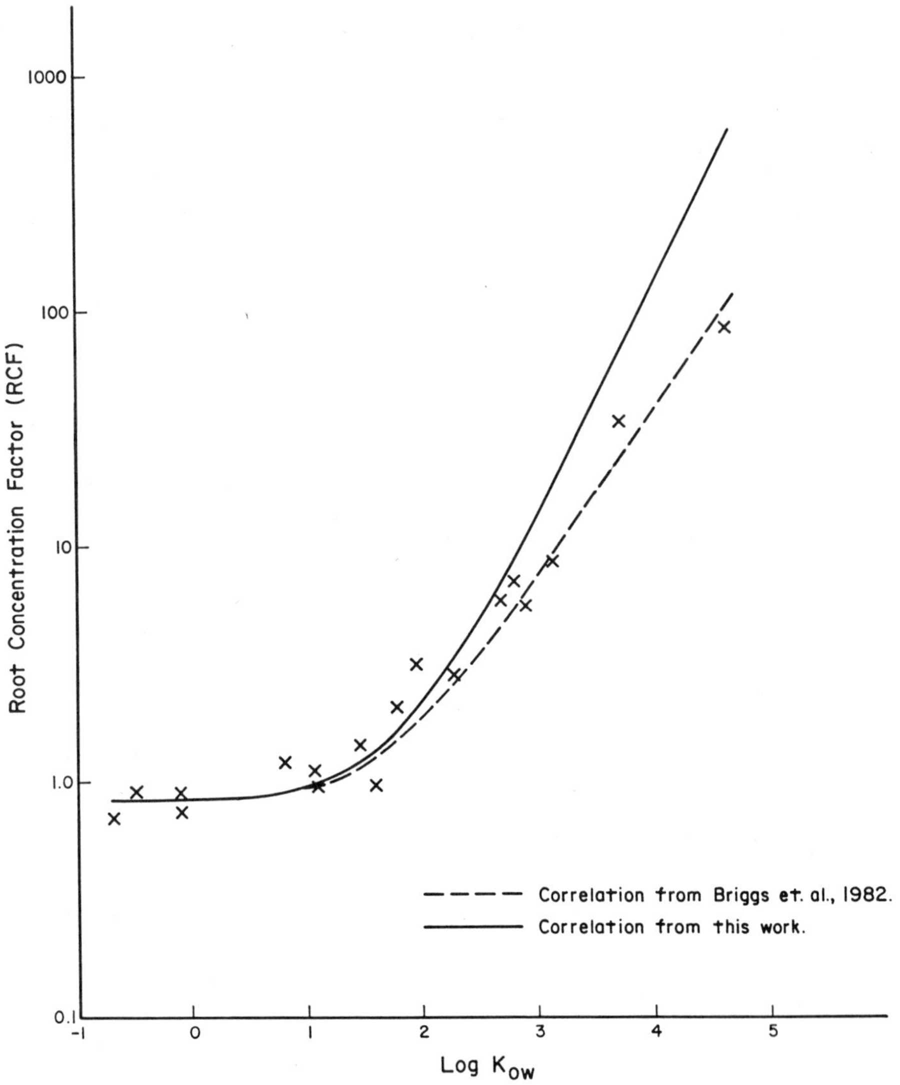

Figure 2. Plot of Root Concentration Factor (RCF) versus log octanol/water partition coefficient (K_{OW}) illustrating correlation from this work (——) and correlation from Briggs et al., 1982[1] (----)

<u>Foliage.</u> A novel approach is used in the calculation of the Z value for foliage. The inner leaf assumed to consist of 50% air, 48% water and 2% lipid has a fugacity capacity Z_{P5} where

$$Z_{P5} = 0.5/RT + 0.48/H + 0.02\ K_{OW}/H$$

Kerler and Schonherr (1988) determined cuticle/water partition coefficients for a number of chemicals, with log K_{OW} between 2 and 8, in leaves of rubber plant and bitter orange. They observed that cuticle/water and octanol/water partition coefficients were similar in magnitude. Here, we assume the cuticle to have the properties of octanol with a fugacity capacity Z_{P6} where

$$Z_{P6} = Z_o = K_{OW} \cdot Z_W = K_{OW}/H$$

By combining the Z values for inner leaf and cuticle with their appropriate volume fractions, a total leaf fugacity capacity, Z_{BL}, can be calculated as

$$Z_{BL} = \phi_{P5} Z_{P5} + \phi_{P6} Z_{P6}$$

These plant Z values are included in Table 1. It must be emphasized that the Z value correlation presented above are hypotheses which are not justified by experimental data. It is clearly important to conduct experiments involving partitioning of a range of stable chemicals into various live, dead and macerated plant tissues to establish the nature of the partitioning properties. We believe that developing these correlations using single phase Z values helps to elucidate the phenomena. This avoids some of the obscuring features of partition coefficients which are inherently properties of chemical in two phases. It remains to be established how these plant-water-air partitioning characteristics are best correlated, and indeed if K_{OW} or Z_O and Z_W are adequate descriptors of the processes.

Chemicals in air may be absorbed into the cuticle with possible subsequent diffusion through the cuticle, or may enter via the stomata or small pores. In addition, chemicals may partition from soil water into the roots. The chemical may then migrate through the xylem and phloem to all other parts of the plant. No attempt is made here to describe these transfer rates, it being assumed for this preliminary model that transport is sufficiently rapid that equilibrium (equal fugacity) is established throughout the entire plant.

For illustrative purposes, the soil, air and all compartments of the plant are assumed to achieve equilibrium and thus have a common fugacity of f (Pa).

Input to the system is in the form of physical chemical properties of the chemical and the fugacity of the system which can be calculated from reported soil or air concentrations as

$$f = C_i/Z_i$$

Z values are calculated for each plant compartment and combined with this common fugacity, f, to calculate the concentration C_i(mol/m^3) in each compartment of the plant

$$C_{pi} = fZ_{pi}$$

The amount M_i (mol) in each compartment is calculated as

$$M_{pi} = C_{pi} V_{pi}$$

Table 3. Physical Chemical Properties and Z values of Five Chemicals[15].

	HCB	p,p'-DDE	p,p'-DDT	α-HCH	γ-HCH
	Z Values (mol/m^3.Pa)				
Molecular Weight (g/mol)	284.8	318.0	345.5	291.0	291.0
Aqueous solubility (g/m^3)	0.005	0.04	0.003	1.0	6.5
Vapor Pressure (Pa)	0.0015	0.001	0.00002	0.003	0.003
Log K_{OW}	5.47	5.7	6.0	3.81	3.8
Z values for Plant (mol/m^3.Pa)					
root	43.5	794	5470	94.0	597
stem	20.2	369	2540	44.1	280
xylem contents	0.012	0.13	0.43	1.1	7.4
phloem contents	0.012	0.13	0.43	1.1	7.4
inner leaf	69	1260	8680	149	943
cuticle	3450	$6.3x10^4$	$4.3x10^5$	7400	$4.7x10^4$
bulk leaf	103	1880	$1.3x10^4$	221	1400
fruit and seeds	-	-			
Organic carbon content of soil - 0.15%					
Air					
Gaseous	$4.04x10^{-4}$	$4.04x10^{-4}$	$4.04x10^{-4}$	$4.04x10^{-4}$	$4.04x10^{-4}$
Aerosols	$1.61x10^6$	$2.42x10^6$	$1.21x10^8$	$8.07x10^5$	$8.07x10^5$
Soil water	0.012	0.13	0.43	1.1	7.4

Table 4. Calculated distribution of five chemicals in plants assuming equal fugacity in all compartments. Reported values at 60 days are given in parentheses (Bacci et al[16]).

Chemical	Concentration in air (mol/m^3)	Fugacity (Pa)	Concentration in Plant Tissues (μg/g) root	stem	foliage
HCB	1.1×10^{-8}	2.7×10^{-5}	0.037	0.17	1.6 (1.3)
p,p'-DDE	3.0×10^{-9}	7.5×10^{-6}	2.1	0.98	8.9 (25)
p,p'-DDT	2.5×10^{-10}	6.2×10^{-7}	1.3	0.60	5.5 (5.4)
α-HCH	2.5×10^{-8}	6.2×10^{-5}	1.9	0.88	7.9 (3.8)
γ-HCH	6.6×10^{-8}	1.6×10^{-4}	32	15	132 (40.6)

Amount (μg) and percentage distribution of p,p'-DDT in various compartments.

roots - 117 (17.1%), stem - 8.5 (1.2%), inner leaf - 372 (54.5%), cuticle - 186 (27.2%)

and the total amount M_T (mol) in the plant as

$$M_T = \Sigma_i M_{pi}$$

Table 3 gives physical chemical properties and estimates Z values for five chemicals which are considered in the next section.

APPLICATION OF THE MODEL

In a study to investigate the significance of root uptake in contamination of foliage Bacci and coworkers[16] experimentally determined foliar concentrations of several pesticides in dwarf bean plants. The plants were grown in containers of both contaminated and uncontaminated soil which were enclosed in a large, covered glass box. They concluded that the foliage concentrations were not directly dependent on the degree of contamination of the soil in which they were grown, but rather that these concentrations were due to vapor uptake of chemical which had volatilized from soil. Their experimental results are used in a preliminary application of the plant model.

Although the Bacci system was not at equilibrium, it was found that levels of each pesticide in the air of the green house remained almost constant over the two month period of observation.

By calculating the fugacity of the air as

$$f_{A1} = C_{A1}/Z_{A1}$$

and assuming this value to be the equilibrium fugacity f, and applicable to all plant parts, the concentration in plant compartments can be calculated as

$$C_{pi} = fZ_{pi}$$

Results of these calculations for hexachlorobenzene, p,p'-DDT, p,p'-DDE, hexachlorobenzene (HCB), α- and γ- hexachlorocyclohexane (HCH) are presented in Table 4 and compared with those measured by Bacci et al[16]. The compartment amounts and percentage distribution of p,p'-DDT in roots, stem and foliage are also given.

DISCUSSION

It is apparent that the model yields reasonable results for the limited number of chemicals tested and the limited plant concentration data available.

Calculation of the percentage distribution of p,p'-DDT shows that this chemical accumulates mainly in the leaf (82%), with 17% in the roots and negligible amounts in the rest of the plant. The other four chemicals show a similar distribution pattern. Partitioning into root and leaf for these five hydrophobic chemicals is controlled by the organic phase, i.e. by the Z_O component of the overall Z value.

We believe that simple models such as this could, with some refinement, give a satisfactory description of the partitioning of chemicals in plants provided that

(i) the chemical can be transported fairly rapidly in the plant, i.e. there are no excessive transport resistances which prevent or delay equilibrium being attained.

(ii) there is insignificant metabolism or other degradation process

(iii) there are no unusual partitioning tendencies attributable to specific chemical interactions between plant components and chemical.

There is a clear need to develop and test the concentration factor expressions, and from them devise more general and reliable methods of calculating Z values. In viewing the plant parts as combinations of water, octanol and air, we have copied the approach which has been used for fish in which the fish is viewed as perhaps 5% octanol 95% water. These "rules" need to be refined and further developed.

The next stage is to include transport rate expressions and address the question of how long it will take to achieve equilibrium. In fish studies this has been done using rate constants. In the fugacity approach D values (with units of mol/Pa.h) are used to express rates of transfer by bulk flow and diffusion and rates of transformation[12]. Fortunately a considerable literature exists on the transport of water through plants (transpiration) and on the gaseous exchange rates of CO_2 and water vapor between leaves and the atmosphere. This literature can be exploited to give transport parameters and ultimately D values, potentially resulting in a complete description of the "toxicokinetics" of chemicals in plants. Such models may be similar to pharmacokinetic models describing the partitioning of chemicals within human organs, tissues and fluids[17].

These D values are essentially conductivities which quantify the ease with which chemical can migrate from one compartment to another. They can also be regarded as reciprocal resistances. The Bacci et al experiments suggest that when DDT is present in soil it has the option of two transport routes to foliage:- (i) uptake by roots and transport in the sap through the xylem, or (ii) evaporation from soil to air and uptake from the air by the leaf surface. The first is primarily an aqueous route and the second a gaseous route. Estimation of the D values corresponding to these routes suggests that the second (air) D value is much larger, thus the DDT experiences a much lower resistance to transfer from soil to leaf by the air route. For a less volatile chemical such as 2,4-D the opposite would apply. It thus becomes possible in principle to ascertain the relative importance of air and soil as sources of contaminant to specific plant parts. Plant models can then be combined with soil transport and transformation models.

Certain difficulties or challenges arise with plants. First, plants, unlike humans and fish, grow very rapidly, thus pseudo steady state models in which the compartment volumes are treated as being constant in the time of the uptake of chemical may not be accurate. In fish studies a "growth dilution" term is often included. How this is best treated in plants is not clear. Plants undergo marked diurnal changes in transport rates associated with changes in photosynthesis. Stomata open and close in accordance with changes in light and water stress altering diffusion rates of water and CO_2. It seems likely that there will be cycling of chemical adsorbed or absorbed to foliage when the foliage falls to the soil and becomes incorporated in litter. The nature of the resistances to chemical transfer offered by root membranes is far from clear.

There is an obvious need for a combination of carefully designed and executed experiments with the development, testing, modification and improvement of models which describe the processes of partitioning, transport and transformation. In this paper we have presented a model which represents only a start on the journey towards comprehensive, validated models of chemical behavior in plants.

ACKNOWLEDGEMENTS

The authors are grateful to Dr. T.C. Hutchinson, Department of Botany, University of Toronto for his critical comments and advice, and are indebted to NSERC for a Strategic Grant, and the Ontario Ministry of the Environment for funding.

REFERENCES

1. G. G. Briggs, R. H. Bromilow and A. A. Evans, Relationships between lipophilicity and root uptake and translocation of non-ionized chemicals in barley, Pest. Sci., 13:495-504 (1982).
2. Y. Iwata and F. A. Gunter, Translocation of the Polychlorinated Biphenyl Aroclor 1254 from soil into carrots under field conditions, Environ. Contam. Toxicol. 4:44-59 (1976).
3. K. Davies, Concentrations and dietary uptakes of selected organochlorines, including PCBs, PCDDs and PCDFs in fresh food composites grown in Ontario, Canada, Chemosphere 17:263-276 (1988).
4. Ontario Ministry of Agriculture and Food (OMAF) and Ministry of the Environment (MOE), Polychlorinated Dibenzo-p-Dioxins and Polychlorinated Dibenzofurans and other Organochlorine Contaminants in Food. Report by Joint OMAF/MOE Toxics in Food Steering Committee.
5. C. Gaggi and E. Bacci, Accumulation of chlorinated hydrocarbon vapors in pine needles, Chemosphere 14:451-456 (1985).
6. E. Bacci, D. Calamari, C. Gaggi, C. Biney, S. Focardi, and M. Morosini, Organochlorine pesticides and PCB residues in plant foliage (Magnifera indica) from West Africa, Chemosphere 17:693-702 (1988).
7. J. P. Villeneuve and E. Holm, Atmospheric background of chlorinated hydrocarbons studied in Swedish lichens, Chemosphere 13:1133-1138 (1984).
8. G. G. Briggs, R. H. Bromilow, A. A. Evans and M. Williams, Relationship between lipophilicity and the distribution of non-ionised chemicals in barley shoots following uptake by roots, Pest. Sci. 14:492-500 (1983).
9. E. Bacci and C. Gaggi, Chlorinated hydrocarbons vapours and plant foliage: kinetics and applications, Chemosphere 16:2515-2522 (1987).
10. F. Kerler and J. Schonherr, Permeation of lipophilic chemicals across plant cuticles: prediction from partition coefficients and molar volumes, Arch. Environ. Contam. Toxicol. 17:7-12 (1988).
11. F. Kerler and J. Schonherr, Accumulation of lipophilic chemicals in plant cuticles: prediction from octanol-water partition coefficients, Arch. Environ. Contam. Toxicol. 17:1-6 (1988).
12. D. Mackay, S. Paterson, B. Cheung, and W. B. Neely, Evaluating the Environmental behaviour of Chemicals with a Level III fugacity model, Chemosphere 14:335-374 (1985).
13. D. Mackay and S. Paterson, Evaluating the regional multimedia fate of organic chemicals: a Level III fugacity model. Report to the Ontario Ministry of the Environment (1988).
14. D. Calamari, M. Vighi and E. Bacci, The use of terrestrial plant biomass as a parameter in the fugacity model, Chemosphere 16:2359-2366 (1987).
15. L. Suntio, W. Y. Shiu, D. Mackay, J. N. Seiber and D. Glotfety, Critical review of Henry's Law constants for pesticies, Rev. Environ. Contam. 103:1-59 (1988).

16. E. Bacci and C. Gaggi, Chlorinated pesticides and plant foliage: translation experiments, <u>Bull. Environ. Contam. Toxicol.</u> 37:850 (1986).
17. S. Paterson and D. Mackay, A steady state fugacity-based pharmacokinetic model with simultaneous multiple exposure routes, <u>Environ. Toxicol. Chem.</u> 6:395-408 (1987).

MULTIPLE PATHWAY EXPOSURE FACTORS (PEFs) ASSOCIATED WITH MULTIMEDIA POLLUTANTS*

Thomas E. McKone

University of California
Lawrence Livermore National Laboratory
Livermore, CA 94550

INTRODUCTION

When pollutants enter the environment, they contact people through air, water, food, and soil in varying amounts throughout a lifetime. Managing the health and environmental risks of multimedia pollutants requires an integrated and comprehensive assessment of environmental transport and human exposure. In response to this need, a number of multimedia monitoring and modelling efforts have emerged. However, the need remains to integrate multimedia concentration data with a comprehensive picture of human exposure.

According to the U.S. EPA (1987) exposure can be defined as "the contact with a chemical or physical agent. The magnitude of the exposure is determined by measuring or estimating the amount of an agent available at the exchange boundaries, i.e., lungs, gut, skin during some specified time." This paper describes methods for addressing several potential exposure pathways and provides a link between human exposure and chemical concentrations in multiple environmental media. This approach links environmental concentrations to human exposure through pathway-exposure factors (PEFs). The PEF incorporates information on human physiology, human behavior patterns, and environmental transport into a term that translates a unit concentration (in mg/m^3, mg/kg, or mg/L) in a specified environmental media (air, soil, or water) into daily exposure in mg/kg-d for a specified route (inhalation, ingestion, or dermal absorption). This process of exploring the data associated with human/environment interactions and proposing exposure models provides insight for risk-management activities.

Table 1 lists the matrix of PEFs that are used to link nine exposure pathways with the five environmental media. The methods used to derive each expression in this table are summarized here and discussed in greater detail in McKone (1988). The procedures used to calculate PEFs are compatible with the multimedia GEOTOX model that was developed to simulate the transport and fate of chemicals in multiple media (McKone and Layton, 1986). The GEOTOX model explicitly calculates concentrations in air, ground and surface water, and soil. Nevertheless, it is equally feasible to obtain the required input concentrations from direct measurement or simulations using media-specific models (e.g., air dispersion, ground-water transport, etc.).

---

*Work performed under the auspices of the U.S. Department of Energy by the Lawrence Livermore National Laboratory under contract number W-7405-ENG-48.

Table 1. Summary matrix of values and expressions for the reference population PEFs derived in this paper[a].

	Environmental Concentrations				
Pathways	Air (gases) C_a PEF Units = m^3/kg-d	Air (particles) C_p m^3/kg-d	Soil C_s kg/kg-d	Ground Water C_g L/kg-d	Surface water C_r L/kg-d
Inhalation	$F_{aa} = 0.38$	$F_{pa} = 0.31$	$F_{sa} = 9.0 \times 10^{-9}$	$F_{wa} = 3.2 \times 10^5 \times \left[2.5/D_{\ell}^{2/3} + RT/\left(D_a^{2/3} H\right)\right]^{-1}$	
Ingestion					
Water	--	--	--	$F_{ww} = 0.032$	
Fruits, vegetables	$F_{av} = \mathrm{Min}(16, 0.0025\ RT/H)$[b]	$F_{pv} = 14$	$F_{sv} = 1.1 \times 10^{-3}\ K_{sp}$	--	--
Grains	$F_{ag} = \mathrm{Min}(26, 0.0039\ RT/H)$	$F_{pg} = 22$	$F_{sg} = 7.9 \times 10^{-4}\ K_{sp}$	--	--
Milk	$F_{ak} = \mathrm{Min}[6800\ B_k, (0.85 + 0.12\ RT/H)B_k]$	$F_{pk} = 6600\ B_k$	$F_{sk} = (0.0028 + 0.12\ K_{sp})\ B_k$	$F_{wk} = 0.33\ B_k$	
Meat	$F_{at} = \mathrm{Min}[2500\ B_t, (0.38 + 0.037\ RT/H)B_t]$	$F_{pt} = 2100\ Bt$	$F_{st} = (0.0012 + 0.038\ K_{sp})\ B_t$	$F_{wt} = 0.14\ B_t$	
Fish	--	--	--	--	$F_{rf} = 3.2 \times 10^{-4}$ BCF
Soil	--	--	$F_{ss} = 1.5 \times 10^{-6}$	--	
Dermal absorption	--	--	$F_{sd} = 2.6 \times 10^{-6}$	$F_{wd} = 0.038$	

[a] Subscripts refer to the source media (a = air (gases), p = air (particles), s = soil, w = potable water and r = surface water) and pathways (a = inhalation, w = water, v = vegetables, g = grain, t = meat, k = milk, f = fish, s = soil ingestion, and d = dermal absorption). Concentrations in potable water C_w are taken as the average of surface, C_r, and groundwater, C_g, concentrations. D_{ℓ} refers to the water diffusion coefficient and D_a refers to air diffusion coefficient.

[b] The function Min (a,b) implies that one takes the minimum of a or b.

The remainder of this paper consists of four sections. The first provides a discussion of exposure routes and PEFs associated with contaminants in ambient outdoor air, including both gases and particles. The next deals with the transfer of contaminants from soils to humans through ingestion and indoor inhalation. The third section addresses PEFs that relate contaminant levels in ground and surface water to human exposure. The final section provides a summary and discussion of the methods presented in this report. This section also provides a sample calculation that is used to illustrate the use of the pathway exposure factors for the contaminants arsenic, benzene, and TNT.

EXPOSURE PATHWAYS ASSOCIATED WITH CONTAMINATED OUTDOOR AIR

This section deals with models used to calculate inhalation and ingestion exposures attributable to gaseous or particulate contaminants in ambient outdoor air. The concentrations C_a and C_p in mg/m^3 of the contaminant as a gas or attached to particles in air are the starting points for the exposure estimates.

Inhalation Exposures

Given the concentrations of contaminant in the gaseous and solid phases of outdoor air, and assuming an adult or a child is active 16 hours per day; resting 8 hours per day; spending 2/3 of their active hours indoors, experiencing 0.75 times as much suspended particulate matter indoors as outdoors (Hawley, 1985); breathing 0.020 m^3/kg-h while active, and breathing 0.0071 m^3/kg-h while resting (McKone, 1988); we calculate the inhalation PEFs, F_{pa} and F_{aa} in Table 1.

Fruits, Vegetables, and Grains

Small atmospheric particles containing contaminants can transfer these contaminants directly to the surfaces of fruits and vegetables by deposition. The pathway exposure factors F_{pv} and F_{pg} account for the ingestion of contaminants as a result of particle deposition on fruits, vegetables, and grains. Contaminants in the gaseous phase of the atmosphere can also be intercepted by the surfaces of fruits, vegetables, and grains. In order to calculate the transfer of atmospheric particles from atmosphere to fruits, vegetables, and grains; we consider the balance between material that deposits on the exposed and edible portion of food crops and material that is removed by weathering and senescence. For example, the PEF for transfer from gases in air to vegetables and fruits is calculated as:

$$F_{av} = \left(\frac{I_v}{BW}\right) \times 0.47\ V_{da}\ f_v/(M_f R_v) \qquad (1)$$

where (I_v/BW) is the human intake of vegetables and fruits per unit body weight, 0.0053 kg (fresh mass)/kg-d; the intake of grains per unit body weight is 0.0040 kg (fresh mass)/kg-d; 0.47 is the fraction of the total mass of ingested fruits and vegetables that consists of unprotected produce or leafy vegetables; V_{da} is the deposition velocity of contaminant gas molecules onto food crops, ~ 600 m/d; the deposition velocity of atmospheric particles onto food crops is ~ 500 m/d; f_v is the fraction of the target population's vegetables and fruits that come from the area with atmospheric concentrations C_a and C_p (set to unity); M_f is the annual average inventory of food crops per unit area, 3.0 kg (fresh mass)/m^2; and R_v is the rate constant for the removal of chemicals from vegetation surfaces as a result of weathering and senescence, day^{-1}.

For particles, R_v is on the order of 0.03 day^{-1}. For gases deposited onto fruits, vegetables, grains, and pasture, we assume that the amount of contaminant incorporated into the plant tissues from the air meets the restriction that $C_a/C_v \leq H/RT \times 10^3\ L/m^3$; where C_a is the air concentration in mg/m^3, C_v is the plant concentration in mg/kg (= mg/L), H is the Henry's law constant in torr-L/mol, R is the gas constant (62.4 torr-L/mol-K), and T is the temperature in kelvins. This makes the removal rate constant R_v for gases dependent on the Henry's law constant and leads to the condition that

$$F_{av} = \text{Min}\ (16,\ 0.0025\ RT/H) \tag{2}$$

$$F_{ag} = \text{Min}\ (26,\ 0.0039\ RT/H) \tag{3}$$

and the function Min (a,b) implies that one uses the minimum of a and b.

Milk and Meat Exposures

Concentrations of contaminants in the solid and gaseous phases of the atmosphere, C_p and C_a, can lead to contamination of meat and milk and thus result in human exposures through the inhalation of atmospheric gases and particles by cattle and the ingestion of contaminants deposited on vegetation consumed by meat- and milk-producing cattle.

The PEF for meat or dairy pathways links the intake by humans of contaminants to atmospheric concentration, inhalation by cattle, deposition onto pasture, uptake by cattle, and biotransfer factors. As an example, the PEF for dairy exposure is defined by:

$$F_{pk} = \frac{I_k}{BW} \left[I_{nc} + (I_{vdc}\ V_{dp}) / (M_p R_v) \right] f_k\ B_k \tag{4}$$

where I_k/BW is the human intake of dairy products per unit body weight, 0.0070 kg (fresh mass)/kg-d; the human intake per unit body weight of meat products is 0.0031 kg (fresh mass)/kg-d; I_{nc} is the inhalation rate for beef and dairy cattle, 120 m^3/d; I_{vdc} is the ingestion rate of pasture grasses by beef cattle, 17 kg (dry mass)/d; the ingestion rate of pasture grasses by dairy cattle is 12 kg (dry mass)/d; M_p is the annual average inventory of pasture crops per unit area, 0.3 kg (dry mass)/m^2; R_v is the weathering and senescence rate constant (= Max[0.03; $(V_{da}H)/(RTM_p)$] day^{-1}); f_k is the fraction of the target population's dairy products that come from an area with contaminant concentrations C_p (set to unity); and B_k is the biotransfer factor from cattle uptake to dairy products, the steady-state contaminant concentration in fresh dairy products divided by the daily contaminant intake by dairy cattle, d/kg i.e. (mg/kg)/(mg/d). Travis and Arms (1988) have proposed correlations for estimating biotransfer factors in meat and milk.

EXPOSURE PATHWAYS ASSOCIATED WITH CONTAMINATED SOILS

This section addresses methods for calculating PEFs that relate contaminant exposure through inhalation, ingestion, and dermal absorption to contaminant concentrations in soil. The concentration C_s (in mg/kg) in surface soil is the starting point for these exposure estimates.

Indoor Inhalation Exposures

According to Murphy and Yocom (1986), several studies around smelters indicate that surface soil and dust tracked into buildings on shoes or clothes, by pets, or other sources is an important source of indoor suspended particles. They have estimated that the concentration of indoor particles attributable to surface soil is on the order of 30 $\mu g/m^3$ or 3.0 x 10^{-8} kg (soil)/m^3(air). Based on this value, and the assumption that both adults and children daily spend 4 out of 16 active hours outdoors, 12 out of 16 active hours indoors, and 8 out of 8 resting hours indoors, we have calculated the soil/indoor inhalation PEF, F_{sa}, to be on the order of 9 x 10^{-9} (mg/kg-d)/(mg/kg).

Fruit, Vegetable, and Grain Exposures

This PEF accounts for the transfer of contaminants in soil to the internal portion of fruits, vegetables and grains. The concentrations in the fresh mass of biota can be estimated using the plant/soil partition coefficient K_{sp}, which is the ratio of the contaminant concentration in biota dry mass per unit concentration in soil, [mg/kg(plant DM)]/[mg/kg(soil)]. The PEFs relating contaminant concentrations in fruits and vegetables (F_{sv}) (or grains, F_{sg}) is the product of K_{sp} and the dry mass intake of fruits and vegetables per unit body weight, kg/kg-d. We assume the dry mass fraction of fruits, vegetables and grains to be on the order of 0.2.

Milk and Meat Exposures

Contaminants in surface soil can enter meat or milk through either the direct ingestion of soil by cattle or the ingestion by cattle of pasture grass that has taken up a contaminant from soil. The PEFs, F_{st} and F_{sk} account for human contaminant exposures attributable to ingestion of contaminated soil and pasture vegetation by beef and dairy cattle. The ingestion uptake by cattle is composed of direct ingestion of a soil contaminant and ingestion of a contaminant that has been taken up from the soil into the edible plant parts. Transfer from cattle uptake to meat and milk is modeled using biotransfer factors. For example, the meat PEF, F_{st}, is:

$$F_{st} = \left(\frac{I_t}{BW}\right) x \left(I_{sc} + I_{vbc} \quad K_{sp}\right) f_t \; B_t \tag{5}$$

where I_t/BW is the human intake of meat per unit body weight; I_{sc} is the soil-ingestion rate for beef and dairy cattle, 0.40 kg/d; and B_t is the biotransfer factor for contaminant uptake to meat per unit intake by cattle, d/kg.

Soil Ingestion Exposures

Assessing human exposures to contaminants in soil through direct ingestion of soil requires an estimate of age-dependent human soil ingestion. LaGoy (1987) has reviewed empirical data on human soil intake and used these data to make preliminary estimates for use in risk assessment. Using LaGoy's data, we calculate the average intake of soil by children to be 4.3 mg/kg-d, and for adults 0.71 mg/kg-d. Combining these values gives a lifetime equivalent PEF of 1.5 x 10^{-6} (mg/kg-d)/(mg/kg).

Dermal Absorption from Soil

Dermal absorption of contaminants from soil occurs through the accumulation of contaminated soil on skin. The amount of soil that accumulates on human skin depends on a number of factors such as age, type of soil, exposed surface area, soil conditions, etc. There is a great deal of variability in these factors, making the estimation of soil dermal absorption a relatively uncertain process. In order to calculate a PEF for dermal absorption from soil, we define the exposure for this pathway in terms of the amount of soil contaminant that passes from the soil matrix on the skin into the underlying tissue. Hawley (1985) has reviewed the soil dermal-exposure pathway. Based on Hawley's model, McKone (1988) has estimated the PEF for dermal absorption from soil to be on the order of 2.6×10^{-6} (mg/kg-d)/(mg/kg).

EXPOSURE PATHWAYS ASSOCIATED WITH CONTAMINATED SURFACE AND GROUND WATER

This section provides procedures for calculating ingestion, inhalation, and dermal absorption exposures attributable to chemicals in ground and surface water. The concentrations C_r and C_g (in mg/L) of the contaminant in surface and ground water are starting points for these exposure estimates.

Fish Ingestion Exposures

In calculating the intake of contaminants by the fish pathway, we assume that fish and other seafood are in chemical equilibrium with surface waters so that the contaminant concentration in fish tissue C_f is equal to the surface-water contaminant concentration C_r times the bioconcentration factor in fish BCF. The daily average exposure attributable to the transfer of contaminants from water to fish is the product of the contaminant concentration in fish C_f and the PEF F_{rf}. The PEF F_{rf} is the product of the daily human intake of fish per unit body weight, 0.0003 kg/kg-d, and BCF.

Drinking Water

For the water-ingestion pathway, the product of the pathway-exposure factor F_{ww} and the water-supply concentration C_w (in mg/L) gives the average population exposure in mg/kg-d. The PEF F_{ww} is the drinking-water intake per unit body weight, 0.034 kg/kg-d.

Inhalation Exposure from Potable Water

McKone (1987) has developed a model that describes the daily concentration profiles of volatile compounds within various compartments of the indoor air environment as a result of home water use. The results of this model provide a basis for calculating the pathway-exposure factor, F_{wa}, which can be multiplied by water-supply concentration to give the daily lifetime average indoor inhalation exposure in mg/kg-d. This model provides an estimate of the pathway-exposure factor for indoor inhalation that relates this PEF to liquid and gas diffusion coefficients and the Henry's law constant for the compound considered:

$$F_{wa} = 3.2 \times 10^5 \left(m^2/s\right)^{-2/3} \left[2.5/D_\ell^{2/3} + RT/D_a^{2/3}H\right]^{-1} \tag{6}$$

where F_{wa} is the water-to-inhalation PEF; D_ℓ is the contaminant diffusion coefficient in water, m^2/s; D_a is the contaminant diffusion coefficient in air, m^2/s; R is the gas constant, torr-L/mol-K; T is the temperature, ~ 293 K; and H is the Henry's constant in torr-L/mol. McKone (1987) estimates the uncertainty in this PEF to be bounded by a factor of 4. For many organic compounds, D_a is on the order of 5×10^{-6} m^2/s and D_ℓ is on the order of 5×10^{-10} m^2/s. This gives $F_{wa} \simeq 0.08$ when $H \geq 160$ torr-L/mol and $F_{wa} \leq 0.007$ when $H \leq 1.6$. In the latter case the indoor inhalation PEF is small compared to the drinking-water ingestion PEF and can be ignored.

Dermal Absorption

We assume that dermal absorption occurs during bathing and showering. The model we use for dermal absorption from potable water is based on a paper by Brown et al. (1984). Assuming that dermal uptake of contaminants occurs mainly by passive diffusion through the stratum corneum; resistance to diffusive flux through layers other than the stratum corneum is negligible; steady-state diffusive flux is proportional to the concentration difference between water on the skin surface and internal body water; children spend approximately the same amount of time bathing, swimming, or showering per week as adults; and the amount of time adults spend in showering, bathing, or swimming is equivalent to 80% immersion of the skin surface for a period of 10 min/d in water containing a contaminant at concentration C_w, McKone (1988) has calculated the PEF for dermal absorption from water to be on the order of 0.038 (mg/kg-d)/(mg/L).

Milk and Meat Exposures

The concentrations C_w of a contaminant in water supplies can result in the contaminantion of fresh meat and milk of beef and dairy cattle that use these supplies. Assuming a steady state partitioning between meat/milk and ingested water for cattle, one can estimate PEFs for people consuming these foods as the product of human intake, cattle intake, and a biotransfer factor. As an example the PEF for water to meat is

$$F_{wt} = (I_t/BW)\ I_{wbc}\ B_t \tag{7}$$

where (I_t/BW) is the human intake of meat per unit body weight, kg (fresh mass)/kg-d; I_{wbc} is the daily intake of water by beef cattle, 44 L/d; the daily intake of water by dairy cattle is 48 L/d; and B_t is the biotransfer factor from cattle intake to meat concentration, (mg/kg)/(mg/d).

SAMPLE CALCULATION AND DISCUSSION

This paper discusses methods for integrating multiple exposure routes from multiple environmental media; air, soil, and water; into a matrix of factors that relate concentrations to human exposure. This section presents an example that illustrates the application of the PEF approach to three environmental contaminants--arsenic, benzene, and TNT. This is followed by a discussion of PEFs. Table 1 provides a matrix that summarizes the values and expressions for the PEFs developed in this paper. The listings in this table represent population average or lifetime equivalent PEFs that combine PEFs for adults and children into a single term.

An Example Calculation

A sample calculation can illustrate the use of PEFs for estimating human exposure to three toxic but chemically different species--arsenic, benzene, and TNT. Arsenic and benzene are known human carcinogens; military-grade TNT (containing TNT isomers and isomers of dinitrotoluene as impurities) has caused malignant tumors in mice and rats in chronic toxicity studies (Layton et al., 1987). Table 2 lists the physicochemical properties of the three contaminants used in this exercise and taken from Layton et al. (1986, 1987). Table 3 lists environmental concentrations in air, soil, and water used for the sample calculations. These values were obtained using the GEOTOX program with steady-state inputs of the contaminants to the soil compartment. The resulting concentrations correspond to the equilibrium distribution of each contaminant among the environmental system modelled in GEOTOX. The landscape data used for this calculation were data compiled by McKone and Layton (1986) to represent the southeastern region of the United States. The results of the calculation were scaled so that the soil concentration for each contaminant equals 1 ppm (mg/kg). The results of the sample calculation for arsenic, benzene, and TNT are summarized respectively in Figure 1. Examination of the results reveals that, for arsenic, ingestion is the dominant route of exposure followed by dermal absorption; for benzene, inhalation is the most important exposure pathway followed by

Table 2. Physicochemical properties of the candidate species.

Property	Arsenic	Benzene	TNT
Molecular weight (g)	74.9	78.1	227
Henry's law constant (torr-L/mol)	1.0×10^{-6}	4.1×10^{3}	2.0×10^{-3}
Octanol/water partition factor (K_{ow})	--	135	39.8
Organic carbon partition factor(K_{oc})	--	76	534
Soil/water sorption coefficient (L/kg)	1300	--	--
Diffusion coefficient (air) (m^2/s)	$\sim 10^{-6}$	5.0×10^{-6}	5.9×10^{-6}
Diffusion coefficient (water) (m^2/s)	$\sim 10^{-10}$	5.0×10^{-10}	5.8×10^{-10}
Soil/plant partition coefficient (dry mass) (K_{sp})	0.11	4.34	6.3
Bioconcentration factor in fish (L/kg)	75	75	10
Meat/diet biotransfer factor in cattle (d/kg)	6.2×10^{-5}	1.3×10^{-4}	7.2×10^{-5}
Milk/diet biotransfer factor in cattle (d/L)	6.2×10^{-5}	1.2×10^{-5}	6.5×10^{-6}

Table 3. Steady-state environmental concentrations used in the sample calculations. Concentrations are predicted using the GEOTOX model with parameter inputs from Table 2.

Description	Arsenic	Benzene	TNT
Concentration of contaminant in air (mg/m^3)	~0	23	2.5×10^{-8}
Concentration of contaminant attach to air particles (mg/m^3)	5.2×10^{-8}	4.2×10^{-8}	5.0×10^{-8}
Concentration of contaminant in soil (mg/kg)	1.0	1.0	1.0
Concentration of contaminant in ground water (mg/L)	1.0×10^{-3}	0.62	0.10
Concentration of contaminant in surface water (mg/L)	1.8×10^{-3}	0.59	0.0043
Concentration of contaminant in potable water (mg/L)	1.4×10^{-3}	0.61	0.052

dermal absorption; and for TNT, overall exposure is dominated by ingestion and dermal absorption, with inhalation exposure expected to be several orders of magnitude lower.

Discussion and Conclusions

One of the fundamental assumptions made in this paper is that all the pathways are fully available. That is, the individual at risk is assumed to

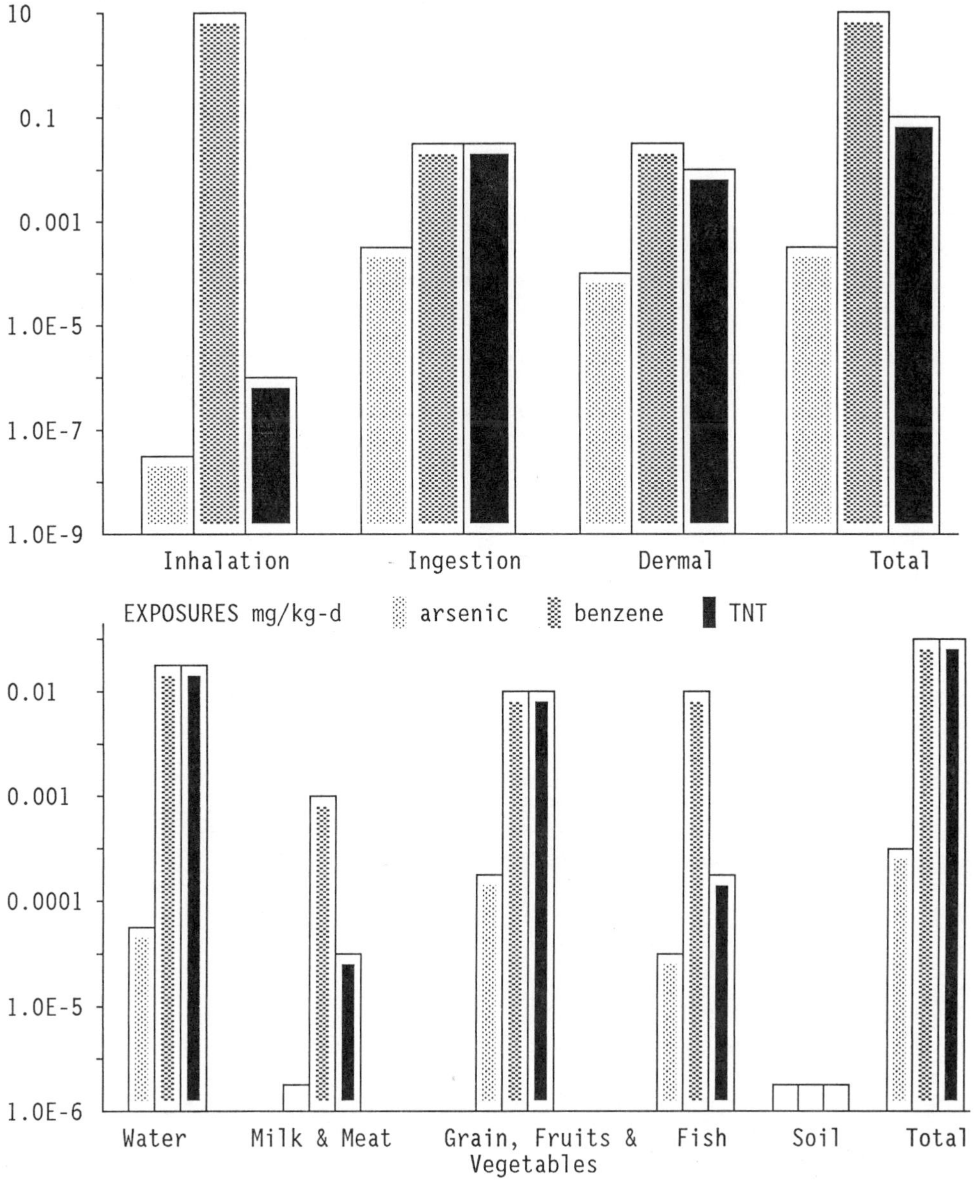

Figure 1. Contribution of each exposure pathway to total exposure for arsenic, benzene, and TNT in the example problem.

obtain all of his or her air, water, and food from an environment having the specified contaminant concentrations in air, water, and soil. Except for contaminants that are uniformly released and spread over a large region, this is not likely to be the case. A more realistic approach to assessing human exposures requires that we now go back to the expression for each exposure route and prescribe methods for adjusting these expressions to account for the variability of individual exposure that is attributable to spatial variation in concentration and the time dependence of pathways. One approach to this problem, which is compatible with the PEF approach, is to specify the concentrations in each media (air, soil, water) as a probability distribution, which characterizes the variation of concentrations as a result of population movement, individual life styles (i.e., food supply, housing, and water-supply choices), and the time and spatial character of the source. The incorporation of uncertainties could result in some reordering of the important pathways. For example the factor F_{wa} has an uncertainty factor of about 2-4.

Although the results in this paper have not been subjected to a rigorous sensitivity or uncertainty analysis, there are some conclusions that can be drawn from the example calculation: dermal absorption of chemicals from soils is not likely to be an important exposure route unless all other soil and water pathways are unavailable; inhalation of volatile chemicals transported from potable water supplies to indoor air has the potential for being as important as the direct ingestion of these compounds from potable water supplies; when addressing the intake of chemical species through milk and meat, it is important to consider all potential intake routes for these chemicals by cattle; for a volatile compound such as benzene, the atmosphere provides a source of contaminant to milk and meat through inhalation and deposition/ingestion; and for TNT and arsenic pasture, soil, and water provide sources of contaminants expected in the milk and meat supply.

REFERENCES

Brown, H. S., D. R. Bishop, and C. A. Rowan, 1984, The role of skin absorbtion as a route of exposure for volatile organic compounds (VOCs) in drinking water, Am. J. Public Health 74, 479-484.

Hawley, J. K., 1985, Assessment of health risk from exposure to contaminated soil, Risk Anal. 5, 289-302.

LaGoy, P. K., 1987, Estimated Soil Ingestion Rates for Use in Risk Assessment, Risk Anal. 7, 355-359.

Layton, D. W., T.E. McKone, C.H. Hall, M.A. Nelson, and Y.E. Ricker, 1986, Demilitarization of Conventional Ordnance: Priorities for Data-Base Assessments of Environmental Contaminants, Lawrence Livermore National Laboratory, Livermore, CA, UCRL-15902.

Layton, D. W., B. Mallon, W. Mitchell, L. Hall, R. Fish, L. Perry, G. Snyder, K. Bogen, W. Malloch, C. Ham, and P. Dowd, 1987, Data-Base Assessment of the Health and Environmental Effects of Conventional Weapons Demilitarization: Explosives and Their Co-Contaminants, Lawrence Livermore National Laboratory, Livermore, CA, UCRL-21109.

McKone, T. E., 1987, Human exposure to volatile organic compounds in household tap water: the indoor inhalation pathway, Environ. Sci. Technol. 21, 1194-1201.

McKone, T. E., 1988, Methods for Estimating Multipathway Exposures to Environmental Contaminants, Lawrence Livermore National Laboratory, Livermore, CA, UCRL-21064.

McKone, T. E., and D. W. Layton, 1986, Screening the potential risks of toxic substances using a multimedia compartment model: estimation of human exposure, Regul. Toxicol. Pharmacol. 6, 359-380.

Murphy, B. L., and J. E. Yocom, 1986, Migration factors for particulates entering the indoor environment, Proceedings of the 79th Annual Meeting of the Air Pollution Control Association, Minneapolis, MN, June 22-27.

Travis, C. C. and A. D. Arms, 1988, Bioconcentration of organics in beef, milk, and vegetation, Environ. Sci. Technol. 22, 271-274.

U.S. Environmental Protection Agency (U.S. EPA), 1987, The Risk Assessment Guidelines for 1986, Office of Health and Environmental Assessment, Washington, DC, Report No. EPA/600/8-87/045.

SUMMARY

It is now recognized that the environment is a complex system of interacting environmental media. Pollutants do not stay in the medium where they originate but move across environmental phase boundaries. Thus, environmental pollution becomes a multimedia problem. The papers in this environmental volume present a unique collection of works that describe various problems relating to intermedia pollutant transport. This book presents the viewpoint that intermedia modeling and monitoring of chemical pollutants are complementary.

Monitoring studies in the Great Lakes Region have shown that rain scavenging and dry deposition can be important processes by which pollutants enter fresh water bodies. Although the acidification of water bodies via acid rain has been widely studied, it is only recently that field studies in measurement of dry deposition velocities were conducted. It is not a trivial task but the proposed dual tracer method appears to be promising. Also, the coupling of aerosol monitoring with that of trace elements in human blood could lead to more accurate estimate of human exposure to chemicals associated with the aerosol phase. A number of papers dealt with the transport of pollutants to the atmosphere from water bodies, soils, surface impoundments, landfills, and even from dredged sediment material, via volatilization, and from aerosol formation and resuspension. It is now apparent that in order to be able to predict the volatilization of organic chemicals from the soil environment, one must account for the diurnal variations of temperature and moisture in the soil matrix. The various discussions have made it clear that there is a need to understand solute adsorption especially as a function of soil moisture content as well as surface chemistry,

Pollutants that enter the soil system also present a potential groundwater contamination problem. Although there are numerous models of solute transport in the unsaturated soil zone, the heterogenous nature of the soil makes it difficult to apply simple models or laboratory results to the field. Thus, the treatment of heterogeneities via modeling approaches appear to be promising.

CONTRIBUTORS

D. T. Allen
Department of Chemical Engineering
5531 Boelter Hall
University of California, Los Angeles
Los Angeles, CA 90024

Richard L. Bell
Department of Chemical Engineering
University of California
Davis, CA 95616

Dermont C. Bouchard
R. S. Kerr Environmental Laboratory
U. S. Environmental Protection Agency
P. O. Box 1198
Ada, OK 74820

Yoram Cohen
Department of Chemical Engineering
5531 Boelter Hall
University of California, Los Angeles
Los Angeles, CA 90024

Steven J. Eisenreich
Department of Civil and Mineral Engineering
CME Building, Room 122
University of Minnesota
Minneapolis, MN 55455

Edward M. Godsy
U. S. Geological Survey
345 Middlefield Road
Mail Stop 458
Menlo Park, CA 94025

Dean S. Jeffries
Environment Canada
National Water Resource Institute
P. O. Box 5050
Burlington, Ontario
Canada L7R 4A6

Walter John
Air and Industrial Hygiene Laboratory
California Department of Health Services
2151 Berkeley Way
Berkeley, CA 94704

Isaac R. Kaplan
Institute of Geophysics and Planetary Physics
5855 Slichter Hall
University of California, Los Angeles
Los Angeles, CA 90024

Hilary H. Main
Department of Chemical Engineering
5531 Boelter Hall
University of California, Los Angeles
Los Angeles, CA 90024

Thomas E. McKone
Lawrence Livermore Laboratory
P. O. Box 5507, L-453
Livermore, CA 94550

Barbara J. Morrison
Department of Chemical Engineering
University of California
Davis, CA 95616

Shyam K. Nair
Bechtel Environmental, Inc.
800 Oak Ridge Turnpike
Oak Ridge, TN 37831-0350

Sally Paterson
Department of Chemical Engineering
University of Toronto
Toronto, Canada M5S 1A4

Danny D. Reible
Department of Chemical Engineering
Louisiana State University
Baton Rouge, LA 70803-7303

Patrick A. Ryan
Department of Chemical Engineering
5531 Boelter Hall
University of California, Los Angeles
Los Angeles, CA 90024

Hiroshi Sakugawa
Institute of Geophysics and
Planetary Physics
5853 Slichter Hall
University of California, Los Angeles
Los Angeles, CA 90024

David R. Shonnard
Department of Chemical Engineering
University of California
Davis, CA 95616

Louis J. Thibodeaux
Hazardous Waste Resource Center
3418 CEBA
Louisiana State University
Baton Rouge, LA 70803-6421

Curtis C. Travis
Oak Ridge National Laboratory
P.O. Box 2008
Bldg. 4500 South - MS 109
Oak Ridge, TN 37831

Martinus Th. van Genuchten
U. S. Department of Agriculture
Agricultural Resource Service
4500 Glenwood Drive
Riverside, CA 92501

Wayne A. Wilford
U. S. Environmental Protection Agency
Great Lakes National Program Office
230 South Dearborn St.
Chicago, Illinois 60604

INDEX

Acid rain, acidification, 41, 45, 55
Adsorption, 127-129, 161, 217
Advection, 214
Air-dry soil, 161
Anaerobic degradation, 222-226
Arsenic, 289
Atmospheric hydrogen peroxide, 53
Atmospheric aerosols, 20, 73, 87

Benzene, 94, 101, 289
Biodegradation, 222
Box model, 74
Bulk phase infiltration, 239

Confined disposal facility (CDF), 22
Convection, 94
Creosote, 213
Crude oil, 167

Deposition
  atmospheric, 20
  dry, 73, 22, 87
  velocity, 74
  wet, 41, 22, 87
Dieldrin, 94, 161
Diffusion, 94
Diffusion experiments in soils, 162, 168
Dispersion, 214
Diurnal cycle, 170
Diurnal variation of H_2O_2, 57
Dredge disposal, 122
Dual tracers, 73

Effects of soil temperature, 163
Emission, 105, 131, 163
Environmental monitoring, 10
  integration with modeling, 11
Equilibrium processes, 123
  between air and water, 124
  between water and sediment, 126
  between sediment and air, 127
Exposure pathways, 6, 87, 283
  inhalation, 285
  ingestion
    beef and milk, 260, 286, 289

Exposure pathways (continued)
  ingestion (continued)
    drinking water, 288
    grains, fruits and vegetables, 285
    fish, 288
  dermal absorption, 287

Field-scale transport processes, 177
Food chain, 259
Fugacity model, 257, 274

Gasoline contamination, 163
Global pollution, 86
Great Lakes, 19
Ground water, 219, 288

Heat transfer model, 168
Henry's Law, 124
Heterogenous oxidation of SO_2, 67
Hydrocarbon, 167
Hydrologic cycle, 20, 177
Hydrophobic chemicals, 122, 269

Incinerators, 262
Industrial sources, 263
Initial soil concentration of contaminant, 165
Intake
  inhalation, 285
  ingestion
    beef and milk, 260, 286, 289
    drinking water, 288
    grains, fruits and vegetables, 285
    fish, 288
  dermal absorption, 287
Intermedia transport model, 5

Lake ecosystem, 19, 27-31, 42
Landfill sites, 107
Lead, 32, 73
Lindane, 94
Low pressure impactor, 73

Mass balance calculations, 27, 94
Moisture, 101, 163
Motor vehicles, 74, 283

Multimedia-compartmental models (MCM), 8
Multimedia exposure, 5, 283
Multimedia transport model, 8, 257
Multiphase chemical transport model, 94

Non-aqueous phase liquids, 189, 196, 237
n-nonane, 165

Oil film, 125
Organic contaminants
  Creosote, 213
  Crude oil, 167
  DDT, 24, 32
  1,2-dichloroethylene, 114
  Dioxin, 24, 257
  Furan, 24
  Lindane, 94
  n-nonane, 165
  PAH, 24
  PCB, 24, 32, 35, 122
  TCDD, 257-262
  TNT, 289
Oxidation 57, 63

Pathway exposure factors (PEFs), 283
Pesticides, 269
Plant model, 271
Plant toxicokinetics, 270
Plants and Vegetation, 259, 271
Porous media, 98, 190
  chemical transport in, 98, 190
Precipitation scavenging, 22
Pulp and paper mills, 264

Relative humidity, 101, 165
Residual non-aqueous phase mobility, 245
Resuspension, 131

Seasonal variation of gaseous H_2O_2, 57
Sediment, 127
Soil, 286
  diffusion coefficient, 95
  moisture, 93, 163
  solute transport in, 177
  structured, 179
  temperature, 163
Solar radiation, 60
Solute transport
  stochastic models in, 180
  nonequilibrium during, 190
Sorption, 189-191, 216
Source depletion model, 75
Spatial-multimedia (SM) models, 10
Spatial-multimedia-compartmental
  model (SMCM), 10
Substrate utilization, 226
Subsurface processes, 189, 214, 237
Sulfate, 42-46
Sulfur dioxide, 53

Temperature profile, 170
TNT, 289
Trace elements
  transport of, 85
  in human blood, 86
  in aerosol, 88
Transport processes, 5, 133
  in dry soil, 95
  in moist soil, 98
  in hazardous waste sites, 105
  in structure soil, 179

Vadose zone transfer processes, 183, 239
Vehicle emissions, 74
Vignette models, 123
Volatile organic compounds (VOCs), 121, 161, 167
Volatilization fluxes, 98
Volatile emissions, 107, 114
  from dredged material disposal, 121
  from land fills, 107
  from motor vehicles, 78, 283

Wind entrainment of contaminants, 105